Advances in Processing Technology

Advances in Processing Technology

Editor-in-Chief
Gopal Kumar Sharma FAFST(I)
Former Additional Director & Head
Grain Science and Technology Division
DRDO-DFRL (Defence Food Research Laboratory)
Ministry of Defence, Government of India
Mysuru, Karnataka

Edited by
Anil Dutt Semwal
Director
DRDO-DFRL (Defence Food Research Laboratory)
Ministry of Defence, Government of India
Mysuru, Karnataka

Dev Kumar Yadav
Scientist 'D'
Grain Science and Technology Division
DFRL-DRDO (Defence Food Research Laboratory)
Ministry of Defence, Government of India
Mysuru, Karnataka

CRC Press is an imprint of the
Taylor & Francis Group, an **informa** business

NEW INDIA PUBLISHING AGENCY
New Delhi-110 034

First published 2022
by CRC Press
2 Park Square, Milton Park, Abingdon, Oxon, OX14 4RN

and by CRC Press
6000 Broken Sound Parkway NW, Suite 300, Boca Raton, FL 33487-2742

CRC Press is an imprint of Informa UK Limited

British Library Cataloguing-in-Publication Data
A catalogue record for this book is available from the British Library

Library of Congress Cataloging-in-Publication Data
A catalog record has been requested

ISBN: 978-1-032-15742-9 (hbk)
ISBN: 978-1-003-24551-3 (ebk)

DOI: 10.1201/9781003245513

Preface

With the increase in socioeconomic status around the world consumers are demanding instant, Ready-to-eat processed foods with better nutritional value and safety. Scientists and industrialists are engaged in developing innovative and newer technologies to cater the immense demand of consumers. As a matter of fact global food processing-technology business has reached to multi trillion dollars. It was reported that approximately, 16 million people work in various food industries. Recent advances in food processing and technology is not only important to meet the increasing productivity demands but to adopt refined automation, control and monitoring methods and techniques. The novel technologies in Food Science enable the Food & Beverages Sector to enhance the quality of the eatables and drinks intern consumers' satisfaction and high profitability. New food products *viz.* Designer foods, Nutraceutical foods have emerged by the application of the newer food processing technologies and adopting challenges in biochemistry, nutrition science, applied biotechnology e.g. genetically modified foods, organic crops balanced protein based diet etc. Hence, the food which we consume every day is the result of extensive food research and systematic investigation into a variety of foods' properties and compositions. The present book is an amalgamation of various topics which are quite relevant to academics pertaining to food science and technology. Sincere attempts have been made to map consumer's perception in terms of sensory evaluation of processed foods and their role on quality determination. To cover food safety, the topic of advancement in the traceability and transparency of food supply chain was also discussed in length. Besides, providing basic nutrition food has become an essential source of health promoting phyto-ingredients too. To take care of concerned population therapeutic foods has also been discussed with their future trends. Similarly, recent trends in functional and Nutraceutical foods were also discussed in detail so as to give an exhaustive overlook of such subject matter. To give impetus to the growing and aged generations the importance of the technology of weaning and geriatric foods was described in detail. Bio-preservation of various food products including fermentation had always attracted researchers for various reasons inclusive of its novel and chemical free approach of preservation which has been aptly covered under current expansions in microbiology for food preservation and also under progression in biotechnology and its application in food processing. The cross linkage of advance technologies inclusive of nano-science was elaborated as technological advances in nano- science for specific food and nutrition delivery.

Oil and spice commerce are two giants pillars in food processing industries and readers would surely be wishing to understand the developments in the technology of oils refineries and condiments. Smart and intelligent packing systems always extend an upper hand as far as shelf life monitoring of any processed food is concerned especially when these are import worthy products. The science and technological approach of these packing innovations was also well covered.

With our experience of working with defence organization gave us an inclination to discuss and write some of the topic related to defence ration which was well covered under heads 'Past, present and future perspective of Army Operational Ration.' The Inception of editing this book came when we thought of compiling most relevant topics of food processing sector while giving nearly three and half decade to this area. There was always a subconscious thrust to provide a comprehensive knowledge to the budding scientists, academicians, entrepreneurs, defense personals and students' at large belonging to this field.

I am very happy to be backed by my efficient editorial team who worked tirelessly but very meticulously in shaping the presenting book. I must record my gratitude with great sense of appreciation to the contributing authors who have given their best to the assigned topics. Our team is confident about the fact that the readers of this will surely have a deep sense of satisfaction after going through the content of each topic covered in the present book entitled 'Advances in Processing Technology'. Kindly do let us know your feedback on this book which will help us in bringing up its better edition in coming times.

Editor-in-Chief

Contents

1

Sensory Evaluation Techniques and Consumer Perception Studies for Food Product Innovations

Shiby V.K.* and *Aisha Tabassum

Freeze Drying and Animal Products Technology Division, Defence Food Research Laboratory, Mysuru, Karnataka, India

Introduction

Quality is the ultimate criterion of the desirability of any food product to the consumer. Overall quality depends on quantity, nutritional and other hidden attributes along with sensory quality. Sensory quality is of great importance to both processors and consumers, since it attracts consumers; it satisfies consumers' aesthetics and gustatory sense. Sensory quality is the combination of sensory senses of perception responsible for choosing and eating a food.

According to Institute of Food Technologists and the American Society for Testing and Materials, sensory evaluation is a scientific method used to evoke measure, analyze, and explain those responses to products as perceived through different senses like sense of sight, smell, touch, taste, and hearing (Stone and Sidel, 2004). Sensory evaluation gives a general rule for the preparation and serving of samples under controlled conditions so that unfair decisions are minimized. People for sensory evaluation are often placed in individual test booths so that the judgments they give are true and do not reflect the opinions of those around them.

Samples with random numbers are given to the evaluators so that they do not form judgments based upon labels, but rather on their sensory experiences. Products may be given in different orders to each evaluator to help measure and to ensure internal validity by controlling the potential confounds created by sequence and order effects. To control unwanted variation and to improve test precision, standard procedures may be established for sample temperature, volume, and spacing in time, as needed.

Sensory evaluation or testing is a quantitative science in which numerical data are collected to establish lawful and specific relationships between product characteristics and human perception. Sensory methods draw heavily from the techniques of behavioural research in observing and quantifying human responses. For example, we can assess the proportion of times people are able to discriminate small product changes or the proportion of a group that expresses liking for one product over another. Another example is having people generate numerical responses reflecting their perception of products taste and smell.

Sensory evaluation comprises a set of test methods with guidelines and established techniques for product presentation, well-defined response tasks, statistical methods, and guidelines for interpretation of results. Discrimination tests, descriptive analysis and affective or hedonic testing are the three primary types of sensory tests which focus on the existence of overall differences among the products. Correct application of sensory technique involves correct matching of method to the objective of the tests, and this requires good communication between sensory specialists and end-users of the test results. Logical choices of test participants and appropriate statistical analyses form part of the methodological mix. Analytical tests such as the discrimination and descriptive procedures require good experimental control and maximization of test precision. Affective tests on the other hand require true consumers; individuals who are pre-screened to be actual users of the product tested and test conditions that enable generalization to how products are experienced by consumers in the real world.

Sensory tests provide useful information about the human perception of product changes due to ingredients, processing, packaging, or shelf life. Sensory evaluation departments not only interact most heavily with new product development groups but may also provide information to quality control, marketing research, packaging, and, indirectly, to other groups throughout a company. Sensory information reduces risk in decisions about product development and strategies for meeting consumer needs. A well functioning sensory program will be useful to a company in meeting consumer expectations and insuring a greater chance of market place success. The usefulness of the information provided is directly proportional to the quality of the sensory measurement.

Classification of sensory tests

The human senses have been used for centuries to assess the quality of foods. We all form judgments about foods whenever we eat or drink "Everyone carries his own inch-rule of taste, and amuses himself by applying it, triumphantly, wherever he travels" (Adams, 1918). This does not mean that all judgments are useful or that anyone is qualified to participate in a sensory test. In the past, production of good quality foods often depended upon the sensory evaluation of a single expert who was in charge of production or made decisions about process changes in order to make sure the product would have desirable characteristics. This was the historical tradition of brew masters, wine tasters, dairy judges, and other food inspectors who acted as the arbiters of quality. Modern sensory evaluation replaced these single authorities with panels of people participating in specific test methods. This change occurred for several reasons. First, it was recognized that the judgments of a panel would in general be more reliable than the judgments of single individual and it entailed less risk since the single expert could become ill, travel, retire, die, or be otherwise unavailable to make decisions. Replacement of such an individual was a nontrivial problem. Second, the expert might or might not reflect what consumers or segments of the consuming public might want in a product. Thus for issues of product quality and overall appeal, it was safer (although often more time consuming and expensive) to go directly to the target population. Although the tradition of informal, qualitative inspections such as bench top "cuttings" persists in some industries, they have been gradually replaced by more formal, quantitative, and controlled monitoring (Stone and Sidel, 2004). The recent sensory evaluation methods comprise a set of measurement techniques with established track records of use in industry and academic research. Much of what we consider standard procedures comes from pitfalls and problems encountered in the practical experience of sensory specialists over the last 70 years of food and consumer product research, and this experience is considerable. The primary concern of any sensory evaluation specialist is to insure that the test method is appropriate to answer the questions being asked about the product in the test. For this reason, tests are usually classified according to their primary purpose and importance. A summary on the types of sensory testing is given in Table 1.

Table 1: Sensory tests and panellist characteristics

Class	Type of Interest	Type of Test	Panelist Characteristics
Descriptive	Are products perceptibly different in any way?	Analytic	Screened for sensory acuity, oriented to test method, sometimes trained
Discrimination	How do products differ in specific sensory characteristics?	Analytic	Screened for sensory acuity and motivation, trained or highly trained
Affective	How well are products liked or which products are preferred?	Hedonic	Screened for products, untrained

Descriptive analyses

One important class of sensory test methods is to quantify the perceived intensities of the sensory attributes of a product. These procedures are called as descriptive analyses. The first descriptive analysis performed by a panel of trained judges was on Flavour Profile by Little consulting group in the late 1940s (Caul, 1957). This group was faced with developing a comprehensive and flexible tool for analysis of flavour to solve problems involving unpleasant off flavours in nutritional capsules and questions about the sensory impact of monosodium glutamate in various processed foods. They formulated a method involving extensive training of panellists that enabled them to characterize specific flavours in a food and the intensities of these flavours using a simple category scale and noting their order of appearance. This advance was noteworthy on several grounds. It replaced the reliance on single expert judges (brew masters, coffee tasters, and such) with a panel of individuals, under the realization that the consensus of a panel was likely to be more reliable and accurate than the assessment by a single individual. Secondly, it helped in characterizing the individual attributes of flavour and provided a comprehensive analytical description of differences among a group of products under development. Several variations and refinement in descriptive analysis techniques were forthcoming. A group at the General Foods Technical Centre in the early 1960s developed and refined a method to quantify food texture, much as the flavour profile had enabled the quantification of flavour properties (Brandt et al., 1963). Texture Profile method, was used to characterize the rheological and tactile properties of foods and how these changed over time with mastication. These characteristics are helpful in evaluating food breakdown or flow. For example, hardness is related to the physical force required to penetrate in to sample and thickness of a fluid or semisolid is related in part to physical viscosity. Texture profile panellists were also trained to recognize specific intensity points along each scale, using standard products or formulated pseudo-foods for calibration. Other approaches were developed for descriptive analysis problems. At Stanford Research Institute, in properties of a food, and not just taste and texture (Stone et al., 1974). This method was termed Quantitative Descriptive Analysis R_ or QDA R_ for short (Stone and Sidel, 2004). QDA R_ procedures borrowed heavily from the traditions of behavioural research and used experimental designs and statistical analyses such as analysis of variance. This insured independent judgments of panellists and statistical testing, in contrast to the group discussion and consensus procedures of the Flavour Profile R_ method. Other variations on descriptive procedures were tried and achieved some popularity, such as the Spectrum Method R (Meilgaard et al., 2006) that included a high degree of calibration of panellists for intensity scale points. Still other researchers have employed hybrid techniques that include some features of the various descriptive approaches (Einstein, 1991). Today many product development groups use hybrid approaches as the advantages of each may apply to the products and resources of a particular company. Descriptive analysis has proven to be the most comprehensive and informative sensory evaluation tool. It is applicable to the characterization of a wide variety of product changes and research questions in food product development. The information can be related to consumer acceptance and to instrumental measures by means of statistical analysis such as regression and correlation. An example of a descriptive ballot for texture assessment of cookie product is shown in Table 2. The cookies were assessed at different time intervals in a systematic, uniform and controlled manner, typical of an analytical sensory test procedure. For example, the first bite may be defined as cutting with the incisors. The panel for such an analysis would consist of perhaps 10–12 well-trained individuals, who were oriented to the meanings of the terms and given practice with examples. Intensity references to exemplify scale points are also given in some techniques. Note the amount of detailed information that can be provided in this example and bear in mind that this is only looking at the product's texture—flavour might form an equally detailed sensory analysis, perhaps with a separate trained panel. The relatively small number of panellists (a dozen or so) is justified due to their level of calibration. Since they have been trained to use attribute scales in a similar manner, error variance is lowered and statistical power and test sensitivity are maintained in spite of fewer observations (fewer data points per product). Similar examples of texture, flavour, fragrance, and tactile analyses can be found in Meilgaard et al., (2006).

Table 2: Descriptive evaluation of cookies texture attributes

Phase	Attributes	Word anchors
Surface	Roughness	Smooth–rough
	Particles	None–many
	Dryness	Oily–dry
First bite	Fracturability	Crumbly–brittle
	Hardness	Soft–hard
	Particle size	Small–large
First chew	Denseness	Airy–dense
	Uniformity of chew	Even–uneven
Chew down	Moisture absorption	None–much
	Cohesiveness of mass	Loose–cohesive
	Toothpacking	None–much
	Grittiness	None–much
Residual	Oiliness	Dry–oily
	Particles	None–many
	Chalky	Not chalky–very chalky

Discrimination testing

The test is also known as difference test. The simplest sensory tests merely attempt to answer whether any perceptible difference exists between two types of products. These are the discrimination tests or simple difference testing procedures. Analysis is generally based on the statistics of frequencies and proportions (counting right and wrong answers). From the test results, the differences are interpreted based on the proportions of persons who are able to choose a test product correctly from among a set of similar or control products. This test method was used by Carlsberg breweries and Sea grams distilleries in 1940s (Helm and Trolle, 1946; Peryam and Swartz, 1950). In this test, two products were from the same batch while a third product was different. Panellists would be asked to choose the odd sample from among the three. Ability to discriminate differences would be inferred from consistent correct choices above the level expected by chance. In breweries, this test served primarily as a means to screen judges for beer evaluation, to insure that they possessed sufficient discrimination abilities. Another multiple-choice difference test was developed at about the same time in distilleries for purposes of quality control (Peryam and Swartz, 1950). In the duo–trio procedure, a reference sample was given and then two test samples. One of the test samples matched the reference while the other was from a different product, batch or process. The participant would try to match the correct sample to the reference, with a chance probability of one-half. As in the triangle test, a proportion of correct choices above that expected by chance is considered evidence for a perceivable difference between products. A third popular difference test was the paired comparison, in which participants would be asked to choose which of two products was stronger or more intense in a given attribute example flavour. Due to the fact that the panellist's attention is concentrated to a specific attribute, this test is very sensitive to differences. These difference tests are shown in Fig 1. A discrimination test will be conducted with 25–40 participants who have been screened for their sensory acuity to common product differences and familiar with the test procedures. This generally provides an adequate sample size for documenting clear sensory differences. In part, the popularity of these tests is due to the simplicity of data analysis. Statistical tables derived from the binomial distribution give the minimum number of correct responses needed to conclude statistical significance as a function of the no. of participants. Analyst merely needs to count answers and refer to a table to give a simple statistical conclusion, and results can be easily and quickly reported.

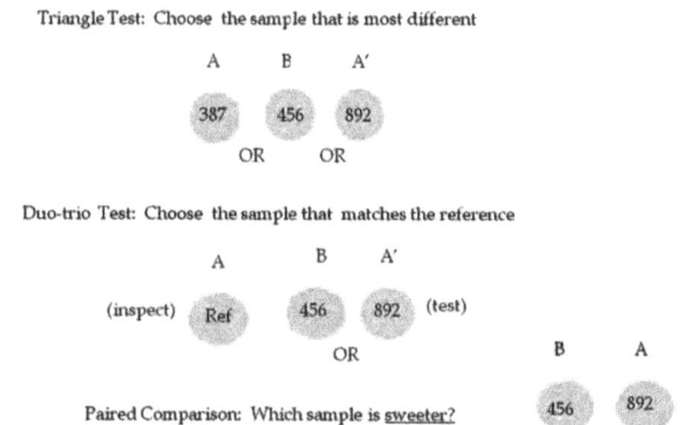

Fig. 1: Common methods for discrimination testing

Affective testing

The third major class of sensory tests is the attempt to find out the degree of liking or disliking of a product, called hedonic or affective test methods. The most appropriate way to solve this problem is to offer people a choice among alternative products and see if there is a clear preference from the majority of respondents. The problem with choice tests is that they are not very informative about the magnitude of liking or disliking from respondents. An historical landmark in this class of tests was the hedonic scale developed at the U.S. Army Food and Container Institute in the late 1940s (Jones *et al.*, 1955). This method provided a balanced 9-point scale for liking with a cantered neutral category and attempted to produce scale point labels with adverbs that represented psychologically equal steps or changes in hedonic tone. In other words, it was a scale with ruler-like properties whose equal intervals would be amenable to statistical analysis. Typically a hedonic test today would involve a 75–150 consumers who are regular users of the product. The test includes several alternative versions of the product and can be conducted in some central location or sensory test facility. In early 1970s, a group proposed a method for descriptive analysis that would be the remedy for some of the apparent shortcomings of the Flavour Profile method. This method provides an opportunity for individual preferences and thus a need for increased numbers of people to insure statistical power and test sensitivity. It also provides an opportunity to look for segments of people who may like products with different colours or flavours. Diagnostic information concerning the reasons for liking or disliking a product can also be generated by using this method. Researchers in the food industry were occasionally in contact with psychologists who studied about sense organs and had developed skills for assessing sensory function (Moskowitz, 1983). The development of 9-point hedonic scale serves as good illustration of what can be realized when there is interaction between psychologists and food scientists. A psychological measurement technique called thurstonian scaling was used to validate the adverbs for the labels on the 9-point hedonic scale. Differences in language, goals, and experimental focus probably contributed to some difficulties. Psychologists focus on the individual person while sensory evaluators focus on the food product (the stimulus). However, since a sensory perception involves the necessary interaction of a person with a stimulus, it should be apparent that similar test methods are necessary to characterize this person–product interaction.

Analytical vs. Hedonic tests

The central principle for sensory evaluation is that the test method should be matched to the objectives of the test. The selection of the test procedure follows from questions about the objective of the investigation. To fulfil this goal, it is necessary to have clear communication between the sensory test manager and end-user. A dialogue is often needed. When it is to know whether or not there is any difference at all among the products a discrimination test is indicated. To analyze whether consumers like the new product better than the previous version a consumer acceptance test is needed. Do we need to know what attributes have changed in the sensory quality of the new product? Then a descriptive analysis procedure is called for. Sometimes there are multiple objectives and a sequence of different tests is required (Lawless and Claassen, 1993). This can present problems if all the answers are required at once or under severe time pressure during competitive product development. One of the most essential task of the sensory specialist in food industry is to insure a clear information needed by the end-users regarding product specifications. Test design require a number of conversations, interviews with many people, or even written tests that can specify why the information is needed and how the results will be predicted in making specific decisions and subsequent actions to be taken. The sensory specialist is the best person to understand the uses and limitations of each procedure and what would be considered appropriate versus inappropriate conclusions from the data. There are two important corollaries to this principle. The sensory test design involves not only the selection of an appropriate method but also the selection of appropriate participants and statistical analyses. For discrimination test the panellists are selected based on having average to good sensory acuity for the critical characteristics (tastes, smells, textures, etc.) of products to be evaluated. They are familiarized with the test procedures and may undergo greater or lesser amounts of training, depending upon the method. In descriptive analysis, they adopt an analytical frame of mind, focusing on particular aspects of the product as mentioned on their questionnaires. Panellists are asked to put personal preferences and hedonic reactions aside, and specify only the attributes which are present in the product and at what levels of sensory intensity, extent, amount, or duration. In contrast to this analytical frame of mind, consumers in an affective test act in a much more integrative fashion. They perceive a product as a whole pattern. Although their attention is sometimes captured by a specific aspect of a product, particularly if the product is bad, unpleasant etc. their reactions to the product are often immediate and based on sensory stimulation it is expressed as liking or disliking. This occurs without a great deal of thought. Consumers are effective in giving opinion based on the integrated pattern of perceptions. In such consumer tests, participants must be chosen carefully to make sure that the results will generalize to the population of interest. Participants should be regular users of the product, since they are most likely to form the target market and know about the similar type of products. They possess reasonable assumptions and a frame of reference within which they can form an opinion relative to other similar products they have tried. The analytic/hedonic distinction gives rise to some highly important rules of thumb and some warnings about matching test methods and respondents. It is unwise to ask trained panellists about their preferences or whether they like or dislike a product. They have been asked to assume a different, more analytical frame of mind and to place personal preference aside. However, they have not necessarily been selected to be regular buyers of the product, so they do not consider as a part of target population to which one would like to generalize hedonic test results. A common analogy here is to an analytical instrument. You would not ask a gas chromatograph or a pH meter whether it liked the product, so why ask your analytical descriptive panel (O'Mahony, 1979).

Conversely, problems arise when consumers are asked to furnish very specific information about product attributes. Consumers not only act in a non-analytic frame of mind but also often have very fuzzy concepts regarding specific attributes, for example confusion in sour and bitter taste of the product. Individuals often differ markedly in their interpretations of sensory attribute words on a questionnaire. While a trained texture profile panel has no trouble in agreeing how cohesive a product is after chewing, we cannot expect consumers to provide precise information on such a specific and technical attribute. In summary, we avoid using trained panellists for affective information and we avoid asking consumers about specific analytical attributes. Related to the analytic–hedonic distinction is the question of whether experimental control and precision are to be maximized or whether validity and generalizability to the real world is necessary. Many a time there is a trade-off between the two and is difficult to maximize both simultaneously. Analytic tests in the lab with specially screened and trained judges are more reliable and lower in random error than consumer tests. However, we give up a certain amount of generalizability to real-world results by using artificial conditions and a special group of participants. Conversely, in the testing of products by consumers in their own homes we have not only a lot of real-life validity but also a lot of noise in the data. O'Mahony (1988) has made a distinction between sensory evaluations Type I and Type II. In Type I sensory evaluation, reliability and sensitivity are key factors, and the participant is viewed much like an analytical instrument used to detect and measure changes in a food product. In Type II sensory evaluation, participants are from the consuming population, and they may evaluate food under more naturalistic conditions. Their emphasis here is on prediction of consumer response. Every sensory test falls somewhere along a continuum where reliability versus actual-life inference are in a potential trade-off relationship. This factor must also be discussed with consumers to see where their emphasis lies and what level of trade-off they find comfortable.

Discrimination tests involve choices and counting numbers of correct responses. The statistics derived from the binomial distribution or those designed for proportions such as chi-square are appropriate. Conversely, for most scaled data, we can apply the familiar parametric statistics appropriate to normally distributed and continuous data, such as means, standard deviations, *t*-tests and analysis of variance. Every time selecting an appropriate statistical test is not so easy therefore, sensory specialists are wise to have thorough training in statistics and to involve statistical and design specialists in a complex project in its earliest stages of planning. Occasionally, these central principles will fail so, they should not be put aside as a matter of bare expediency or cost savings and never without a logical analysis. For instance use of a discrimination test before consumer acceptance. Although our ultimate interest may lie in whether consumers will accept a new product variation, we can conduct a simple difference test to see whether any change is perceivable at all. The logic in this sequence is the following: if a screened and experienced discrimination panel cannot tell the difference under carefully controlled conditions in the sensory laboratory, then a more heterogeneous group of consumers are unlikely to see dissimilarity in their less controlled and more variable world. If no difference is perceived, there can logically be no systematic preference. So a more time consuming and costly consumer test can sometimes be avoided by conducting a simpler but more sensitive discrimination test first. The added reliability of the controlled discrimination test provides a "safety net" for conclusions about consumer perception. Of course, this logic is not without its pitfalls—some consumers may interact extensively with the product during a home use test period and may form stable and important opinions that are not captured in a short duration laboratory test, and there is also always a possibility of a false negative result (the error of missing a difference). MacRae and Geelhoed (1992) describe an interesting case of a missed difference in a triangle test where a significant preference was then observed between water samples in a paired comparison. The sensory professional must be aware that these anomalies in experimental results will sometimes arise, and must also be ready for the reasons why they occur.

Sensory data and human perception of food

Human perceptions regarding foods and consumer products are based on the results of complex sensory and interpretation processes. Perceptions of such multidimensional stimuli as conducted by the parallel processing of the human nervous system are difficult or impossible to predict from instrumental measures. In many cases instruments lack the sensitivity of human sensory systems (eg. smell). Instruments lack the ability to mimic the types of peri-receptor filtering that occur in biological fluid like saliva or mucus that can cause chemical partitioning of flavour materials. Instrumental assessments give values that miss an important perceptual activity that is the interpretation of sensory responses by the human brain before responding. The brain lies interposed between sensory stimuli and generation of responses that form data. Only human sensory data provide the best models for how consumers are likely to perceive and respond to food products in real life. Sensory data is analyzed and interpreted to form predictions about how products have changed during a product development program. In the food and consumer products industries, these changes arise from three important factors: ingredients, processes, and packaging. A fourth consideration is often the way a product ages, in other words its shelf life, but we may consider shelf stability to be one special case of processing, albeit usually a very passive one (but also consider products exposed to temperature fluctuation light-catalyzed oxidation, microbial contamination, and other "abuses"). Ingredient changes occur due to various reasons. They can be included to improve product quality, to reduce costs of production, or simply because a certain supply of raw materials has become unavailable. Any change in the processing arise the problems in quality parameters like sensory, nutritional, microbiological stability factors, to reduce costs or to improve manufacturing productivity.

Packaging changes arise from considerations of product stability or other quality factors, e.g., a certain amount of oxygen permeability may insure that a fresh beef product remains red in colour for improved visual appeal to consumers. Packages function as carriers of product information and brand name. Packaging and print ink may cause changes in flavour or aroma of the product and sometimes transfer of off-flavour into the product. Package also functions as an important barrier to oxidative changes, to the potentially deleterious effects of light-catalyzed reactions, and to microbial infestations and other nuisances.

Consumer sensory tests and market research services

Another challenge to the effective communication of sensory results concerns the resemblance of sensory data to those generated from other research methods. Problems may arise as a result of apparent similarity of some sensory consumer tests to those conducted by marketing research services. However, some important differences exist. Sensory tests are almost always conducted on a blind-labelled basis. That is, product identity is usually obscured other than the minimal information that allows the product to be selected in the proper category (e.g., cold breakfast cereal). In contrast, marketing research tests often deliver explicit concepts about a product—label claims, advertising imagery, nutritional information, or any other information that may enter into the mix designed to make the product conceptually appealing (e.g., bringing attention to convenience factors in preparation).

In a sensory test all the important biasing factors are removed in order to collect the information based on sensory properties only. In a scientific inquiry, isolation of the variables of interest (ingredients, processing, packaging changes) and assessing sensory properties as a function of these variables, and not as a function of conceptual influences is important. This is done in order to minimize the effects of a larger cognitive load of expectations generated from complex conceptual information. There are many potential response biases and task demands that are entailed in "selling" an idea as well as in selling a product. Participants often like to please the experimenter and give results as per the expectations of the consumers. Plenty of literature is available on the effect of factors such as brand label on consumer response. Product information interacts in complex ways with consumer attitudes and expectancies.

Sensory evaluation is conducted to inform product developers about whether they have met their sensory expectations in terms of perception of product characteristics. This information can only be obtained when the test method is as free as possible from the influence of conceptual positioning. The product developer has a right to know if the product meets its sensory goals just as the marketer needs to know if the product meets its consumer appeal target in the overall conceptual, positioning, and advertising mix. However, in case of product failures, strategies for improvement are never clear without both types of information. Sometimes the two styles of testing will give apparently conflicting results (Oliver, 1986). But the situation is never that one is "correct" and the other is "wrong." These are simply different types of evaluations and are even conducted on different participants. For example, taste testing in market research tests may be conducted only on those persons who previously express a positive reaction to the proposed concept. This seems reasonable, as they are the likely purchasers, but bear in mind that their product evaluations are conducted after they have already expressed some positive attitudes and people like to be internally consistent. However, a blind sensory consumer test is carried out on a sample of regular product user, with no pre-screening for conceptual interest or attitudes. So they are not necessarily the same sample population in each style of test and differing results should not surprise anyone.

Apparent similarity of sensory testing and quality grading

A second arena of apparent comparability to sensory evaluation is with the traditional product quality grading systems that use sensory criteria. The grading of agricultural commodities has a significant influence on the movement to assure consumers of quality standards in the foods they purchase. Such techniques were widely applicable to simple products such as fluid milk and butter (Bodyfelt et al., 2008), where an ideal product could be largely agreed upon and the defects that could arise in poor handling and processing gave rise to well-known sensory effects. Further impetus brought from the fact that competitions could be held to examine whether novice judges-in-training were meeting the opinions of experts. Traditionally it is applied more in case of livestock grazing where a young person could judge a cow and receives awards at a state fair for acquiring to apply the same criteria and critical eye as the expert judges. There are noteworthy differences in the ways in which sensory testing and quality judging are performed. The commodity grading and the inspection tradition have severe limitations in the current era of highly processed foods and market segmentation. There are fewer and fewer "standard products" relative to the wide variation in flavours, nutrient levels (e.g., low fat), convenience preparations, and other choices that line the supermarket shelves. Also, one person's product defect may be another person's marketing boon, which is, the glue that did not work so well that gave us the ubiquitous post-it notes. Quality judging methods are poorly suited to research support programs. These techniques have been widely questioned on a number of scientific grounds. The defect identification in quality grading emphasizes root causes (e.g., oxidized flavour) whereas the descriptive approach uses more elemental singular terms to describe perceptions rather than to infer causes. In case of oxidized flavours, the descriptive analysis panel would use a number of terms (oily, painty, and fishy) since oxidation give rise to a number of qualitatively different sensory effects. Another notable difference with respect to mainstream sensory evaluation is that the quality judgments combine an overall quality scale (presumably reflecting consumer dislikes) with diagnostic information about defects, which is mostly like a descriptive analysis focusing on the negative aspects of products. In mainstream sensory evaluation, the descriptive function and the consumer evaluation would be clearly separate in two distinct tests with different respondents. Whether the response of a single expert can effectively represent consumer opinion is highly questionable at this time in history.

Significance of experimental design in sensory evaluation

Pineau et al., (2009) through his studies found that experimental design is a powerful tool to assess the impact of various parameters at a time while reducing the number of trials and samples to be analysed.

When sensory profiles (samples × panellists × attributes) are the outcome measures, it is recommended to analyze the data based on a two-steps approach building on the use of Fisher's least significant difference (LSD) as an intuitive and easy-to-interpret descriptive tool. This technique evaluates the impacts of the analysed parameters on the multivariate sensory profiles and compares these impacts with relevant LSD.

To illustrate the approach, a fractional factorial design with 32 experiments has been used to assess the potential of 11 parameters of a vending machine to modulate sensory properties for various coffee types and dosages. With minimal effort, this study shows that machine parameters influence appearance; modulate specific odour and flavour attributes, but only marginally impact taste/after taste. This study illustrates the high action-ability of a two-step data analysis approach, which has proven the efficiency to drive innovation and renovation in hundreds of cases covering multiple food and beverage categories.

This technique has been applied for many years for studies where in design of experiment has been implemented with monadic sensory profiling as an outcome. The application ranges from various product categories to various levels of design complexity (2–15 parameters, with 2–8 levels each) and has found to be very efficient to drive innovation in products and renovation while optimizing the use of limited resources. Apart from sensory, other types of data could also be analyzed similarly as long as it is possible to get a relevant estimate of the measurement variability and that this variability is considered important in comparison with the production (pilot/kitchen) variability. The replicates of (some of) the samples or external data can possibly give the estimate of the variability (e.g., a method validation procedure could provide information about the measurement variability of a given instrument).

Acevedo et al., (2018) developed a methodology to optimize mixtures of natural, non-caloric sweeteners—with the highest sweetness and the lowest bitterness—for carbonated soft drinks. To this end, and with the aid of a trained sensory panel, researchers first determined the most suitable mixtures of tagatose, sucrose, and stevia in a soft drink matrix, using a three-component simplex lattice mixture design. Then they developed a multi-objective thermodynamically-based decision model to this purpose. Results indicated that both, sucrose as well as tagatose, were effective in reducing stevia's bitterness. However, an increase of bitterness intensity was found above 0.23 g/L of stevia (sucrose equivalency or SE >5). Both, sensory analysis and multi-objective decision modelling identified similar optimal mixtures, corresponding to 23–39 g/L sucrose, 0.19–0.34 g/L stevia, and 34–42 g/L tagatose, depending on the desired sweetness/bitterness balance. Within this constrained area, a reduction of almost 60% of sucrose is possible to achieve in both approaches, keeping bitterness intensity low.

Current demand of low-calorie beverages has significantly risen as a result of consumer concerns on the negative effects of refined sugars present in carbonated soft drinks. Consequently, natural sweeteners, and their mixtures, are being increasingly used for these product developments. This study provides a methodology to optimize mixtures of natural, non-caloric sweeteners for preparing carbonated soft drinks with the lowest possible caloric content, while maintaining the tastiness—high sweetness and low bitterness—of full caloric ones, containing the bulk sweetener tagatose and the high-intensity sweetener stevia.

Effect of carriers on sensory profile

Kwak et al., (2019) investigated the effect of wheat origin (domestic versus imported) on the sensory profile, consumer acceptability and physicochemical properties of pan bread. The pan bread made from imported wheat flours were similar in terms of physicochemical and textural properties of those made with domestic wheat flours. The sensory profiles of breads showed difference depending on wheat origin. The sensory profile of pan bread made from imported wheat flours were similar to domestic wheat flour from large milling companies.

However, domestic wheat flour containing some bran originating from mid-size companies showed a strong cereal odour/flavour, yeasty flavour, powdery mouth feel aftertaste, and crumbliness. In consumer test, there was an increment of purchase intent (PI) for the highly accepted domestic samples between the blind and informed tests, but there was no change in PI for imported samples. Also consumers seemed to value the country of origin for domestic wheat flour only for the highly accepted samples. Therefore, high quality domestic wheat flour can meet the consumer expectations with increase PI.

It is found that the country of origin is one of the marketing points for the food industry to attract consumers. Consumer acceptance should be a major concern for the Food manufacturers. Although they advertise the use of domestic ingredients in their products, consumers would not change their PI dramatically for products with low consumer acceptability. Consumers prefer to give additional value to food products made from domestic ingredients only if these products meet certain level of acceptability. Unconditional ethnocentrism for food products made from domestic ingredients does not exist. Therefore for foods with domestic ingredients, the industry should consider that consumer acceptance of products are meeting consumer expectations in order to maximize the advantage of the country of origin

Myers et al., (2018) conducted study designs on non-nutritive sweeteners (NNS), sensory characteristics of NNS combined with different carriers should be better understood. The objective of the research was to determine an appropriate carrier (water or applesauce) for a high dose of specific NNS types (aspartame, sucralose, or stevia) to inform future metabolic and controlled feeding study designs. Adult participants ($n=67$) sampled six sweetener-carrier combinations (water and applesauce containing high concentrations of aspartame [30 g/oz], sucralose [8.25 g/oz], and stevia [6.75 g/oz]). Participants completed Check-All-That-Apply, affective attribute questionnaires, emotional terminology questionnaires and paired preference questionnaires. Applesauce was preferred (aspartame = 79.1%; sucralose = 83.6%; and stevia = 74.6%) significantly more than water for all sweetener types ($p < .001$) also the applesauce samples were found to have a higher mean acceptability scores. Participants preferred NNS delivered in applesauce rather than water. The study also found Applesauce likely as a more appropriate and tolerable carrier for high-dose NNS, which may contribute to designing effective intervention, studies with greater participant compliance.

This investigation is useful to understand which carrier (applesauce or water) is more suitable for participants consuming high doses of NNS also helps in designing future research studies to determine health outcomes associated with NNS consumption. In research studies, Applesauce is likely to be a more appropriate and tolerable carrier for high-dose NNS. Using an acceptable carrier for delivery of high doses of NNS may contribute to designing effective intervention studies, with greater participant compliance. An acceptable method for delivering NNS to research participants will be useful for designing intervention studies aimed at examining impact of NNS consumption on health outcomes.

Word co-occurrence network analysis for consumer perception

Yano et al., (2018) gave importance to the opinions of potential consumers of cosmetic products containing plant-derived ingredients and analysed the impact of personal characteristics on those opinions. A word association method was used to collect opinions about plant-derived cosmetics from 901 Japanese females. In order to identify the main themes in responses, including sensory properties and efficacy of plant-derived cosmetics, word co-occurrence network analysis was used. Multivariate probity analysis was conducted using the themes identified as dependent variables and personal attributes of respondents as independent variables. The results specified that while respondents viewed plant-derived cosmetics to be gentle to the skin and safe to use, they also viewed such products to be expensive and to have a slow effect. Some respondents, especially older women, were concerned about potential negative effects. However, older respondents, full-time workers, and respondents with high household income were less likely to view plant-based products as expensive.

This research demonstrates that word co-occurrence network analysis along with a community detection approach can be a useful tool for systematically identifying the main themes in responses to open-ended survey questions. By combining this method with discrete choice analysis it is possible to evaluate the influence of characteristics of respondents on the occurrence of the themes identified. The use of these two approaches in the study provided valuable insights on the opinions of potential Japanese consumers about plant-derived cosmetics also the impact of demographic and socioeconomic characteristics of individuals on those opinions. Similar methods can be used to study the effect of demographic and socio economic factors that affect consumers liking and decision making in purchasing food products also.

Rocha et al., (2018), evaluated consumers' perception regarding frankfurter sausages with different healthiness attributes. Descriptions of different frankfurter sausages were used (traditional, with natural antioxidants, with sodium reduction, with fat reduction, source of dietary fibre, and with omega 3) as stimuli. All consumers evaluated the six stimuli and conducted a test of word association prior to the evaluation of the acceptance of each concept, along with the familiarity level, consumers' emotional responses, and their perceptions of food with healthiness attributes. The category "healthiness" presented the highest frequency mentions for the stimuli with healthier attributes. The negative associations related to the traditional frankfurter sausages were unhealthy, fat/cal, and high blood pressure. Regarding the emotional evaluations, significant differences were observed among different types of sausage. It is possible to arrive at a conclusion that different types of frankfurter sausage promoted distinctive associations and emotions when recurring categories were considered, suggesting that the development and promotion of these types of products must consider all the particularities obtained through these results.

There is a trade potential for the development of healthier meat products, because of the exposure of traditional meat products in the media regarding possible adverse health effects. The information and uniqueness obtained in this study on the perception of consumers with respect to sausage with various attributes of healthiness can direct the industry in the development and promotion of these products. Besides that, these results indicate the importance of carrying out research with product concepts to evaluate consumer perception in order to develop and launch those products in the consumer market.

Rojas-Rivas et al., (2019) shows demographic changes in contemporary societies has promoted sectors of consumers being informed about the food they buy for health reasons, as functional foods, whose demand has grown in diverse countries as Mexico, identified as one of the emergent market with the highest consumption of these products in Latin America. However, there is little knowledge on the attitudes of Mexican consumers toward functional foods. The aim was to research on the perception of functional foods by Mexican urban consumers using the free word association technique. A total of 610 persons were asked the three first words that came to their minds for the stimulus "functional food." Twenty-three categories were grouped in nine dimensions, the most crucial were: Health, Nutrition, and Foods and Nutrients. Differences due to age and schooling level were found in dimensions and categories. Consumers are pretty familiar with idea of "functional food" that explains growth in this market. The promotion of functional foods should take into consideration the demands that specific groups of consumers have for these products.

This work is a first approach on the perception of Mexican urban consumers toward functional foods, with results that may be useful for public or private bodies. As indicated in works conducted in other parts of the world, demographic and socio-economic characteristics are recognized as significant determinants; the understanding toward functional foods is positively associated with educational achievement, age and women. Taking into consideration the diabetes and overweight problems of the Mexican population, research results indicate an opportunity for producers of functional foods for an adequate development, improvement, and promotion of these products to meet specific perceptions of each target group of consumers.

Importance of developing and validating lexicon for discriminate analysis

He et al., (2018) stated the development of lexicon for sufu products, a subtype of fermented soybean curd products that are red coloured. Information available regarding the sensory properties of red sufu products is very limited. Hence, the objectives of this investigation were to develop and validate a lexicon in order to profile the sensory properties of red sufu products. In this study, 10 qualified panellists were selected to develop a list of descriptors using 8 randomly selected commercial red sufus available from local market. Principal component analysis (PCA) was used as a method to reduce the number of descriptors from the initial 117 terms to final 15 descriptors. Furthermore, 12 red sufu samples were selected to confirm the red sufu lexicon. Both PCA and agglomerative hierarchical cluster analysis validated that the newly developed lexicon could discriminate among red sufu samples.

The developed lexicon in this study for red sufus will be useful for researchers and manufacturers to identify and describe similarities and differences of different commercial red sufus to help in their red sufu product development. Moreover, it will be helpful to correlate with the instrumental data to strengthen the flavour quality control in the red sufu products.

Correlating the different quality characteristics with consumer hedonic perception

Kim et al., (2018) determined the quality parameters of Doenjang samples prepared by different methods and established the relationship with consumer acceptability. Doenjang is fermented bean paste product popular in Korea. Consumer acceptance testing on these Doenjang samples was conducted in different locations of South Korea (north and southeast). No differences in consumer preference were found based on the testing location. But differences were found in physiochemical quality parameters of the samples from different regions. Consumer preference in the northern region was driven by color-related attributes while consumer preference in the south-eastern region was driven by the sweetness of the Doenjang samples. This study provided the potential to use objective physiochemical quality control-related data to predict the consumer preferences of Doenjang samples and also reported the different quality attributes that drive consumer preferences, based on the regions. Consumer preferences of the Doenjang samples were predicted according to the quality-related attributes of traditional Doenjang and provided information to strategically design the Doenjang products according to the consumers' preferences in different regions of Korea. This study gives valuable inputs on regional preferences for food and the sensory perception being affected by physiochemical characteristics of the product.

Correlating product rheology with sensory testing

Zhong et al., (2018) established the correlation between measured and sensory viscosities of thickened water and thickened milk which are used for dysphagia management through rheological studies and sensory tests. Magnitude estimation (ME) and triangle tests were used for the sensory evaluation. The perception of thickened water and thickened milk both obeyed the Stevens' power law according to the results of ME given by the participants. It was found in the triangle assessments that within the proposed nectar- and honey-thick viscosity ranges, participants were able to correctly discriminate increase in shear viscosity of thickened water and thickened milk. Results of both ME and triangle tests suggested that human abilities in detecting the viscosity changes of thickened water and thickened milk by finger tactile sensory were similar within nectar- and honey-thick viscosity range in spite of their different compositions and rheological properties. These results show how to correlate rheological data with sensory data during development of beverages and liquid foods.

How results of one type can be translated to another

Hautus et al., (2018) explained A-Not-A design involves two stimuli presented multiple times in a block of trials: a reference stimulus (*A*) and a comparison stimulus (*B*). The combined A-Not-A design employs *A* and several levels of *B* in a block of trials. Both designs were compared in ice tea with five levels of sucrose. Six judges were assessed for sensitivity, including their overall sensitivity, their average sensitivity across four replicated blocks, and the variability in sensitivity across those blocks. The pair wise design gave higher mean sensitivity, but also higher variability, than the combined design. A secondary analysis considered fewer trials in the combined design, such that sensitivity assessments were based on the same number of trials as for the pair wise design. The pattern for sensitivity within each design did not change, but variability was now comparable. This suggests the combined design yields lower variation for an equivalent expenditure of resource, or that the combined design yields similar levels of variation with reduced resourcing. The trade-off is getting slightly lower estimates of sensitivity. Additionally, the combined design produced sensitivity estimates significantly above chance performance levels when test stimuli were identical. However, the magnitudes of these estimates were small.

The A-Not-A test is used in research and development for a range of purposes. The results, expressed in sizes of difference between products, *d'*, and the variance of *d'*, give guidance for business decisions on whether products are similar enough to be distinguishable, or if they are perceivably different. The variance of *d'* is typically predicted from theoretical models. To study the difference between the real-life variance and the variance predicted from the theoretical models, an empirical study was conducted in which many A-Not-A data were collected from a small group of sensory panellists. According to results obtained the real-life variance is larger than predicted, so it might be that small but significant differences in tests might not really be significant anymore if we take into account this real-life variance. The comparison of paired versus combined designs generated judges the effect of test design on the outcome. The combined design, which is more efficient when there is more than one prototype to be compared, produces smaller *d'* values and variance than the paired design for the same number of collected trials. In this sense the study gives insights on how results from one test design can be translated to another.

Sensory evaluation of iron fortified products using fuzzy logic analysis

Shiby, (2008) conducted sensory study using fuzzy logic analysis for formulation of iron fortified drink mixes. Although dairy products are widely consumed, they are deficient in iron. With increase in food production, there is also requirement of advanced technologies for the effective shelf life enhancement of food. The value added convenience dairy products have got an increased demand. Drying is one of the superior preservation processes, which helps in increasing the shelf life of the product, but this leads to loss of texture and taste of products such as *dahi,* (Indian yoghurt) which can be a potential drink mix also as part of the space food. Research supports the prospects of dry *dahi* powder as a vehicle for iron fortification in the food beverage mix industry to modify the iron status of the population without compromising the nutritional and sensory properties, and also increases the scope of use of *dahi* mix as a drink for military and in the preparation of drinks while travelling. Further, it encourages the sensory evaluation using fuzzy analysis of other fermented special products which need to be explored and verified.

A plain reconstituted *dahi* drink and one standard brand of *dahi* drink available in the market were selected for the sensory evaluation of the three samples of iron fortified health drinks prepared using recirculatory convective air dried (RCAD) *dahi* powder.

The sweetened plain reconstituted *dahi* powder (containing *dahi* powder, sucrose and water) was marked as S1, iron fortified health drinks containing *dahi* powder, sucrose, ascorbic acid , water and varying levels of ferrous sulphate, were labelled as S2, S3 and S4 and the market *dahi* drink (control) was marked as S5. The primary objective of the study was to evaluate the sensory scores of the various *dahi* drink samples also to rank different samples according to their sensory qualities, using fuzzy analysis and to find out the optimum ferrous sulphate level that can be used in the formulation of iron fortified *dahi* drink. The ranking of the general sensory attributes color, flavor, homogeneity and taste were also done according to their importance in the overall acceptability of the *dahi* drinks. Buffalo milk *dahi* was prepared in laboratory using a standard procedure, which was subjected to thin layer drying in a recirculatory convective air dryer, and used for further analysis.

Sensory evaluation of reconstituted *Dahi* drinks

After obtaining agreement about the attributes of good quality *dahi* and the meaning of different terminologies used in sensory evaluation, a panel of 12 judges (five females, seven males) was created based on good health, interests in sensory assessment, product knowledge and their willingness to participate on a regular basis (Jaya and Das 2003). Before actual sensory evaluation, judges were familiarized with quality attributes of *lassi* (*dahi* drink) and advised to rinse their mouth with water between testing of consecutive samples. The judges were also asked to take two short sniffs of samples before tasting them and give the score for aroma first in score cards. The samples were rated as "Not satisfactory,""Fair,""Medium,""Good" and "Excellent." The set of observations were analyzed using Fuzzy comprehensive modelling of sensory scores. This sensory method has been successfully applied for mango drinks (Jaya and Das, 2003) and instant green tea powder (Sinija and Mishra, 2011).

Fuzzy comprehensive modelling of sensory scores

This method utilizes linguistic data obtained by subjective evaluation, and also accurate and precise data obtained using objective evaluation. Ranking of the *dahi* powder-based drinks was done using triangular fuzzy membership distribution function, which has been explained in detail by Das (2005). Sensory scores of the drinks' samples were obtained using fuzzy score card (Fig. 1), which were converted to triplets and used for estimation of similarity values used for ranking of samples. The important steps involved in the fuzzy modelling of sensory evaluation were: (1) calculation of overall sensory scores of *dahi* based drinks in the form of triplets; (2) estimation of membership function on standard fuzzy scale; (3) computation of overall membership function on standard fuzzy scale; (4) estimation of similarity values and ranking of the *dahi* based health drink samples; and (5) quality attribute ranking of *dahi* based iron fortified health drinks in general. Set of 3 numbers known as "triplet" is constituted to represent triangular membership function distribution pattern of sensory scales and the distribution pattern of 5 point sensory scales consists of "Not satisfactory/ Not at all important,""Fair/Somewhat important,""Medium/Important,""Good/Highly important" and "Excellent/Extremely important

On comparison of highest similarity values for all samples, their ranking was done as S5 > S1 > S2 > S3 > S4 showing the maximum acceptability for market *dahi* drink and decreasing acceptability of the *dahi* powder based health drinks with increasing levels of ferrous sulphate. The sweetened health drinks based on *dahi* powder containing 10-20 ppm of ferrous sulphate are of acceptable quality without addition of artificial flavour. The ranking for quality attributes for iron fortified health drinks in general is Colour> Taste> Flavour> Homogeneity.

Sensory evaluation of low sodium foods

High intake of salt has been associated with diseases like hypertension, cardiovascular disease and stroke (WHO, 2012). The main source of Na in diets is sodium chloride (NaCl) Quilaqueo *et al.*, (2015). In general, consumption of more than 6 g of NaCl per person per day is associated with health risks, such as high blood pressure levels, thereby contributing to an increased risk of cardiovascular diseases in the population (Choi *et al.*, 2014; Desmond, 2006; Ruusunen and Puolanne, 2005, WHO, 2003).

It has been recommended to adopt new methods for salt reduction, especially in meat products as its content is relatively higher in processed products such as 2% in sausages and other emulsion products and up to 6% in cooked cured products (Jimenez *et al.*, 2001). Some of the approaches to reduce salt content in meat products are substitution of all or part of NaCl with other chloride salts like potassium chloride (KCl), Lithium chloride (LiCl), and calcium chloride ($CaCl_2$) etc. With any of the above mentioned suggestions, use of protein and non protein binders, substitute of part of the NaCl with non chloride salts like phosphates, mineral salt mixture, ascorbic acid and citrates are also used.

The perception of saltiness has the greatest influence on sensory acceptability in meat products with low sodium content (Fellendorf *et al.*, 2015); JAR scales have been used to assess the saltiness intensity perceived by the consumer, in terms of being above or below ideal levels for that attribute. Additionally, reduction of acceptability with saltiness perception was associated with penalty analysis (Popper, 2014). According to Agudelo *et al.*, (2015), the penalty analysis helps to identify possible improvements in the sensory characteristics of products, by identifying ideal levels of attributes, and thereby increase their acceptability.

OAA and saltiness acceptability have been evaluated in restructured low sodium chicken nuggets and low sodium chicken jerky using hedonic scale and JAR scale to study the effect of NaCl replacement with potassium chloride and potassium lactate (Luckose *et al.*, 2015). In addition to hedonic scale rating test saltiness was evaluated using a five point Just-About-Right (JAR Scale) where 1=extremely too little salty than ideal, 2=slightly less salty than ideal, 3=just about right, 4=slightly saltier than ideal, 5=extremely too much salty than ideal). The samples were coded with 2 digit numbers and randomly presented to the sensory panellists.

Consumer study for low sodium salami

Sensory analysis of salami has been evaluated by Marcio Aurelio *et al.*, 2015. The sensory analyses were designed in accordance with ISO 8589 (ISO, 2007). The regular consumers of salami were placed in individual tasting booths, where they received instructions on the use of the scale, the nature of the products and the type of evaluation to be carried out. The seven salami samples were served to the panellist in monadic way and were evaluated under white light on disposable white plates coded with random three-digit numbers. During the sensory evaluation, an interval of two hours was allowed between the samples to avoid sensory fatigue. Mineral water and unsalted crackers were used to clear the palate between samples.The sensory analysis was carried out using 9 point hedonic scale and salt content of the product was evaluated using JAR scale.

Finally, consumers assessed their purchase intent to the tested products by using a five-point structured scale (1 = certainly will not buy, 3 = may or may not buy and 5 = certainly will buy). This study was approved and has been registered by the Ethics in Research Committee of ESALQ-USP under protocol No.104/2012.

Consumer acceptability and salt perception of various foods with low sodium content was studied by Mariloux Malherbe, 2003. Consumer sensory evaluation was carried out as per procedures described by Stone and Sidel (1993). Consumers rated food items in evaluation venues as per the methods given by Lawless and Heymann (1998). Acceptability was evaluated according to a 9-point Hedonic scale and salt perception was rated using a numerical rating 5-point hedonic scale. The study involved a series of twelve separate evaluation sessions. The evaluation sessions were scheduled for the late morning and the mid afternoon. When the consumers arrived at the evaluation venues, test instruction was given and they were asked to complete a short socio demographic questionnaire, which included questions regarding preparation of food and the addition of salt.

Each panellist evaluated two dishes with three sodium variations. Between the evaluation of each sample, water was used for the cleansing of the palate, and between the evaluations of each dish a break of fifteen to twenty minutes was given. Each sample was evaluated for both acceptability and salt perception. Samples (30g) were kept in transparent plastic containers on white plates and served. The serving temperature of 50-55°C for porridge, 60-65°C for mashed potatoes, 67-71°C for beef stew, and 61-64°C for vegetable soup was maintained. Samples were coded by means of three digital codes, to eliminate any assumption in respect to the order of the samples.

A study on sensory properties of two types of Asian foods using difference-from-control test has been carried out by Leong et al 2016. This study examined the effects of sodium reduction and flavour enhancers on the sensory profile of chicken rice and *mee soto* broth, the foods commonly consumed in Singapore. The study involved 24–29 trained panellists for this testing.

Training of panels

About 44 staff members of Singapore Polytechnic (SP), aged between 21 and 60 years, took part in the sensory screening tests. All the panellists were untrained when recruited. Prior to participation, the panellists were screened based on their ability to properly identify the basic tastes (sweet, salty, sour, bitter, and umami). Panellists were given six samples namely, water and solutions of sugar (16 g/L), sodium chloride (5 g/L), caffeine (0.5 g/L), citric acid (1 g/L), and monosodium glutamate (1.9 g/L). The panellists were also asked to rank the solutions in order from saltiest to least for the series of coded salt solutions with concentrations of 0.40%, 0.60%, 0.80%, 1.00%, and 1.50% (w/v) given to them. After the screening test, about 29 panellists were selected. The panellists were then asked to generate descriptors that well defined the two products under the direction of the panel leader. Through consensus, a set of agreed terms were generated by the panellists describing differences amongst the products. Descriptors were referenced and all panellists were trained on those descriptors (Table 3).

Table 3: Definitions of the sensory attributes of chicken rice and *mee soto* broth

Sensory attribute		Interpretation
Chicken rice	Overall flavor	Overall flavor associated with cooked chicken rice[1]
	Chicken flavor	Sensations associated with cooked chicken[1]
	Herbs/spices flavor	Sensations associated with garlic, shallot, ginger and pandan leaf [1]
	Salty	Tastes like sodium chloride
	Umami	The associated Monosodium glutamate[2] taste on the tongue
	Mouthfeel	Full flavor sensation in the mouth[1]
Mee soto broth	Overall flavor	Overall flavor associated with the cooked product
	Chicken flavor	Sensations associated with cooked chicken[1]
	Herbs/spices flavor	Sensations associated with garlic, shallot, ginger and pandan leaf[1]
	Salty	Taste associated with sodium chloride[2]
	Umami	Taste associated with monosodium glutamate[2]
	Mouthfeel	Full flavor sensation in the mouth[1]
	Sweet	Taste associated with sucrose[2]
	Sour	Taste on the tongue associated with citric acid[2]
	Bitter	Taste on the tongue associated with caffeine[2]

[1]Definitions as developed by the panelists.

[2]Definitions of Meilgarrd et al., (2007).

Tasting procedure

The testing was conducted in in the sensory laboratory at Food Innovation & Resource Centre (FIRC) at the Singapore Polytechnic. The requirements like individual booths, standard lightings, and temperature were fulfilled according to BSISO8589 (2010). To determine the sensory differences between the control and the test samples, Directional difference control test (DFC) was conducted.

For the sensory evaluation of the chicken rice, a total of 29 panellists and for *mee soto* broth, 24 panellists were selected. Both groups of panellists participated in 6 training sessions, followed by 6 taste-testing sessions. Each training sessions was conducted for 1 1/2 h. During the training, panellists were given references related to sensory attributes of the chicken rice and *mee soto*.

For taste evaluation of chicken rice, each panellist was given three sets of DFC samples of 50 g each. And for *mee soto* broth, each sample was about 35 g. Among these, one set was a comparison between a control and a blind control while the other two sets were a comparison between a control and one of the six reduced sodium samples. One low-sodium sample and the three blind controls were presented to sensory panellists, using a three-digit coded sample while the control sample was presented as "Control" in three separate sessions with randomised order. All the tests were conducted in duplicate. In order to check the homogeneity of the sample preparation and to monitor the performance of the panel, three blind controls were used. The panellists were requested to assess each coded sample, comparing it to the control (C), and to assess the degree of difference using a 11-point scale where 1 = much less intense than control, 6 = no difference from control, 11 = extremely intense than control. Before the commencement of the experiment, Ethics approval from Singapore Polytechnic Ethics Review Committee and written informed consent from all subjects were obtained.

Challenges in the sensory evaluation of reduced-salt products

As the previous examples have illustrated, sensory evaluation and consumer liking of salt-reduced products are a great challenge (Ulla Hoppu 2017). As the results are found to be related to other sensory characteristics of the product, different salt reduction options may work in different products—such as bread and meat products Jaenke *et al.*, (2017). Because the salt content may also affect, for example, texture characteristics and color (Silow *et al.*, 2016), the evaluation of salty taste intensity is usually not sufficient, but other sensory characteristics may be included.

The use of a trained panel or consumer tests should be carefully planned by considering the practical aspects of sensory evaluation. Trained panels can evaluate small differences in sensory quality and are needed in the initial phase of screening salt reduction options. However, consumer evaluations and preferences are important in order to obtain an overview of the market potential. A better approach would be both trained panel and consumers together. Different consumer groups should be represented in the tests—for example, gender, age, and smoking may be associated with salty food preferences Lampure *et al.,* (2015). The challenge currently is that consumers have so many different lifestyle variables, values, attitudes, and motives, characteristics that might affect their food and product preferences (Pohjanheimo *et al.,* 2010, De Boer and Schosler 2016).

The obtained results from sensory laboratory may also differ from those obtained under real tasting conditions, such as home experiments. Romagny *et al.,* (2016) employed different methodologies and combined home and laboratory evaluations of pleasantness and willingness to pay and evaluate the effect of reducing fat, salt and sugar in commercial food products. They found that, in most cases, the reformulated products maintained consumer acceptance. Willems *et al.,* (2014) reported on repeated in-home consumption on the liking of reduced-salt soups (regular-salt soup compared with 22% and 32% salt-reduced soups) and found no difference in liking the soups when consumed at home (twice weekly for five weeks). Also it was found that the initial liking cannot be considered as predictive of liking after repeating the in-home period.

Herbert *et al.* (2014) introduced an innovative approach to quantify memory for the sensory characteristics of a recently consumed food. And it was found that most people recalled a reduced-salt soup as having a higher salt concentration and suggested that remembered saltiness can be influenced due to ideal saltiness representations. They concluded that concentrations of salt could be reduced to a greater extent than might be predicted by a direct comparison between a regular and a reduced-salt product Herbert *et al.,* (2014).

Conclusion

Sensory evaluation is a quantitative science where numerical data are collected to establish lawful and specific relationships between human perception and product characteristics. It includes a set of test methods with guidelines and established techniques for product presentation, well-defined response tasks, statistical methods, and guidelines for interpretation of results. It gives essential information about human perception of product changes due to ingredients, processing, packaging, or shelf life. Researchers in the past have studies the correlations of sensory parameters with various physic- chemical and rheological properties of products. The development of lexicon for discriminate analysis ,sensory evaluation using fuzzy logic analysis are methods where linguistic descriptions of product attributes play a major role. Even though there are methods for using big number of untrained panellists, the vivid description of product quality require trained panellist services in sensory evaluation. The perceived intensities of the sensory characteristics of a product can be quantified using various sensory evaluation methods. Descriptive analysis has proven to be the most comprehensive and informative sensory evaluation tool. It is applicable to the characterization of a wide variety of product changes and research questions in the area of food product development. The information obtained can be related to consumer acceptance information and to instrumental measures using statistical techniques such as regression and correlation. The progress in sensory studies open possibilities for evaluating quality and consumer perceptions on various new categories of functional foods claiming health benefits to different target groups. In the arena of developing terrain specific foods for Armed forces, the challenges related to physiological and psychological changes in extreme environments need to be evaluated in relation to sensory perception of soldiers. State of the art in sensory evaluation needs to be explored in order understand the preferences while developing food products of different categories and age groups of consumers.

References

Acevedo, W. Ramírez-Sarmiento, C. A. and Agosin, E. (2018). Identifying the interactions between natural, non-caloric sweeteners and the human sweet receptor by molecular docking. *Food Chem.* 264, 164–171. doi: 10.1016/j.foodchem.2018.04.113

Adams, H. (1918). The Education of Henry Adams. The Modern Library, New York.

Agudelo, A. Varela, P. and Fiszman, S. (2015). Fruit fillings development: a multiparametric approach. *LWT - Food Science and Technology* 61: 564-572

Bodyfelt F.W. and Dave Potter. (2009). The sensory evaluation of dairy products. 167-190 Creamed Cottage Cheese.

Bodyfelt, F. W., Drake, M. A. and Rankin, S. A. (2008). Developments in dairy foods sensory science and education: from student contests to impact on product quality.

Bodyfelt, F. W., Tobias, J. and Trout, G. M. (1988). Sensory Evaluation of Dairy Products. Van Nostrand/AVI Publishing, New York.

Bodyfelt, F.W. Drake, M.A. Rankin S.A. (2008). Developments in dairy foods sensory science and education: From student contests to impact on product quality. *International Dairy Journal,* 18(7), 729-734

Brandt, M. A. Skinner, E. Z. and Coleman, J. A. (1963). Texture profile method. *Journal of Food Science*, 28, 404–409.

Brinberg, D. and McGrath, J. E. (1985). Validity and the Research Process. Sage Publications, Beverly Hills, CA

BS ISO 8589 (2010). Sensory Analysis – General guidance for the design of test rooms.

Caul, J. F. (1957). The profilic method of flvor analysis. *Advances in Food Research,* 7, 1–40.

Choi, Y.M. Jung, K.C. Jo, H.M. Nam, K.W. Choe, J.H. Rhee, M.S. and Kim, B.C. (2014). Combined effects of potassium lactate and calcium ascorbate as sodium chloride substitutes on the physicochemical and sensory characteristics of low-sodium frankfurter sausage. *Meat Science,* 96: 21-25.

Das, H. (2005). Food processing operations analysis. New Delhi: Asian Books.Faergestad, E. M., Molteberg, E. L., & Magus, E. M. (2000). Interrelationship ofprotein composition, protein level, baking process and characteristics of hearthbread and pan bread.*Journal of Cereal Science*, 31, 309e320.

De Boer, J. and Schosler, H. (2016). Food and value motivation: Linking consumer affinities to different types of food products. *Appetite*, 103, 95–104. doi: 10.1016/j.appet.2016.03.028. Dekker, New York, pp. 317–338.

Desmond, E. (2006). Reducing salt: a challenge for the meat industry. *Meat Science,* 74, 188-196.

Edgar, Rojas, R. Angelica Espinoza, O. Humberto, Thome,O. Sergio, Moctezuma, P. (2019). "Consumers' perception of amaranth in Mexico: A traditional food with characteristics of functional foods", *British Food Journal*, Vol. 121 Issue: 6, pp.1190-1202

Einstein, M. A. (1991). Descriptive techniques and their hybridisation. In Lawless H. T. and Klein B. P. *Sensory science, theory and applications in foods*. New York: Dekker.

Emily, A.M. Erin, M.P. and Valisa, E.H. (2018). The Reproducibility and Comparative Validity of a Non-Nutritive Sweetener Food Frequency Questionnaire. *Nutrients,*10(3), 334-340

Fellendorf, S. Sullivan, M.G. and Kerry, J.P. (2015). Impact of varying salt and fat levels on the physicochemical properties and sensory quality of white pudding. *Meat Science* 103, 75-82.

Hautus, M.J. van Hout, D. Lee, H.S. Stocks, M.A. Shepherd, D. (2018). Variation of *d'* estimates in two versions of the A-Not A task, 2018 VOL page

He, W. Chen, Y.P. and Chung, H.Y. (2018). Development of a lexicon for red sufu. *Journal of Sensory Studies* , e12461.

Helm, E. and Trolle, B. (1946). Selection of a taste panel. *Wallerstein Laboratory Communications*, 9, 181–194.

Herbert, V. Bertenshaw, E.J. Zandstra, E.H. and Brunstrom, J.M. (2014). Memory processes in the development of reduced-salt foods. *Appetite*, 83, 125–134. doi: 10.1016/j.appet.2014.08.019.

International Organization for Standardization [ISO]. 2007. ISO 8589. Sensory Analysis: General Guidance for the Design of Test Rooms. International Organization for Standardization, Geneve, Switzerland.

Jaenke, R. Barzi, F. McMahon, E. Webster, J. and Brimblecombe, J. (2017). Consumer acceptance of reformulated food products: A systematic review and meta-analysis of salt-reduced foods. *Food Science and Nutrition*, 57, 3357–3372. doi: 10.1080/10408398.2015.1118009.

Jasmine, L. Chinatsu, K. Evelyn, O. Jia, T. Hoi, and Mann, N.L. (2016). A study on sensory properties of sodium reduction and replacement in Asian food using difference-from – control test. *Food Science and Nutrition*, 4(3), 469–478.

Jaya, S. and Das, H. (2003). Sensory evaluation of mango drinks using fuzzy logic. *Journal of Sensory Studies,* 18(2): 163-176.

Jimenez-Colmenero, F. Carballo, J.C. and Cofrades, S. (2001). Healthier meat and meat products: their role as functional foods. *Meat Science*, 59, 5-13.

Jones, L. V., Peryam, D. R. and Thurstone, L. L. (1955). Development of a scale for measuring soldier's food preferences. *Food Research*, 20, 512–520.

Kwak, H.S. Kim, M.J. Kim, S.S. (2019). Sensory profile, consumer acceptance, and physicochemical properties of pan bread made with imported or domestic commercial wheat flour. *Journal of Sensory Studies*, 34(2),

Lampure, A. Schlich, P. Deglaire, A. Castetbon, K. Peneau, S. Hercberg, S. and Mejean C. (2015). Sociodemographic, psychological, and lifestyle characteristics are associated with a liking for salty and sweet tastes in French adults. *Journal of Nutrition*, 145, 587–594. doi: 10.3945/jn.114.201269.

Lawless, H.T. and Claassen, M.R. (1993). The central dogma in sensory evaluation. *Food Technology*, 47(6), 139–146.

Lawless, H.T. and Heymann, H. (1998). Sensory evaluation of food: Principles and practices. 2nd ed. New York. International Thomson Publishing.

Lawless, H.T. and Clalssen. M.R. (1993). Validity of descriptive and defect-oriented terminology systems for sensory analysis of fluid milk. *Journal of Food Science*, 58(1), 108-112

Luckose, F. Pandey, M.C. and Abhishek, V. (2015). Effect of Sodium Chloride Replacement on the Sensory and Physico-Chemical Properties of Restructured Chicken Jerky. *Asian-Australas Journal of Animal Science*. doi: 10.5713/ajas.15.0573.

MacRae, R.W. and Geelhoed, E. N. (1992). Preference can be more powerful than detection of oddity as a test of discriminability. *Perception and Psychophysics*, 51, 179–181.

Marcio, A. Almeida, N. Doris, M.V. Jair, S. Silva, P. Erick, S. Carmen, J. Contreras, C. (2015). Sensory and physicochemical characteristics of low sodium salami. *Scientia Agricola*, 347. (http://dx.doi.org/10.1590/0103-9016-2015-0096)

Mariloux, M. Corinna, M. Walsh, and Cay, A. Merwe, T. Gesinsekologie, V. (2003). Consumer acceptability and salt perception of food with a reduced sodium content. Journal name 31,

Meilgarrd, M. Civille, G. V. and Carr, B. T. (2007). Sensory evaluation techniques, 4th edn CRC Press Inc, Boca Raton, FL.

Mina, K.K, Hyun, J.C. Woosuk, B. (2018). Correlating physiochemical quality characteristics to consumer hedonic perception of traditional *Doenjang* (fermented soybean paste) in Korea. *Journal of Sensory Studies.*

Moskowitz, H. R. (1983). Product Testing and Sensory Evaluation of Foods. *Food and Nutrition*, Westport, CT.

O' Mahony, M. Thiemel, U. Goldstein R. (1988). The warm-up effect as a means of increasing the discriminability of sensory difference tests.

O'Mahony, M. (1979). Psychophysical aspects of sensory analysis of dairy products: a critique. *Journal of Dairy Science*, 62, 1954–1962.

O'Mahony, M. (1988). Sensory difference and preference testing: The use of signal detection measures. Chpater 8 In: H. R. Moskowitz (ed.), *Applied Sensory Analysis of Foods*. CRC, Boca Raton, FL, pp. 145–175.

O'Mahony, M. (1979). Shortcut signal detection measures for sensory analysis. *Journal of Food Science*, 44(1), 302-303

Oliver, T. (1986). The Real Coke, The Real Story. Random House, New York.

Peryam, D. R. and Swartz, V. W. (1950). Measurement of sensory differences. *Food Technology*, 4, 390–395.

Pineau, N. Schlich, P. Cordelle, S. Mathonniere, S. and Issanchou, A. (2009). Temporal dominance of sensations: Construction of the TDS curves and comparison with time-intensity. *Food Quality and Preference*, 20, 450-455

Pohjanheimo, T. Paasovaara, R. Luomala, H. and Sandell, M. (2010). Food choice motives and bread liking of consumers embracing hedonistic and traditional values. *Appetite*, 54, 170–180. doi: 10.1016/j.appet.2009.10.004.

Popper, R. (2014). Use of Just-About-Right in consumer research. In: Varela, P.; Ares, G. Novel techniques in sensory characterization and consumer profiling. CRC Press, Boca Raton, FL, USA.

Quilaqueo, M. Duizer, L. and Aguilera, J.M. (2015). The morphology of salt crystals affects the perception of saltiness. *Food Research International*, 76, 675-681.

Rocha, Y. J. P. Judite, L.G. Regina, L. F. and Marco, A. T. (2018). Evaluation of consumers' perception regarding frankfurter sausages with different healthiness attributes. *Journal of Sensory Studies*, 33(6) 1-14, e12468

Romagny, S. Ginon, E. and Salles C (2017). Impact of reducing fat, salt and sugar in commercial foods on consumer acceptability and willingness to pay in real tasting conditions: A home experiment. *Food Quality*, 56, 164–172. doi: 10.1016/j.foodqual.2016.10.009.

Ruusunen, M. and Puolanne, E. (2005). Reducing sodium intake from meat products. *Meat Science*, 70, 531-541.

Shiby, V.K. and Mishra, H.N. (2008). Modelling of acidification kinetics & textural properties in buffalo milk curd (Indian yoghurt). *International Journal of Dairy Technology*, 61(3): 284-289

Silow, C. Axel, C. and Zannini, E. (2016). Arendt E.K. Current status of salt reduction in bread and bakery products—A review. *Journal of Cereal Science*, 72, 135–145. doi: 10.1016/j.jcs.2016.10.010.

Sinija, V.R. and Mishra, H.N. (2011). Fuzzy analysis of sensory data for quality evaluation and ranking of instant green tea powder and granules. *Food and Bioprocess Technology*, 4(3): 408-416.

Stone, H. and Sidel, J.L. (1993). Sensory Evaluation Practices. 2nd ed. San Diego, California. Academic Press.

Stone, H. and Sidel, J. L. (2004). Sensory Evaluation Practices, 3rd ed.. Academic, San Deigo.

Stone, H. Sidel, J.L. Oliver, S. Woolsey, A. and Singleton R.C. (1974). Sensory evaluation by Quantitative Descriptive Analysis. *Food Technology*, 28 (1), 24-33

Szczesniak, A. S., Loew, B. J. and Skinner, E. Z. (1975). Consumer texture profil technique. *Journal of Food Science*, 40, 1253–1257.

Ulla, H. Anu, H. Terhi, P. Minna, R.P. Sari, M. Anne, P. and Mari, S. (2017). Effect of Salt Reduction on Consumer Acceptance and Sensory Quality of Food. *Foods*, 6(12), 103-110.

Willems, A.A. Hout, D.H. Zijlstra, N. and Zandstra, E.H. (2014). Effects of salt labelling and repeated in-home consumption on long-term liking of reduced-salt soups. *Public Health Nutrition*, 17, 1130–1137. doi: 10.1017/S1368980013001055.

World Health Organization [WHO]. (2003). Diet, Nutrition and the Prevention of Chronic Diseases. World Health Organization, Geneva, Switzerland. (WHO Technical Reports Series, 916).

World Health Organization [WHO]. (2012). Guideline: Sodium Intake for Adults and Children. World Health Organization, Geneva, Switzerland.

Yuki, Y. Eri, K.Y.O. and David, B. (2018). Examining the opinions of potential consumers about plant-derived cosmetics: An approach combining word association, co-occurrence network, and multivariate probit analysis. *Journal of Sensory Studies.*

2

Therapeutic Foods: An Overview

Farhath Khanum

Nutrition, Biochemistry and Toxicology Division
Defence Food Research Laboratory, Mysuru, Karnataka, India

Introduction

Therapeutic foods are products, which are multiutility food products that may be used as nutrition as well as medicine. A nutraceutical may be defined as a substance, which has physiological benefit or provides protection against chronic disease. Nutraceuticals are used to improve health, delay the aging process, prevent chronic diseases, increase life expectancy, or support the structure and/or function of the body. Oflate nutraceuticals are receiving considerable interest due to potential nutritional, safety and therapeutic effects. The examples of therapeutic foods are the foods used for feeding of malnourished children on an emergency basis or to supplement the diets of persons with special nutrition requirements, such as the elderly. Therapeutic foods can also form a part of treatment/prevention/reduction of other health conditions such as diabetes, cardiovascular disease, obesity, anaemia and any other lifestyle disorders.

The incidences of diet-related diseases are progressively increasing due to greater availability of calorie-rich foods and a sedentary lifestyle. Obesity, diabetes, atherosclerosis, and neurodegeneration are major diet-related diseases. Nutraceuticals and functional foods represent a novel therapeutic approach to prevent or attenuate diet-related diseases.

There is always a requirement for cheaper formulations using locally available ingredients that are processed in a safe, reliable, and financially sustainable methods. Some of the literature available on health foods for various diseases is summarized below. However, since the burden of diabetes in India is staggering. In 2017, 72 million cases of people suffering from diabetes were reported and the figures are expected to reach 134 million by 2025, this chapter more emphasis is given on the safe, natural treatment/reduction of diabetes by herbal means.

Nutraceuticals and Diseases

Cardiovascular diseases: Cardiovascular diseases (CVD) is the name for the cluster of disorders of the heart and blood vessels including hypertension (high blood pressure) coronary heart disease (heart attack), cerebrovascular disease (stroke), heart failure peripheral vascular disease, etc. CVD has become the leading cause of death in the world. Majority of the cardiovascular diseases are preventable and controllable. It has been reported that high mortality in cardiovascular disease is associated with a low intake of vegetables and fruits (Rissanen et al 2003, Temple et al 2003). A protective role of diets rich in fruits and vegetables against CVD has been identified in many studies (Hu and Willett 2002). Nutraceuticals in the form of antioxidants, dietary fibers, omega-3 fatty acids (n-3 PUFAs), vitamins, polyphenols and minerals are recommended/suggested along with physical exercise for prevention and treatment of CVD (Magrone et al 2013). It has been demonstrated that the polyphenols present in grapes and wine modulate cellular metabolism and cell signaling, which coincides with reducing arterial disease. Nutraceuticals with calcium channel blocking activity (leading to antihypertensive activity) include α-Lipoic acid, magnesium, pyridoxine (vitamin B6), Vitamin C, N-acetyl cysteine, Hawthorne, Celery, ω-3 fatty acids etc. Fruits, spices etc have been shown to be excellent in reducing/preventing cardiovascular diseases (Balasuriya and Rupasinghe 2012, Chen et al 2019,. Majewska-Wierzbicka and Czeczot 2012, Kulczyński and Gramza-Michałowska 2016, Rahimlou et al 2019, Mao et al 2019, Tsui et al 2018).

Obesity

Obesity is defined as an unhealthy amount of body fat and is a well-established risk factor for many diseases like hypertension, hyperlipidemia, angina pectoris, congestive heart failure, respiratory disorders, renal vein thrombosis, osteoarthritis, cancer and reduced fertility. Obesity is now a global health problem, with around 315 million people estimated to fall into the WHO-defined obesity categories. One of the primary causes for this rapid rise in obesity rates is the increased availability of high-fat and energy-dense foods. Excessive consumption of energy-dense foods (like snacks, processed foods and drinks) can lead to weight gain, which calls for a limit in the consumption of saturated and trans fats apart from sugars and salt in the diet. Nutraceuticals that can increase energy expenditure and/or decrease caloric intake are appropriate for reduction of body weight. Buckwheat seed proteins have beneficial effect in obesity and constipation acting similar to natural fibers present in foods. 5-hydroxytryptophan and green tea extract have also been found to promote weight loss, while the former decreases appetite, the latter increases energy expenditure. Current status of nutraceuticals in obesity, a blend of glucomannan, chitosan, fenugreek, G sylvestre, and vitamin C in the dietary supplement significantly reduced body weight and promoted fat loss in obese individuals. However, further studies are required to establish a long term efficacy and adverse effects.

An effective nutraceutical is the one that can increase energy expenditure and/or decrease caloric intake and is desirable for body weight reduction. Stimulants, such as caffeine, ephedrine, chitosan, ma huang-guarana, and green tea have been shown to be effective in facilitating body weight loss. However, their use is controversial due to their side-effects. A number of spices, herbs and dietary fibre have been shown to promte weight loss (Yang et al 2018, Darooghegi Mofrad et al 2019a,b, Jane et al 2019, Salehi et al 2019).

Cancer

Cancer is a major public health problem in developing countries Cancer rates are increasing and there would be 15 million new cases in the year 2020 . according to the World Cancer Report that is, a rise in 50%. A healthy lifestyle and proper diet can definitely help in prevention of cancer.

Plants rich in carotenoids, daidzein, biochanin, isoflavones and genistein have been shown to posses antioxidant activities and effective in cancer prevention (Delphi and Sepehri 2016, Mariadoss et al 2019, Fu et al 2019, Ko et al 2018, Kumar et al 2018, Micheli et al 2018, Wang et al 2018, Alsherbiny et al 2019, Muhammad et al 2018, Sharma et al 2017, Bassiri-Jahromi 2018, Salehi et al 2019).

Allergies

Allergies also known as allergic diseases, is a condition in which the body has an exaggerated response to either a drug or food due to hypersensitivity of the immune system to typically harmless substances. These diseases include hay fever, food allergies, atopic dermatitis, allergic asthma and anaphylaxis.

Flavonoids /polyphenolic compounds especially quercetin are natural antihistamine and oppose the actions of the histamine. Histamines are the root cause for allergic and inflammatory reactions. Flavonoids can help reduce the inflammation that are caused due to hay fever, bursitis, gout, arthritis, and asthma. Herbs and their extracts, such as garlic, (which contain allicin), green tea (containing catechins and bioflavonoids such as quercetin, hesperidin, rutin) are effective antioxidants (Maleki et al 2019, Goldoni et al 2019, Tanaka et al 2019).

Alzheimer's Disease (AD)

Alzheimer's disease is characterized by progressive dementia with memory loss as the major clinical manifestation. In 1996, almost 4 million people in the United States were clinically diagnosed with AD; which is expected to triple in the next 50 years. Women are more affected than men at a ratio of almost 2:1 could be due in part to the larger population of women who are over 70. Evidence strongly suggest that oxidative stress is causal factor related to a number of neurodegenerative disorders including Alzheimer's disease. There's growing evidence that what you eat can affect the health of your brain. Nutraceutical antioxidants like β-Carotene, curcumin, lutein, lycopene, turmerin etc may have positive effects on these specific diseases by neutralizing the negative effects oxidative stress, mitochondrial dysfunction, and different forms of neural degenerations. A great deal of research has pointed to deleterious roles of metal ions in the development of Alzheimer's disease, by the augmentation of oxidative stress by metal ion. The growing trend in nutraceutical intake is a result of the belief that they delay the development of dementias like the Alzheimer's disease (Kim et al 2018, Botchway et al 2018, Solfrizzi et al 2017, Wesselman et al 2019, Simunkova et al 2019, Zhang et al 2019).

Parkinson's disease

Parkinson's disease is a disorder of the brain that is caused by nerve damage in certain regions of the brain resulting in muscle rigidity, shaking, and difficult walking, usually occurring in mid to late adult life. Vitamin E, creatine, glutathione etc. in food have been shown to be protective against Parkinson's disease. Nutritional supplements have also shown some promising results in preliminary studies, it is important to remember that there is no sufficient scientific data to recommend them for Parkinson's disease at present (Perez-Pardo et al 2017, Zhao et al 2019, Ciulla et al 2019, Wang et al 2016).

Diabetes

Diabetes is a metabolic syndrome characterized by deranged carbohydrate metabolism due to malfunctioning of β- cells of pancreas, impaired insulin secretion, increased oxidative stress and insulin resistance in the peripheral tissues leading to abnormally high blood sugar level generally known ashyperglycemia) and high lipid levels, therefore it is generally referred to as a metabolic disorder. The causes could be hereditary, increasing age, poor diet, imperfect digestion, obesity, sedentary lifestyle, stress, drug-induced, infection in pancreas, hypertension, high serum lipid and lipoproteins, less glucose utilization and other factors. It is estimated that the diabetic patients in India will increase by 195% in the near future.

Normal glucose absorption and metabolism

In brief, the food we consume is broken down to monomer units, glucose in case of carbohydrates and gets absorbed absorbed into the blood stream.

Absorption of glucose from the small intestine: pathways and mechanisms

Absorption from the small intestine involves diffusion, active transport and co-transport. Glucose is absorbed from the small intestine by a mechanism that involves the co-transport (simultaneous transport) of sodium ions. The transport protein carries a glucose molecule with a sodium ion into the epithelial cell by facilitated diffusion. Facilitated diffusion only works if there is a concentration gradient. This is maintained by the active transport of sodium ions out of the epithelial cell; glucose molecules pass out by facilitated diffusion. Absorption is efficient because; microvilli increase the surface area of the epithelial cells for absorption, the distance from the lumen of the intestine to the capillaries is short and there is always a concentration gradient between the lumen of the small intestine and the epithelial cells.

Sensing of glucose by intestinal cells

Epithelial cells lining the inner surface of the intestinal epithelium are in direct contact with a luminal environment that keeps varying with the type of diet. It has been said that the intestinal epithelium can sense the nutrient composition of lumenal contents. It is only recently that the nature of intestinal nutrient-sensing molecules and underlying mechanisms has been elucidated (Tolhurst et al 2012, Breer et al 2012). There are a number of nutrient sensors expressed on the luminal membrane of endocrine cells that are activated by various dietary nutrients. It has been shown that the intestinal glucose sensor, T1R2+T1R3 and the G-protein, gustducin are expressed in endocrine cells (Shirazi-Beechey et al 2011).

Glucose transport

Facilitated diffusion

Molecules/particles upto a molecular weight of 100 can readily diffuse through the membranes but molecular weight of glucose is 180 yet it passes through membranes easily through a facilitated diffusion, a type of passive transport supported by the integral membrane proteins, which span the width of the membrane. There are specific transmembrane proteins for different ions and molecules. For transporting glucose there are different types of facilitators called glucose transporters (GLUT 1-10). GLUT1 and GLUT3 are located in the plasma membrane of cells throughout the body, and are responsible for maintaining a basal rate of glucose uptake. Basal blood glucose level is approximately 5mM. The Km value (an indicator of the affinity of the transporter protein for glucose molecules; a low Km value suggests a high affinity)of the GLUT1 and GLUT3 proteins and is 1mM; that means GLUT1 and GLUT3 have a high affinity for glucose and uptake from the bloodstream is constant.

GLUT2 in contrast has a high Km value (15-20mM) and therefore a low affinity for glucose. They are located in the plasma membranes of hepatocytes and pancreatic beta cells, and their high Km allows for glucose sensing; rate of glucose entry is proportional to blood glucose levels.

GLUT4 transporters are insulin sensitive, and are found in adipose tissue and muscles. As muscle is one of the principle storage site for glucose and adipose tissue for triglyceride (into which glucose can be converted for storage), GLUT4 is important in post prandial uptake of excess glucose from the bloodstream. Moreover, several recent papers show that GLUT 4 is also present in the brain. The drug metmorphin phosphorylates GLUT4, thereby increasing its sensitivity to insulin.

Secondary active transport

Facilitated diffusion can occur between the bloodstream and cells as the concentration gradient between the extracellular and intracellular environments are such that no ATP hydrolysis is required.

However in the kidney, glucose is reabsorbed from the filtrate in the lumen of the renaltubule, where it is at a relatively low concentration, passes through thesimple cuboidal epithelia lining the kidney tubule, and into the bloodstream where glucose is at a comparatively high concentration. Therefore the concentration gradient of glucose opposes its reabsorption, and the secondary active transport of glucose in the kidney is Na^+ linked; therefore a Na^+ gradient must be established. This is achieved through the action of the Na^+/K^+ pump, the energy for which is provided through the hydrolysis of ATP. Three Na^+ ions are extruded from the cell in exchange for two K^+ ions entering through the intra membrane enzyme Na^+/K^+ ATPase ; this leaves a relative deficiency of Na^+ in the intracellular compartment. Na^+ ions diffuse down their concentration gradient into the columnar epithelia, co- transporting glucose. Once inside the epithelial cells, glucose re enter the bloodstream through facilitated diffusion through GLUT2 transporters.

Hepatic portal circulation

Blood from the spleen, pancreas, stomach and intestines enters the liver via the hepatic portal vein. Blood from the intestines is very high in glucose. Liver cells remove excess glucose, amino acids, toxins, bacteria. Blood leaves the liver by means of the hepatic vein, now with normal levels of glucose.

Role of pancreas

The pancreas has two major types of tissues: the *acini* (secrete digestive juices to duodenum) and the *islets of Langerhans* (secrete insulin and glucagon directly into blood). Human pancreas has 1 – 2 million islets of 0.3 mm in diameter, although the islets account for only about 1% of the pancreatic cell population Islets of Langerhans contain 3 types of hormone-secreting cells *Alpha* cells (25%), *Beta* cells (60%), *Delta* cells (10%) and F cells (PP cells) (5%).

- The islets are organized around small capillaries into which the cells secrete their hormones
    - *Alpha* cells secrete *glucagon that* elevates blood glucose concentrations and *Beta* cells secrete *insulin that* reduces blood glucose concentrations
    - *Delta* cells secrete *somatostatin*- slows the rate of food absorption and digestive enzyme secretion
- F cells/PP cells secrete pancreatic polypeptide

- Close interrelation among various cell types allow regulation of secretion of some hormones by other hormones: *insulin inhibits glucagon* secretion, *somatostatin inhibits insulin and glucagon* secretion

Secretion of insulin

Insulin is a peptide hormone consisting of 51 amino acids and is produced in islets of Langerhans in the pancreas. High concentrations of blood sugar trigger the release of insulin. The main role of insulin is transport of glucose from blood to cells by increasing the rate of the glycolytic pathway; the process where glucose is converted to other carbohydrates that are used in the urea cycle or for fatty-acids metabolism and ATP production. Glucose initially enters the β- cells of pancreas with the help of transporters known as GLUT 2 that carry the sugar across the cell membrane, so the action of glucose seems to be responsible for most of the changes that follow. Initially, an increase in the ATP/ADP ratio (Cook and Hales 1984) is observed. This increase selectively affects ATP-sensitive K^+ ion channels by closing them and thus preventing K^+ passage across the cell membrane (Figure below). This decreases the charge difference that already exists between the inside and outside of the membrane -an effect known as *depolarization*. At the same time, an increasing electrical conductivity is observed, driving the opening of voltage-sensitive calcium-ion channels (Cook and Hales 1984, Gauthier and Wollheim 2008). The opening of this channel allows the entrance of Ca^{2+} ions into the membrane which initially bind certain proteins such as calmodulins or synaptotagmins. As a result, this drives the densely packed vesicles of insulin to open and release insulin, by emerging through the membrane of the β-pancreatic cells (Gauthier and Wollheim 2008).

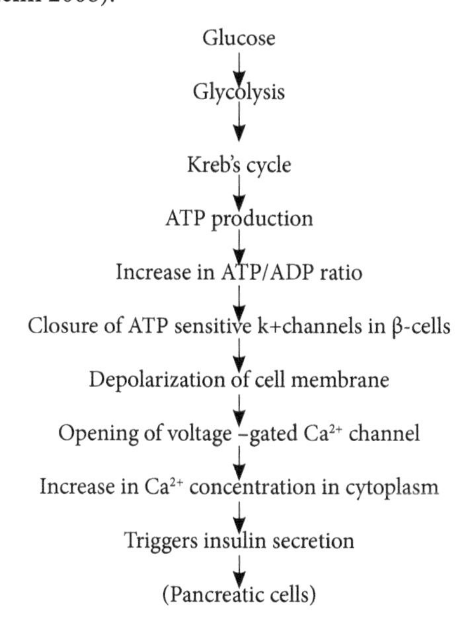

Insulin actions peak after feeding, whereas on starving, the hormone's main role is glucagon production. There are three main sites to consider for insulin's activities, those being the muscle, liver and adipose tissue. Starting from blood glucose levels upon the fed state, insulin has the main role to increase the rate of glucose oxidation (glycolysis) in the liver and muscle while at the same time it converts glucose concentrations to glycogen (the storage form of glucose in the body).

All actions of insulin are performed via insulin receptors, known as *tyrosine kinase receptors*. These are 2-subunit receptors and contain both an extracellular domain for insulin to bind as a ligand as well as an intracellular part, insulin protein kinase (Jangsoon & Pilch 1994) where all phosphorylation events take place. As with most polypeptide hormones, upon binding, conformational changes are undertaken on the two subunits and a series of phosphorylation events proceed through, leading to more actions of the hormones. Internalization of insulin upon binding keeps it in the cell followed by degradation at the end. In addition, it is worth mentioning that all of these actions and signalling pathways are accompanied by the formation of second messengers such as cyclic adenosine phosphate, cAMP. Second messengers are molecules that disperse information around tissues. The decrease in the concentration of cAMP is the main cause for insulin's activities and this is because of insulin suppressing the precursor molecule, adenylate cyclase, from which the second messenger is synthesized (Dzeja & Terzie 2009).

Mechanism of Action of Insulin

The insulin receptor is a combination of 4 subunits held together by disulfide linkages: two a-subunits lying outside the cell membrane and two b-subunits protruding into the cell cytoplasm. When insulin binds to the a-subunit in target tissues, the b-subunits in turn become activated.

Activation of the b-subunits triggers a series of reactions that draw the glucose transporter to the cell membrane. Cells (liver, muscle, adipose, *but not* brain) are now able to increase their uptake of glucose. The cell membrane also becomes more permeable to many amino acids.

Only enough ATP for immediate cellular requirements is made at any one time. Glucose that is not needed for ATP is anabolized into glycogen and stored for later use in the liver and in muscles.

Glucose Metabolism

Glycogenesis: synthesis of glycogen from glucose molecules. Insulin stimulates glycogenesis (glycogen anabolism) and inhibits glycogenolysis (glycogen catabolism)

Why is glucose stored as glycogen?

Glucose is in liquid form. With increase in the number of glucose molecules, the pressure inside the cell increases. Converting glucose to glycogen (in solid form) relieves pressure inside the cell.

Insulin promotes the conversion of all excess glucose in liver that cannot be stored as glycogen into fatty acids. Fatty acids are packaged as triglycerides in low density lipoproteins transported by blood to adipose tissue. Insulin activates lipoprotein lipase in the capillary walls of adipose tissue, which splits triglycerides into fatty acids. This enables them to be absorbed into adipose cells where they are converted again to triglycerides and stored. As an end result decreases high blood glucose levels"

The regulation of blood glucose

Humans consume meals twice or thrice daily according to a set pattern, however their cells and tissues and organs must be supplied with nutrients constantly from which they can derive energy in the form of ATP, for which glucose is the main source. It is therefore important to maintain a constant level of glucose in the blood. This maintenance is a complex interplay of hormones.

Insulin is secreted into blood when food is consumed and glucose level in the blood increases, insulin stimulates uptake, storage and use of glucose by the tissues of the body such as liver cells, muscle and fat cells. These activities reduce blood glucose levels. Insulin in turn is broken down rapidly; this prevents the blood glucose levels from continuing to drop. Several other hormones such as glucogen promote the release of stored energy reserves into the blood and thereby increasing blood glucose levels.

Diabetes (Type 2)

The name of the disease's *Diabetes* was derived from– Greek for siphon or fountain for the characteristic frequent urination sweet as honey. In 1679, a physician tasted the urine of a person with diabetes and described as sweet like honey.

Type 2 diabetes mellitus is a disorder where either there is a reduced insulin secretion or insulin resistance or both. The problems therefore, are without insulin, glucose transport into the cells will be insufficient, lacking glucose, and cells will have to rely on protein and fat catabolism for fuel. Also, when there is not enough insulin, excess glucose cannot be stored in the liver and muscle tissue. Instead, glucose accumulates in the blood above normal levels.

The American Diabetes Association suggests the following targets for most nonpregnant adults with diabetes. More or less stringent glycemic goals may be appropriate for each individual.

- A1C: 7%
- A1C may also be reported as eAG: 154 mg/dl
- Before a meal (preprandial plasma glucose): 70–130 mg/dl
- 1-2 hours after beginning of the meal (Postprandial plasma glucose)*: Less than 180 mg/dl

The high concentration of glucose in the blood (resulting from the lack of insulin) is called hyperglycemia, or high blood sugar. Excess of blood glucose exerts high osmotic pressure in extracellular fluid, causing cellular dehydration and excess of glucose begins to be lost from the body through urine (glycosuria). Dehydration stimulates hypothalamic thirst centers, causing polydipsia or excessive thirst. Polyuria, polydypsia, and polyphagia are the 3 cardinal signs of diabetes. That is, although plenty of glucose is available, it cannot be used, and the cells begin to starve. Without fuel, cells cannot produce energy leading to fatigue and weight loss. A deficiency of insulin will accelerate the breakdown of the body's fat reserves for fuel. Free fatty acids become the main energy substrate for all tissues except the brain. Increased lipolysis results in the production of organic acids called ketones (ketogenesis) in the liver. The increased ketones in the blood lower the pH of blood, resulting in a form of acidosis called ketosis, or ketoacidosis. Ketones are excreted in the urine, a condition known as ketonuria.

Complications of diabetes

Ketosis

Serious electrolyte losses also occur as the body gets itself rid of excess ketones. Ketones are negatively charged and carry positive ions out with them. Sodium and potassium are also lost from the body; because of the electrolyte imbalance, the person get abdominal pains and may vomit, and the stress reaction could increase further and can result in coma and death.

Effects of insulin deficiency on metabolic use of fat

Excess fat metabolism leads to an increase in plasma cholesterol which in course of time could lead to increased plaque formation on the walls of blood vessels. This leads to atherosclerosis and other cardiovascular problems; cerebrovascular insufficiency, ischemic heart disease, peripheral vascular disease, and gangrene. Degenerative changes in cardiac circulation can lead to early heart attacks. Heart attacks are three to four times more likely in diabetic individuals than in nondiabetic individuals. The most common cause of death with diabetes mellitus is myocardial infarction.

Other complications of diabetes

- A reduction in blood flow to the feet can lead to tissue death, ulceration, infection, and loss of toes or a major portion of one or both feet.
- Damage to renal blood vessels can cause severe kidney problems. (Nephropathy). Diabetic nephropathy progresses in three stages; stage one: increased glomerular filtration and kidney hypertrophy; stage two; excretion of albumin in urine (< 30mg/24h) and Stage three: micro albuminuria (30 – 300 mg/24h).
- Damage to blood vessels of the retina can also cause blindness. (Retinopathy)
- Blood vessels in the retina leak and hemorrhage. Patient may notice a decrease in vision if the swelling and hemorrhage affect the macula (Non proliferative).

- Edema of macula is the most common cause of visual loss in diabetic retinopathy.
- New blood vessels grow in the eye (Proliferative).
- These new blood vessels tend to bleed and leak causing vision loss.
- These new blood vessels may also pull on the retina causing retinal detachment.
- High blood glucose also helps bacteria in the mouth to grow, making tooth and gum problems worse.
- Infections: increased susceptibility to infection in general.
- Gingivitis: bacteria grow in the shallow pocket where the tooth and gum meets; gum begins to pull away from the tooth.
- Periodontitis: infection causes actual bone loss, teeth begin to pull away from the jaw itself.

Damage to the nerves

- Numbness and tingling in feet and night leg cramps may result from nerve damage due to prolonged high glucose levels that cause changes in the nerves and "neuron starvation" from lack of cellular glucose.
- Nerve damage can also lead to a loss of the ability to feel pain in the lower extremities (feet), leading to undue pressure leading to calluses and ulcerations (Neuropathy). Open ulcers can lead to infections and could become a cause of amputations.
- Neuropathy can result in two sets of what appear to be contradictory problems. Most patients who have neuropathy have one of these problems but some can be affected by both:

1. Symptoms of pain, burning, pins and needles or numbness which lead to discomfort
2. loss of ability to feel pain and other sensation which leads to neuropathic ulceration.

Regulation of glucose uptake

A major metabolic defect linked with diabetes (type 2) is the failure of peripheral tissues in the body to properly utilize glucose, thereby resulting in chronic hyperglycemia.

Pancreatic α-amylase is a key enzyme in the digestive system and catalyses the hydrolysis of starch to a mixture of smaller oligosaccharides consisting of maltoses, maltotrioses, and a number of α-(l-6) and α-(1 - 4) oligosachharides. These are then degraded by α-glucosidases and further degraded to glucose which gets absorbed and enters the blood-stream. Degradation of dietary starch proceeds rapidly and leads to elevated post-prandial hyperglycemia (PPHG). It has been shown that activity of human pancreatic α-amylase (HPA) in the small intestine correlates well to an increase in post-prandial glucose levels, controlling this is therefore an important aspect in treatment of type

– 2 diabetes. Hence, retardation of starch digestion by inhibition of enzymes such as α-amylase plays a key role in the control of diabetes. Inhibitors of pancreatic α-amylase delay carbohydrate digestion causing a reduction in the rate of glucose absorption and lowering the post-prandial serum glucose levels. Some inhibitors currently in clinical use are acarbose and miglitol which inhibit glycosidases such as α-glucosidase and α-amylase while others such as and voglibose inhibit α-glucosidase. However, many of these synthetic hypoglycemic agents have their own limitations, some are non-specific, can produce serious side effects and fail to alleviate diabetic complications. The main side effects of these inhibitors are insomnia, gastrointestinal bloating, abdominal discomfort, diarrhea and flatulence. Herbal medicines are getting more importance in the treatment of diabetes as they are free from side effects and less expensive when compared to synthetic hypoglycemic agents.

Glycosidases enzymes are widespread in microorganisms, plants, and animals. They are a very important class of enzymes, which catalyze a hydrolytic cleavage of glycosidic bonds in oligosaccharides or glycoconjugates. Among these glycosidases, α-glucosidase catalyzes the cleavage of glycosidic bonds (1-4 linked) involving terminal glucose connected at the site of cleavage through α-linkage at the anomeric center to release a single alpha- glucose molecule.

Glucose transport and the role of insulin

The major consumer of glucose is skeletal muscle. Glucose is the main fuel for contracting muscle, and normal glucose metabolism is vital for health. Glucose enters the muscle cell via facilitated diffusion through the GLUT4 glucose transporter which translocates from intracellular storage depots to the plasma membrane and T-tubules upon muscle contraction (Richter and Hargreaves 2013). Glucose is cleared from the bloodstream by a family of facilitative transporters (GLUTs), which catalyze the transport of glucose down its concentration gradient and into cells of target tissues, primarily striated muscle and adipose. Currently, there are five established functional facilitative glucose transporter isoforms (GLUT1-4 and GLUTX1), with GLUT5 being a fructose transporter (Watson and Pessin 2001).

Role of AMP kinase

AMP activated protein kinase is a member of energy sensing enzymes that are activated by cellular stresses resulting in depletion of ATP, thus working as a fuel gauge, upon activation, AMPR functions to restore cellular ATP by both inhibiting ATP consumption processes as well as promoting /accelerating ATP generating processes (Misra and Chakrabarti 2007).

Sorbitol: When excess of unused glucose remains in circulationthen it enters the polyol pathway where glucose is reduced by the enzyme aldose reductase to sorbitol. The enzyme utilizes NADPH as co factor and oxidizes it to $NADP^+$. Sorbitol is converted to fructose by the sorbitol dehydrogenase which in the process utilizes

NAD^+. Thus activation of the polyol pathway leads to a reduction in the availability of reduced NADPH and oxidized NAD^+ which are necessary cofactors in redox reactions throughout the body. The decreased concentration of these NADPH leads to decreased synthesis of glutathione (reduced), nitric oxide, myo inositol, andtaurine. Myo-inositol is particularly required for the normal function of nerves. Sorbitol may also glycate nitrogens on proteins, such as collagen, and the products of these glycations are referred-to as AGEs – advance glycation products. AGEs are thought to cause disease in the human body.

Chronic Somogyi rebound, a contested explanation of phenomena of elevated blood sugars in the morning. Also called the Somogyi effect and posthypoglycemic hyperglycemia, it is a rebounding high blood sugar that is a response to low blood sugar. When managing the blood glucose level with insulin injections, this effect is counter-intuitive to insulin users who experience high blood sugar in the morning as a result of an overabundance of insulin at night. The Somogyi effect can occur when a person takes long-acting insulin for diabetes. If the blood sugar level drops too low in the early morning hours, hormones (such as growth hormone, cortisol, and catecholamines) are released. These help reverse the low blood sugar level but may lead to blood sugar levels that are higher than normal in the morning. An example of the Somogyi effect is:

- A person who takes insulin doesn't eat a regular bed time snack and the person's blood sugar level drops during the night.
- A person's body responds to the low blood sugar in the same way as in the dawn phenomenon, by causing a high blood sugar level in the early morning.

Dawn phenomenon

The Dawn phenomenon occurs when

- Growth hormone, cortisol, and catecholamines produced by the body cause the liver to release large amounts of glucose into the bloodstream. These hormones are released in the early morning hours. These hormones also may block the effect of insulin partially, whether it's insulin produced by the body or insulin from the last injection.
- If the body doesn't produce enough insulin (which occurs in people with type 1diabetes and a few people with type 2 diabetes), blood sugar levels may rise. This may cause high blood sugar in the morning before the person eats.

Treatment of diabetes with synthetic drugs is costly and chances of side effects are high. For example, long-term use of Exenetide (Byetta) has led to several side effects such as nausea, vomiting, diarrhoea, dizziness, headache, jittery feeling and acidity. Sulfonylureas cause abdominal upset, headache and hypersensitivity, while Metformin causes diarrhoea, nausea, gas, weakness, indigestion, abdominal discomfort and headache. Thiazolidinediones has side effects like, upper respiratory infections and sinusitis, headache, mild anaemia, retention of fluid in the body which may lead to heart failure and muscle pain. Ayurveda, Unani and other traditional system for the treatment of diabetes describe a number of plants used as herbal drugs. Hence, they play a major role as alternative medicine due to less side effects and low cost. The active principles present in medicinal plants have been reported to possess pancreatic beta cells regenerating, insulin releasing and fighting the problem of insulin resistance *Aloe vera* juice stimulates the release of insulin from the beta-cells in human, *Acacia catechu* wood extract enhances the regeneration of pancreatic beta cells in rabbits, *Momordica charantia* fruit extract enhances insulin secretion by the islets of Langerhans etc. A significant proportion of these plants have been observed to possess potent antioxidant activity, which may contribute to anti-diabetic property in streptozotocin/alloxan, induced animal model not only in Ayurveda, but also in several other traditional systems of medicine such as Unani, it is described that plants useful in diabetes also possess strong antioxidant/free-radical scavenging properties. In Ayurveda, diabetes is described as 'Madhumeha'and diabetes in Unani. Ayurvedic and Unani preparations contain active hypoglycemic constituents that are isolated from the plant and herb species of India. These active principles are dietary fibres, flavonoids, alkaloids, saponins, amino acids, steroids, peptides and others. These have produced potent hypoglycemic, anti-hyperglycemic and glucose suppressive activities (Saxena and Vikram, 2004, Gupta et al 2008). The above effects achieved by either insulin release from pancreatic ß-cells, inhibited glucose absorption in gut, stimulated glycogenesis in liver or increased glucose utilization by the body (Grover et al., 2002; Saxena and Vikram, 2004). These compounds also exhibited their antioxidant, hypolipidemic, anticataract activities, restored enzymatic functions, repair and regeneration of pancreatic islets and the alleviation of liver and renal damage (Mukherjee et al., 2010). Some active constituents have been obtained from plants possess insulin like activity and could be provide alternate for insulin therapy.

Anti diabetic plants

Table 1. Antidiabetic plants and their active componenets

Sl.No.	Botanical name	Family	Local (Hindi)/ common name	Plant parts used	Active compounds	Proposed mechanism of action	Reference
1.	*Abelmoschus esculentus (L) Moench*	Malvaceae	Ladies Finger, Bhendi, okra	Seed, fruit and peel	quercetin glucosides, pentacyclic triterpene ester	regulating DPP-4 signaling, type 1 glucagon-like peptide receptor (GLP-1R), attenuates insulin resistance , reduce oxidative stress and renal fibrosis	Huang *et al.* 2017a, b, Sabitha *et al.* 2011, Mishra *et al.* 2016, Peng *et al.* 2019, Dubey and Mishra 2017, Anjani, *et al.* 2018
2.	*Abelmoschus moschatus*	Malvaceae	Musk mallow, rose mallow, muskdana	Aerial portion	Myricetin	Stimulates glucose uptake, promotes glycogen synthesis, Increases sensitivity to insulin Increases insulin action on GLUT4 expression, Reverses the defect in expression of insulin receptor substance -1 (IRS-1) and p^{85} regulatory sub unit of PI 3-kinase. Increased phosphorylation of IR and IRS-1 as well as Akt.	Liu *et al.* 2005, 2007a,b, 2010, Rawat and Parmar 2013
						Improves insulin sensitivity in soleus muscle through enhancement of GLUT4, Increase insulin sensitivity through increased post-receptor insulin signaling mediated by enhancements in IRS-1-associatedPI3-kinase and GLUT 4 activity.	

3	*Acacia arabica*	Fabaceae	babul, kikar or Indian gum Arabic	bark extract, leaves,	phenolic, flavonoid	Decreases insulin resistance. Works as antioxidant. Reduces total cholesterol, triglycerides, LDL cholesterol, malondialdehyde and significant increase in HDL-cholesterol and coenzyme Q10 levels.	Hegazy *et al.* 2013, Aadil *et al.* 2012
4	*Acacia nilotica*	Fabaceae	gum arabic tree, babul, thorn mimosa, Egyptian acacia or thorny acacia	leaves	phenolic, flavonoid	Antioxidant activity	Roozbeh *et al.* 2017, Asad et al .2015
5	*Acacia tortilis, Vachellia-tortilis*	Fabaceae	umbrella thorn and Isra-eli-babool	Leaves, gum exudates	Tannins	Normalizes carbohydrate and lipid metabolism	Mukhtar *et al.* 2017, Bhateja and Dahiya 2014
6	*Achyranthes aspera L.*	Amaranthaceae	Chirchira; Latjira,prickly chaff flower, devil's horsewhip	Root, whole plant	Oleolenic acid, Ursolic acid	Normalizes carbohydrate metabolism	Zambare *et al.* 2011 a, Zambare *et al.* 2011 b,
7	*Adiantum capillus-veneris L.*	Adiantaceae / Pteridaceae	Venus hair fern	Seed	flavonoids, triterpenoids, oleananes, phenylpropanoids, carbohydrates, carotenoids, and alicyclics	presence of flavonoids and tannins responsible for the anti-diabetic effect, Lowered plasma cholesterol and triglyceride concentrations as well as HMG-CoA reductase activity	Ibraheim *et al.* 2011; Ahmed *et al.* 2012
8	*Aegle marmelos Linn.*	Rutaceae	Bael Holy fruit tree	Fruit, root bark leaf	oxazoline, coumarins, alkaloids & carotenoids	helps in delaying the progression of diabetic cataract	Compose 2003, Sankeshi *et al.* 2013, Augusti 1994, Kumar *et al.* 2013, Anusha *et al.* 2013

9	*Allium cepa Linn.*	Liliaceae	Pyaz, onion	bulb	flavonoids, phenolic acids, organosulfur compounds	hypoglycaemic agents capable of lowering blood glucose level, Stimulating effects on glucose utilization and antioxidant enzyme	Mootoo-samy&Mahmood-ally 2014
10	*Allium sativum L.*	Amarylli-daceae	Lahsun, garlic	whole plant, bulb	organosulfur com-pounds & phenolic acids	Increases the pancreatic secre-tion of insulin from the β-cells or release of bound insulin, Antihyperglycemic and antino-ciceptive effect	Ahmad & Ahmad 2006, Poonam *et al.* 2011,
11	*Aloe vera*	Liliaceae	Gheekuar Aloe	Gel/pulp	amino acids, anthra-quinones, enzymes minerals, vitamins, lignin's, monosac-charide's, polysac-charides, salicylic acid, saponins, and steroids	Stimulating synthesis and/or release of insulin, reduce fasting blood glucose and triglyceride levels in type 2 diabetic patients	Kim *et al.* 2009; Yagi *et al.* 2009
12	*Andrographis paniculata Wall. ex Nees*	Acantha-ceae	Kalmegha, King of bitter	Leaf, whole plant	andrographolide	Increases glucose metabolism	Fasola *et al.* 2010; Xu *et al.*, 2012
13	*Annona squamosa Linn.*	Annona-ceae	Sharifa, Custard apple	leaf	flavonoids, phenols, phenolic glycosides, saponins and cyano-genic glycosides	stimulation of surviving β – cells to release more insulin, inhib-iting the action of intestinal α glucosidase	Tomarand Sisodia 2012; Panda *et al.* 2013
14	*Asparagus racemosus L.*	Aspar-agaceae	Satavar, Shatavari	Root	steroidal saponins	inhibition of carbohydrate digestion and absorption, and enhancement of cellular insulin action	Hannan *et al.* 2011; Somani *et al.* 2012

15	*Azadirachta-indica A. Juss*	Meliaceae	Neem, Indian Lilae tree	Bark, leaf, seed	Quercetin and ß-sitosterol, polyphenolic flavonoids	reduced plasma cholesterol and triglycerides significantly and increased their hepatic glucokinase activity probably by enhancing the insulin release from pancreatic islets	Bhat *et al.* 2011; Atangwho *et al.* 2012
16	*Buniumpersi-cumBois.*	Apiaceae	Kali Jeera	Seed, whole plant	Cuminaldehyde, γ-terpinene, c-Terpinen-7-al, p-Cymene, limonene		Mohammadi, *et al.* 2010
17	*Calamus erectus*	Arecaceae	Cane/ Bet Rattan	Fruit	Flavanoids and phenolic acids	α-glucosidase and αamylase enzyme inhibition	Ghosal and Mandal 2013
18	*Capparis decidua*	Cappara-ceae	ker-da, kair, karir, kirir,	bark	Sterols, diterpine alcohol, Stachydrine, β-carotene, Rutin, Isothiocynate, Glucosides, Hydrocarbons and Fatty acids	Inhibit alpha-glucosidase and decrease glucose transport through the intestinal epithelium	Sharma *et al.* 2010; Zia-Ul-Haq *et al.* 2011
19	*Centellaasiat-ica L.*	Apiaceae	Brahmi booti, Indian pennywort	Leaf	phenolic compounds and triterpene saponins	Normalizes carbohydrate and lipid metabolism	Chauhan *et al.* 2010; Ramachandran and Saravanan, 2013
20	*Cinnamo-mumtamala (Hamm.) Nees. &Eberm.*	Lauraceae	Tejpat, Bayberry	Leaf	Monoterpenoides,	glucose lowering activity, pancreatic (enhancement of insulin secretion), extra pancreatic (peripheral utilization of glucose) mechanism	Bhist2011,Chandola *et al.* 1980, Soni *et al.* 2013
21	*Cocciniaindi-ca W.&A.*	Cucurbi-taceae	gherkins	Fruit, leaf, root, whole plant	steroids, alkaloids, flavonoids, glycosides, tannins and Phenolic compounds	Stimulates the release of insulin and blocks the formation of glucose in the bloodstream	Balaraman *et al.* 2010; Gunjan *et al.* 2011, Ramakrishnan *et al.* 2011, Shibib *et al.* 2011

22	*Coccini-agrandis (Linn.) Voigt*	Cucurbi-taceae	Kundri, Ivy guard	Fruit, leaf, root, whole plant	Flavaoids, triter-penoids, saponins and sterols	enhance secretion of insulin from the b-cells of Langerhans or through extrapancreatic mechanism	Munasinghe *et al.*2011; Al-Amin *et al.* 2013
23	*Cuscutare-flexa*	Cuscuta-ceae	Amar bel, Giant Dodder	whole plant	Flavanoids, ter-penoids, quinines, tannins	Suppressed the glucose level, reduced plasma cholesterol and triglycerides significantly and increased their hepatic glucokinase activity probably by enhancing the insulin release from pancreatic islets	Rahmatullah *et al.* 2010
24	*Cynodon-dactylon (L.) Pers.*	Poaceae	Bermuda grass, Bahama grass, In-dian *doab*	Leaf, whole plant	flavanoids, alkaloids, glycosides, terpenoi-des, triterpenoids steroids, saponins, tannins, resins, phy-tosterols, reducing sugars, carbohy-drates, proteins, volatile oils and fixed oils	enhancing the glycolytic flux and a concomitant decrease in gluconeogenesis	Rai *et al.* 2010; Singh *et al.* 2009
25	*Datura stra-monium L.*	Solana-ceae	Dhatura	Seed	atropine	β-cells secreted insulin when treated with seed extract	Neeraj *et al.* 2013
26	*Dilleniain-dica*	Dillena-ceae	Elephant ap-ple, Chulta/Chal-ta or Ouu	Leaf	triterpenoids, flavo-noids, tannins	stimulation of insulin secretion,-stimulation of glucose uptake by peripheral tissue, inhibition of endogenous glucose production, or activation of gluconeogenesis	Kumar *et al.* 2011a,b.

27	*Diplazium esculentum*	Athyria-ceae	Edible fern	Young frond	steroids, triter-penoids, tannins and phenolic substances, flavonoids, carbo-hydrates, gum and mucillages	Stimulates the release of insulin and blocks the formation of glucose in the bloodstream	Tag *et al.* 2012
28	*Eclipta alba L.*	Astera-ceae	false daisy, ,*bhrin-graj*	Leaf	coumestans, alka-loids, flavonoids, glycosides, polyacet-ylenes, triterpenoids	Normalizes carbohydrate and lipid metabolism	Ananthi *et al.* 2003; Jaiswal *et al.* 2012
29	*Eugenia jambolana*	Myrta-ceae	Jamun,black berry	Fruit, pulp and seed	tannins, alkaloids, steroids, flavonoids, terpenoids, fatty acids, phenols	Normalizes carbohydrate and lipid metabolism	Achrekar *et al.* 1991
30	*Ficus amplis-sima*	Moraceae	kal-itchch	Bark, leaf	triterpenoids, phen-ylpropanoids, chala-cone , flavonoids and phenolic acids	Suppressed the glucose level, reduced plasma cholesterol and triglycerides significantly and increased their hepatic glucokinase activity probably by enhancing the insulin release from pancreatic islets	Arunachalam & Parimelazhaghan 2013; Khan *et al.* 2011
31	*Ficus bengha-lensis L.*	Moraceae	Bargah, Indian Baugar tree	Leaf	utin, friedelin, taraxosterol, lupeol, β-amyrin along with psoralen, bergapten, β-sisterol and quer-cetin-3-galactoside	glycogenesis and enhanced peripheral uptake of glucose are the probable mechanisms	Augusti *et al.* 1994; Khan *et al.* 2011
32	*Ficus glom-erata*	Moraceae	Indian Fig Tree or Goolar	Bark, fruit	flavonoids, triter-penoids (basically lanosterol), alkaloids, and tannins	possible mechanism includes -cells and subsequent release of insulinthestimulation of and activation of the insulin receptor	Sharma *et al.* 2010; Khan *et al.* 2011

33	*Ficus racemosa L.*	Moraceae	Indian Fig Tree or Goolar	Bark, fruit	β-sitosterol, glauanol acetate	glucose homeostasis	Ahmed *et al.* 2010; Veerapur *et al.* 2012; Khan *et al.* 2011
34	*Ficus relegiosa*	Moraceae	sacred fig, *pīpal*	Leaf	phenols, tannins, steroids, lanosterol, stigmasterol, lupen-3-one	pancreatic secretion of insulin from existing -cells of islets	Pandit *et al.* 2010; Khan *et al.* 2011
35	*Gymnemasylvestre R.*	Asclepiadaceae	Gudmar, Periploca of the wood	Whole plant	alkaloids terpenoids, tannins, saponins, flavanoids, phenols, steroids, gymnemic acid	Lowers plasma glucose level	Ahmed *et al.* 2010; Kang *et al.* 2012
36	*Heliotropiumindicum L.*	Boraginaceae	Hathsura, Indian turnsole	Leaf	pyrrolizidine alkaloids, indicine (Principal), echinitine, supinine, heleurine, heliotrine, lasiocarpine, its N-oxide, acetyl indicine, indicinine and antitumour alkaloid, indicine-n-oxide. The plant also contains rapone and lupeol and an ester of retronecine	Inhibit alpha-glucosidase and decrease glucose transport through the intestinal epithelium	Francois *et al.* 2014
37	*Hemidesmus indicus L.*	Apocynaceae	Kshirini,Indian sarsaparilla	Root	phenols, glycosides, flavonoids and steroids	Block the entry of glucose from the intestine	Gayathri *et al.* 2010; Gayathri *et al.* 2012.

38	*Lagerstroemia speciosa(L.) Pers*	Lythraceae	Giant Crape-myrtle, Queen's Crape-myrtle, Pride of India	Leaf	Triterpenes, tannins, ellagic acids, glycosides and flavones	suppression of gluconeogenesis and stimulation of glucose oxidation using the pentose phosphate pathway	Hou *et al.* 2009; Shareef *et al.* 2013
39	*Lectucagracilis*	Asteraceae		Leaf			Tag *et al.* ⊠2012
40	*Mangiferaindica L.*	Anacardiaceae	Aam, Mango	Seed	Phytochemical, carotenoids, polyphenols, beta-carotene, lutein, alphacarotene,polyphenols, xanthonoid, mangiferin	Reduction of intestinal absorption of glucose	Teja *et al.* 2011; Alim *et al.* 2012
41	*Millingtonia-hortensis*	Bignoniaceae	Neem chameli, Indian cork tree, tree jasmine,	Flower	flavonoids	Normalizes carbohydrate and lipid metabolism	Tag *et al.* ⊠2012
42	*Mimosa pudica L.*	Fabaceae	Chui-mui Touch-me-not	Whole plant	phytosterol, amino acids, alkaloids, flavonoids, tannins, glycosides and fatty acids	Suppressed the glucose level, reduced plasma cholesterol and triglycerides significantly and increased their hepatic glucokinase activity probably by enhancing the insulin release from pancreatic islets	Sutar *et al.* 2009; Bashir *et al.* 2013
43	*Momordica charantia L*	Cucurbitaceae	Karela, bitter guard	Fruit, leaf, whole plant	triterpene glycosides, saponin, cucurbitane triterpenoids	saponin-rich fraction stimulates insulin production	Keller *et al.* 2011; Choudhary *et al.* 2012
44	*Morindacitrifolia*	Rubiaceae	(Noni)	fruit	anthraquinones, flavonolglycosides,iridoidglycosides,lipid glycosides, triterpenoids	reduction of glucose uptake into muscle cells	Carvalho *et al.*2013; Lee *et al.* 2012; Nayak *et al.* 2010

45	*Musa sapientum L*	Musaceae	Kela, banna	Fruit	saponins, flovonoids, glycosides, steroids and alkaloids	stimulation of insulin production and glucose utilization	Kumar *et al.* 2013; Choudhari *et al.* 2012
46	*Ocimum sanctum L.*	Lamiaceae	Tulsi, holy basil	Whole plant, leaves	triterpenoid with hydroxyl or carboxylic or ketonic groups , The 3-carboxylic acid or hydroxyl or ketonic substituted triterpenoid	increasing the pancreatic secretion of insulin from existing ⊠ cells of islets of Langerhans or its release from bound insulin, and/or (b) enhanced glucose utilization by peripheral tissues	Patil *et al.* 2011; Nim *et al.* 2013
47	*Oxalis griffithii*	Oxalidaceae	wood sorrel	Leaf	phytochemicals, alkaloids	Normalizes carbohydrate and lipid metabolism	Tag *et al.* ⊠2012
48	*Phyllanthus emblica L.*	Phyllanthaceae	Indian gooseberry	Fruit, seed, whole plant	phytochemicals oils, phosphatides, essential xed oils, phosphatides, essential oils, tannins, minerals, vitamins, amino acids, fatty acids, glycosides	effects against lipid peroxidation by scavenging the free radicals and reducing the risk of diabetic complications	Sultana *et al.* 2014
49	*Psidium guajava*	Myrtaceae	Amrud, guava	leaf	carbohydrates, alkaloids, flavones, tannins, steroidal glycosides, phenols, coumarin	Inhibition of alpha- amylase and alpha-glucosidase enzymes	Basha and Kumari 2012; Manikanandan *et al.* 2013
50	*Pterocarpus marsupium Roxb.*	Fabaceae	Vijaysar, India malabar	bark	phenolics,marsupsin, pterosupin, pterostilbene, isoflavone	possible Pancreatic beta-cell regeneration,Upregulation of Glut-4 and PPAR gamma by isoflavone	Mohankumar *et al.* 2012; Gupta *et al.* 2009

51	*Saccharum spontaneum*	Poaceae	Kams, Kans grass	Young shoot	alkaloids, flavonoids, tannins, steroids, terpenoids, glycosides and phenolic constituents	Lowers plasma glucose level	Tag *et al.* 2012
52	*Salacia oblonga*	Celastraceae	Saptarangi, banaba	Roots and stem	Salalcinol, kotalanol, phenolic compounds, sesquiterpenes and triterpenes	increases GLUT4 expression and translocation in muscle cells mediated through two independent pathways that are related to 5⊠-AMP-activated protein kinase and PPAR-γ	Girón *et al.* 2009; Bhat *et al.* 2012
53	*Swertiachirata L.*	Gentianaceae	KirayatChinata, Bitter stick	Root	Xanthones, flavonoids, mono or poly methyl ethers, phytochemical swerchirin	a-glucosidase inhibitory potential to manage hyperglycemia and phenolic-linked antioxidant activity to reduce micro vascular complication	Kavitha and Dattatri 2013; Phoboo *et al.* 2013; Verma *et al.* 2013
54	*Swietenia mahagoni Jacq.*	Meliaceae	Indian Mahagony	Seed	tannins, alkaloids, saponins, terpenoids, anthraquinone, cardiac glycosides, saponins, and volatile oils	hypoglycaemic activity of is mediated by agonistic activity to PPAR γ receptor which after activation improves insulin sensitivity,	De *et al.* 2011; Hasan *et al.* 2013
55	*Syzygium cumini (L.) Skeels (Syn. E. jambolana)*	Myrtaceae	Jamun, black berry	Leaf, seed	anthocyanins, glucoside, ellagic acid, isoquercetin, kaemferol and myrecetin,alkaloid, jambosine, and glycoside jambolin or antimellin	increased insulin release from b-cells, inhibits adenosine deaminase activity and reduces glucose levels	Bopp *et al.* 2009; De Bona *et al.* 2011

56	*Tamarindus indica L.*	Fabaceae	Imli, Tamarind	Seed	sulfur amino acids, phenolic antioxidants, proanthocyanidins and epicatechin.	selectively increases glucose transporter-2, glucose transporter-4, and islets' intracellular calcium levels and stimulates β-cell proliferation resulting in improved glucose homeostasis	Agnihotri & Singh 2013, Sole and Srinivasan 2012; Parvin *et al.* 2013
57	*Terminalia arjuna W.&A*	Combretaceae	Arjuna,	Seed	Alkaloids, flavonoids , tannins, saponins	augmenting endogenous antioxidant mechanisms	Biswas *et al.* 2011; Morshed *et al.* 2011
58	*Terminalia bellirica L.*	Combretaceae	*Behada,* Bahera or Beleric	Seed	Gallic acid	stimulates basal insulin output and potentiated glucose-stimulated insulin secretion concentration-dependently in the clonal pancreatic β-cells	Kasabri *et al.* 2010; Latha and Daisy, 2011
59	*Terminalia chebula Retz.*	Combretaceae	Harad, Harra	Seed	Phenolic acid, tannic acid,Benzoic acid, Flavonoid	Increased release of in vitro insulin levels from pancreatic β-cells through direct insulinotropic effect	Mukherjee *et al.* 2010; Huang *et al.* 2012
60	*Tinospora cordifolia L.*	Menispermaceae	Geloy, guruch	Bark, leaf, root, whole plant	alkaloids, diterpenoid lactones, glycosides, steroids, sesquiterpenoid, phenolics, aliphatic compounds and polysaccharides	hypoglycemic effects via mechanisms of insulin releasing and insulin-mimicking activity and thus improves postprandial HYPERLINK "https://www.sciencedirect.com/topics/medicine-and-dentistry/post-prandial-hyperglycemia"hyperglycemia.	Abdel Berry *et al.* 1997, Puranik *et al.* 2010

61	*Trigonellafoenum graecum L.*	Fabaceae	Methi, Fenugreek	Seed, whole plant	saponins and alkaloids, hydroxyisoleucine	delay in gastric emptying time and glucose absorption rate; reducing the glucose uptake in the small intestine by its high-fiber content that slows carbohydrate metabolism and lowered blood glucose; restoring the function of pancreatic tissues,	Baquer *et al.* 2011, Ganeshpurkar 2013; Shankar *et al.* 2012, Mooventhan and Nivethitha 2017
62	*Vernonia anthelmintica Willd*	Asteraceae	bakchi, baksi, kalijira, somraj, bukchi, ghrajiri. Iron Weed	Whole plant, seeds	flavonoids, carotenoids, glycosides, steroids, phenols and tannins	stimulates insulin secretion from the remnant b cells of islets of langerhans	Fatima *et al.* 2010
63	*Vinca rosea L*	Apocynaceae	Sadabahar, rose periwinkle	Leaf	alkaloids	Regeneration of β-cells of pancreas	Ghosh and Suryawanshi 2001; Ahmed *et al.* 2010
64	*Vitex negundo L.*	Lamiaceae	Nirgundi, Sephali, Sambhalu Five-Leaved Chaste Tree	Leaf	Phenols and flavonoids	Normalizes carbohydrate metabolism and liver functions	Manikandan *et al.* 2011; Sundaram *et al.* 2012, Falguni *et al.* 2017
65	*Withania somnifera (L.) Dunal*	Solanaceae	Ashwagandha, Indian ginseng	Leaf, root, whole plant	Flavonoids	Normalizes carbohydrate metabolism	Udayakumar *et al.* 2009; Udayakumar *et al.* 2010

Conclusion

Metabolic imbalance causing diabetes mellitus has become a characteristic of the fast moving, materialistic world. Differences in social structure, psychological stress, obesity, hormonal imbalance and heredity are optimizing the growth of this pandemic. Increasing population with diabetes has a huge requirement of effective remediation. The Indian flora has a huge variety of medicinal plants, which are used traditionally for their anti-diabetic property. However, careful assessment including sustainability of such herbs, ecological and seasonal variation in activity of phyto-constituents, metal contents of crude herbal anti-diabetic drugs, thorough toxicity study and cost effectiveness is required for their popularity. These efforts may provide treatment for all and justify the role of novel traditional medicinal plants having anti-diabetic potentials.

All the studies reviewed have used crude extracts of the plant materials, nobody has reported the amount of the plant material to be consumed (in case of edible plants) as extracting and then using may not be possible for the general population. In case of non edible plants, whether it is safe to consume directly? Another question that arises is how long one should consume at a stretch, should there be a gap?

References

Aadil R, Barapatre A, Rathor N, Pottam S and Jha H. (2012). Comparative study of *in vitro* antioxidant and antidiabetic activity of plant extracts of Acacia Arabica, Murraya koeingii, Catharanthus roseus and Rawolfia serpentine. *International Journal of Phytomedicine* 4 (4) 543-551.

Achrekar, S.; Kakliji, G. S.; Pote, M. S. and Kelkar, S. M. (1991). Hypoglycemic activity of Eugenia jambolana and Ficus bengalenesis: Mechanism of action. *In Vivo.*, 5: 143-147.

Agrawal N. K and Uma Gupta. (2013). 1Evaluation of hypoglycemic and antihyperglycemic effects of Acacia tortilis seed extract in normal and diabetic rats. *International Journal of PharmTech Research* Vol.5, No.2, pp 330-336.

Agnihotri, A. and Singh, V. (2013). Effect of Tamarindus indica Linn. and Cassia fistula Linn. Stem bark extracts on oxidative stress and diabetic conditions. *Acta. Pol. Pharm.*, 70(6): 1011-1019.

Ahmad, M. S. and Ahmad, N. (2006). Antiglycation properties of aged garlic extract: possible role in prevention of diabetic complications. *J. Nutr.*, 136: 796-799.

Ahmed, A.; Jahan, N.; Wadud, A.; Imam, H.; Hajera, S. & Bilal, A. (2012). Physicochemical And Biological Properties Of Adiantum Capillus-Veneris Linn: An Important Drug of Unani System of Medicine. *Int. J. Cur. Res. Rev.*, 4(21) 70-75.

Ahmed, A., Rao, A.S. and Rao, M.V. (2010). *In vitro* callus and in vivo leaf extract of Gymnema sylvestre stimulate β-cells regeneration and anti-diabetic activity in Wistar rats. *Phytomedicine.*, 17(13): 1033-1039.

Ahmed, M. F.; Kazim, S. M.; Ghori, S. S.; Mehjabeen, S. S.; Ahmed, S. R.; Ali, S. M. and Ibrahim, M. (2010). Antidiabetic activity of Vinca rosea extracts in alloxan-induced diabetic rats. *Intl. J. Endocrinol.*

Akhtar MS, Iqbal J. (1991) Evaluation of the hypoglycaemic effect of Achyranthes aspera in normal and alloxan-diabetic rabbits. *J Ethnopharmacol.* 31(1):49-57.

Akhtar MS, Khan MA, Malik MT. (2002) Hypoglycaemic activity of Alpinia galanga rhizome and its extracts in rabbits. *Fitoterapia.* 73(7-8):623-8.

Akhtar MT, Bin Mohd Sarib MS, Ismail IS, Abas F, Ismail A, Lajis NH, Shaari K. Anti-Diabetic Activity and Metabolic Changes Induced by Andrographis paniculata Plant Extract in Obese Diabetic Rats. Molecules. 2016 Aug 9;21(8). pii: E1026. doi: 10.3390/molecules21081026.

Al-Amin, M.; Uddin, M. M. N.; Rizwan, A.; & Islam, M. (2013). Effect of Ethanol Extract of Coccinia grandis Lin leaf on Glucose and Cholesterol Lowering Activity. *Brit. J. Pharmaceut. Res.*, **3**(4): 1070-1078.

Alim, A.; Sharmin, R.; Hossain, M. S.; Ray, D. N.; Anisuzzaman, M. & Ahmed, M. (2012). Evaluation of hypoglycemic effect of compound (s) from petroleum ether fraction of ethanol extract of Mangifera indica red leaves. *IOSR J. Pharm.*, 2(6): 14-19.

Alharbi W. D. M. and A. Azmat. (2011). Hypoglycemic and hypocholesterolemic effects of Acacia tortilis (Fabaceae) growing in Makkah. *Pakistan Journal of Pharmacology* 28(1):1-8.

Alsherbiny MA, Abd-Elsalam WH, El Badawy SA, Taher E, Fares M, Torres A, Chang D, Li CG. (2019) Ameliorative and protective effects of ginger and its main constituents against natural, chemical and radiation-induced toxicities: A comprehensive review. *Food Chem Toxicol.* 123:72-97.

Ananthi, J.; Prakasam, A.; and Pugalendi, K. V. (2003). Antihyperglycemic Activity of Eclipta alba Leaf on Alloxan-induced Diabetic Rats. Yale *J. Biol. Med.*, 76: 97-102.

Anusha, C., Sarumathi, A., Shanmugapriya, S., Anbu, S., Ahmad R. S., Saravanan N. (2013). The effects of aqueous leaf extract of Aegle marmelos on immobilization-induced stress in male albino Wistar rats. *Int. J. Nutr. Pharmacol. Neurol. Dis.*, 3: 11-6.

Arunachalam, K. and Parimelazhagan, T. (2013). Antidiabetic activity of Ficus amplissima Smith. bark extract in streptozotocin induced diabetic rats. *J. Ethnopharmacol.*, 147(2): 302-10.

Asad M, Munir TA, Farid S, Aslam M, Shah SS. (2015) Duration effect of Acacia nilotica leaves extract and glibenclamide as hypolipidaemic and hypoglycaemic activity in alloxan induced diabetic rats. *J Pak Med Assoc.* 65(12):1266-70.

Atangwho, I. J.; Ebong, P. E.; Eyong, E. U.; Asmawi, M. Z. and Ahmad, M. (2012). Synergistic antidiabetic activity of *Vernonia amygdalina* and *Azadirachta indica* Biochemical effects and possible mechanism. *J. Ethnopharmacol.*, 141(3): 878-887.

Augusti, K. T.; Cherian, R. D. S.; Sheeta, C. G. and Nair, C. R. (1994). Effect of leucopelargonin derived from *Ficus bengalensis* (Linn.) on diabetic dogs. *Ind. J. Med. Res.*, 99: 82-86.

Azantsa B. G. K., G. R. Takuissu, E. J. Tcheumeni, M. Fonkoua, Dibacto E. R. K, Ngondi J. L, Oben J. E. (2019) Antihyperglycemic Mechanisms of Allium sativum, Citrus sinensis and Persea americana extracts: Effects on Inhibition of Digestive Enzymes, Glucose Adsorption and Absorption on Yeast Cells and Psoas Muscles. *Diabetes Res Open J.* 5(2): 26-34

Balaraman, A. K.; Singh, J.; Dash, S.; & Maity, T. K. (2010). Antihyperglycemic and hypolipidemic effects of *Melothria maderaspatana* and *Coccinia* indica in Streptozotocin induced diabetes in rats. Saudi. *Pharmaceut.* J., 18(3): 173-178.

Balasuriya N, Rupasinghe HP. (2012) Antihypertensive properties of flavonoid-rich apple peel extract. *Food Chem.* 135(4):2320-5.

Baquer, N. Z., Kumar, P., Taha, A., Kale, R.K., Cowsik, S.M. and McLean, P. (2011). Metabolic and molecular action of *Trigonella foenum*-graecum (fenugreek) and trace metals in experimental diabetic tissues. *J. Biosci.*, 36(2):383-96.

Barry, A.J.A., Hassan, A.I.A. and Al-Hakiem, M.H. (1997). Hypoglycaemic and antihyperglycaemic effect of *T. foenum* graecum leaf in normal and alloxan induced diabetic rats. *J. Ethnopharmacol.*, 58: 149-155.

Basha, S. K. and Kumari, V. S. (2012). In vitro antidiabetic activity of Psidium guajava leaves extracts. *Asian. Pac. J. Trop. Dis.*, 98-100.

Bashir, R.; Aslam, B.; Javed, I.; Muhammad, F.; Sarfraz, M. and Fayyaz, A. (2013). Antidiabetic efficacy of Mimosa pudica (Lajwanti) root in Albino rabbits. *Intl. J. Agric. Biol.*, 15(4): 782-786.

Bassiri-Jahromi S. *Punica granatum* (Pomegranate) activity in health promotion and cancer prevention. *Oncol Rev*. 2018 Jan 30;12(1):345.

Bhat, B. M.; Raghuveer, C. V.; D'Souza, V. & Manjrekar, P. A. (2012). Antidiabetic and Hypolipidemic Effect of Salacia Oblonga in Streptozotocin Induced Diabetic Rats. *J. Clin. Diagnostic. Res.* 6(10): 1685-1689.

Bhat, M., Kothiwale, S. K., Tirmale, A. R., Bhargava, S. Y., & Joshi, B. N. (2011). Antidiabetic properties of Azardiracta indica and Bougainvillea spectabilis: *in vivo* studies in murine diabetes model. Evid. Based. Complement. *Alternat. Med.*,561-625. doi: 10.1093/ecam/nep033

Bhateja P K and R. S. Dahiya. (2014). Antidiabetic Activity of *Acacia tortilis* (Forsk.) *Hayne* ssp. raddiana Polysaccharide on Streptozotocin-Nicotinamide Induced Diabetic Rats. *BioMed Research International*. Volume 2014, Article ID 572013.

Biswas, M., Kar, B., Bhattacharya, S., Kumar, R.S., Ghosh, A.K. & Haldar, P.K. (2011). Antihyperglycemic activity and antioxidant role of Terminalia arjuna leaf in streptozotocin-induced diabetic rats. *Pharmaceut. Biol.*, 49(4): 335-340.

Bopp, A., De Bona, K.S., Bellé, L. P., Moresco, R. N. & Moretto, M. B. (2009). Syzygium cumini inhibits adenosine deaminase activity and reduces glucose levels in hyperglycemic patients. Fundamental. *Clin. Pharmacol.*, 23(4): 501-507.

Botchway BOA, Moore MK, Akinleye FO, Iyer IC, Fang M. Nutrition: (2018) Review on the Possible Treatment for Alzheimer's Disease. *J Alzheimers Dis.* 61(3):867-883. doi: 10.3233/JAD-170874

Breer, H., Eberle, J., Frick C., Haid D., Widmayer P. (2012). Gastrointestinal chemosensation: chemosensory cells in the alimentary tract. Histochem. *Cell. Biol.*, 138(1): 13-24.

Campos, K. E., Diniz, Y. S., Cataneo, A. C., Faine, L. A., Alves, M. J. and Novelli, E. L. (2003). Hypoglycemic and antioxidant effect of onion, Allium cepa: dietary onion addition, antioxidant effect and hypoglycemic effects on diabetic rats. *Int. J. Food Sci. Nutr.*, 54: 241-246.

Carvalho, M. C., Paulo, D. B., de Castro e Silva, P. A. N., Nunes, H. M. and Martins, M.C.C. (2013). Investigation of anti-hyperglycemic effect in juice and lyophilized forms of Morinda citrifolia. *Emir. J. Food Agric.*, 25(10): 767-771.

Chandola, H.M., Tripathi, S. N. and Udupa, K. N. (1980). Hypoglycemic response of Cinnamonum tamala in patients of maturity onset (NIDDM) Diabetes. *J. Res. Ayurv. Sidha.*, 1: 275-290.

Chaturvedi P. (2012) Antidiabetic potentials of Momordica charantia: multiple mechanisms behind the effects. *J Med. Food.* 15(2):101-7.

Chauhan, P. K.; Pandey, I. P. & Dhatwalia, V. K. (2010). Evaluation of the anti-diabetic effect of ethanolic and methanolic extracts of Centella asiatica leaves extract on alloxan induced diabetic rats. *Adv. Biol. Res.*, 4: 27-30.

Chen MH, Lin CH, Shih CC (2014). Antidiabetic and Antihyperlipidemic Effects of Clitocybe nuda on Glucose Transporter 4 and AMP-Activated Protein Kinase Phosphorylation in High-Fat-Fed Mice. *Evid Based Complement Alternat Med.* ; 981046. doi: 10.1155/2014/981046.

Chen C, Zhang Q, Wang FQ, Li CH, Hu YJ, Xia ZN, Yang FQ. (2019) In vitro anti-platelet aggregation effects of fourteen fruits and vegetables. *Pak J Pharm Sci.* 32(1):185-195.

Choudhari, S. A., Khatwani, P. F., & Kulkarni, S. R. (2012). Assessment of antidiabetic potential of flowers of musa sapientum and development of tablet dosage form. *Inventi Rapid: Ethnopharmacology,* Inventi:pep/733/12

Choudhary, S.K., Chhabra, G., Sharma, D., Vashishta, A., Ohri, S. & Dixit, A. (2012). Comprehensive Evaluation of Anti-hyperglycemic Activity of Fractionated Momordica charantia Seed Extract in Alloxan-Induced Diabetic Rats. Evid. Based. Complement. *Alternat. Med.*, 293650.

Chul-Won Lee, Hyung-Seok Lee, Yong-Jun Cha, Woo-Hong Joo, Dae-Ook Kang and Ja-Young Ciulla M, Marinelli L, Cacciatore I, Stefano AD. (2019). Role of Dietary Supplements in the Management of Parkinson's Disease. *Biomolecules.* 10;9(7). pii: E271. doi: 10.3390/biom9070271

Cook, D. L. and Hales, C. N. (1984). Intracellular ATP directly blocks K^+ channels in pancreatic β-cells. *Nature.*, 311(5983): 271-3.

Darooghegi Mofrad, M., Rahmani, J., Varkaneh, H.K., Teymouri, A., Mousavi, S.M. (a). The effects of garlic supplementation on weight loss: A systematic review and meta-analysis of randomized controlled trials. *Int J Vitam Nutr Res.* 2019 Jul 30:1-13.

Darooghegi Mofrad, M., Mozaffari, H., Mousavi, S.M., Sheikhi, A., Milajerdi, A.(b). (2019). The effects of psyllium supplementation on body weight, body mass index and waist circumference in adults: A systematic review and dose-response meta-analysis of randomized controlled trials. *Crit Rev Food Sci Nutr.* 18:1-14.

Dey, A. (2011). Achyranthes aspera L: Phytochemical and pharmacological aspects. *International Journal of Pharmaceutical Sciences Review and Research* Volume 9, Issue 2, 2011; Article-013 Volume 9, Issue 2, Article-013 ISSN 0976 – 044X

Delphi, L., Sepehri, H. Apple pectin: (2016). A natural source for cancer suppression in 4T1 breast cancer cells in vitro and express p53 in mouse bearing 4T1 cancer tumors, *in vivo. Biomed Pharmacother.* 84:637-644.

De Bona, K.S., Bellé, L. P., Bittencourt, P. E. R., Bonfanti, G., Cargnelluti, L. O., Pimentel, V. C. & Moretto, M. B. (2011). Erythrocytic enzymes and antioxidant status in people with type 2 diabetes: Beneficial effect of Syzygium cumini leaf extract *in vitro. Diab. Res. Clin. Practice.*, 94(1): 84-90.

De, D.; Chatterjee, K.; Ali, K. M.; Bera, T. K. & Ghosh, D. (2011). Antidiabetic potentiality of the aqueous-methanolic extract of seed of *Swietenia mahagoni* (L.) Jacq. in streptozotocin-induced diabetic male Albino rat: A correlative and evidence-based approach with antioxidative and antihyperlipidemic activities. Evid. Based. Complement. *Alternat. Med.*, 8: 1-11.

Dzeja, P. and Terzic. (2009). Adenylate Kinase and AMP Signaling Networks: Metabolic Monitoring, Signal Communication and Body Energy Sensing. *Int. J. Mol. Sci.*, 10: 1729-1772.

Falguni F. Z, Md. A. Islam, Md. M. Hasan, S. M. M. M. H. Mousum, Md. Ashraduzzaman, S. Khatun. (2017). Antioxidant and Antidiabetic Properties of *Vitex nigundo* L. Leaves Most. *American Journal of Life Sciences* 5(1): 21-26.

Fasola, T. R., Ayodele, A. E., Odetola, A. A., & Umotok, N. E. (2010). Foliar epidermal morphology and anti-diabetic property of Andrographis paniculata (Burm. f.) Wall ex. Nees. Ethnobotanical. *Leaflets.*, 14: 593-598.

Fatima, S. S., Rajasekhar, M. D., Kumar, K. V., Kumar, M. T. S., Babu, K. R., & Rao, C. A. (2010). Antidiabetic and antihyperlipidemic activity of ethyl acetate: Isopropanol (1: 1) fraction of Vernonia anthelmintica seeds in Streptozotocin induced diabetic rats. *Food. Chem. Toxicol.*, 48(2): 495-501.

François, M. G.; Tehoua, L.; Ouattara, H.; Yapi, A. (2014). Comparative of the antihyperglycemic activity of Sclerocarya birrea, Khaya senegalensis, Heliotropium indicum and Ocimum gratissimum to rats wistar. *Amr. J. Biosci.*, 2(2): 60-63.

Fu S, Zhang Y, Shi J, Hao D, Zhang P. (2019). Identification of gene-phenotype connectivity associated with flavanone naringenin by functional network analysis. *Peer J.* 2019 Mar 19;7:e6611. doi: 10.7717/peerj.6611. eCollection.

Ganeshpurkar, A.; Diwedi, V. and Bhardwaj, Y. (2013). *In vitro* α -amylase and α-glucosidase inhibitory potential of Trigonella foenum-graecum leaves extract. *Ayu.*, 34(1): 109-12.

Gauthier, B. R. and Wollheim, C. B. (2008). Synaptotagmins bind calcium to release insulin. Am. *J. Physiol. Endocrinol. Metab.*, 295(6): 1279-86.

Gayathri, M., & Kannabiran, K. (2010). Hypoglycemic effect of 2-hydroxy 4-methoxy benzoic acid isolated from the roots of Hemidesmus indicus on streptozotocin-induced diabetic rats. *Pharmacology*. online., 1: 144-54.

Gayathri, M., & Kannabiran, K. (2012). Effect of 2-hydroxy-4-methoxy benzoic acid isolated from Hemidesmus indicus on erythrocyte membrane bound enzymes and antioxidant status in streptozotocin-induced diabetic rats. *Ind. J. Pharmaceut. Sci.*, 74(5): 474-478.

Geetha G, K. Gopinathapillai P, Sankar V. Anti diabetic effect of Achyranthes rubrofusca leaf extracts on alloxan induced diabetic rats. *Pak J Pharm Sci*. 2011 Apr;24(2):193-9.

Ghosal, M. and Mandal P. (2013). *In-vitro* antidiabetic and antioxidant activity of calamus erectus roxb. Fruit: a wild plant of darjeeling himalaya. *Intl. J. Pharma & Bio. Sci.*, 4(2): 671-684.

Ghosh, S., & Suryawanshi, S. A. (2001). Effect of Vinca rosea extracts in treatment of alloxan diabetes in male albino rats. *Ind. J. Exptl. Biol.*, 39(8): 748-759.

Girón, M. D., Sevillano, N., Salto, R., Haidour, A., Manzano, M.; Jiménez, M. L. & López-Pedrosa, J. M. (2009). Salacia oblonga extract increases glucose transporter 4-mediated glucose uptake in L6 rat myotubes: Role of mangiferin. *Clin. Nut.*, 28(5): 565-574.

Goldoni FC, Barretta C, Nunes R, Broering MF, De Faveri R, Molleri HT, Corrêa TP, Farias IV, Amorin CK, Pastor MVD, Meyre-Silva C, Bresolin TMB, de Freitas RA, Quintão NLM, Santin JR. (2019). Effects of Eugenia umbelliflora O. Berg (Myrtaceae)-leaf extract on inflammation and hypersensitivity. *J Ethnopharmacol.* 1;244:112133

Hu FB and Willett WC. 2002). Optimal diets for prevention of coronary heart disease. *JAMA*. 288: 2569-2578.

Grover, J. K.; Yadav, S. And Vats, V. (2002). Medicinal plants of India with hypoglycemic potentials. *J. Ethnopharmacol.*, 81: 81–100.

Gunjan, M.; Goutam.; Jana, K.; Jha, A. K.; Mishra, U. (2010). Pharmacognostic and antihyperglycemic study of Coccinia indica. *Intl. J. Phytomed.*, 2: 36-40.

Gupta, R., and Gupta, R. S. (2009). Effect of Pterocarpus marsupium in streptozotocin-induced hyperglycemic state in rats: comparison with glibenclamide. *Diabetol. Croat.*, 38: 39-45.

Gupta, R., Bajpai, K. G. and Saxena, A. M. (2008). An Overview of Indian Novel Traditional Medicinal Plants with Anti-Diabetic Potentials. *African J. Traditional.* Complementary. *Alternative. Med.*, **5**(1): 1-17.

Hardikar MR, Varma ME, Kulkarni AA, Kulkarni PP, Joshi BN. Elucidation ofhypoglycemic action and toxicity studies of insulin-like protein from Costus igneus. *Phytochemistry*. 2016;124: 99-107.

Hegazy G A., , A M. Alnoury, , H G. Gad, (2013). The role of Acacia arabica extract as an antidiabetic, antihyperlipidemic, and antioxidant in streptozotocin-induced diabetic rats. *Saudi Med J.* 34 (7): 727-733

Hannan, J. M.; Ali, L.; Khaleque, J.; Akhter, M.; Flatt, P. R. & Wahab, A. Y. H. (2011). Antihyperglycaemic activity of Asparagus racemosus roots is partly mediated by inhibition of carbohydrate digestion and absorption, and enhancement of cellular insulin action. *Br. J. Nutr.*, 8: 1-8.

Hasan, S. M.; Khan, M.; Umar, B. U.; Sadeque, M. Z. (2013). Comparative study of the effect of ethanolic extract of Swietenia mahagoni seeds with rosiglitazone on experimentally induced diabetes mellitus in rats. Bangladesh. *Med. Res. Counc. Bull.* 39(1): 6-10.

Hegde PK, Rao HA, Rao PN. (2014). A review on Insulin plant (*Costus igneus* Nak). *Pharmacogn Rev.* 8(15):67-72.

Hou, W.; Li, Y.; Zhang, Q.; Wei, X.; Peng, A.; Chen, L. & Wei, Y. (2009). Triterpene acids isolated from Lagerstroemia speciosa leaves as α-glucosidase inhibitors. *Phytother. Res.*, 23(5): 614-618.

Huang, Y. N.; Zhao, D. D.; Gao, B.; Zhong, K.; Zhu, R. X.; Zhang, Y. & Gao, H. (2012). Anti-hyperglycemic effect of chebulagic acid from the fruits of Terminalia chebula *Retz. Intl. J. Mol. Sci.*, 13(5): 6320-6333.

Huang CN, Wang CJ, Lee YJ, Peng CH. (2017). Active subfractions of Abelmoschus esculentus substantially prevent free fatty acid-induced β cell apoptosis via inhibiting dipeptidyl peptidase-4. *PLoS One.* 17;12(7):e0180285.

Huang CN, Wang CJ, Lin CL, Lin HT, Peng CH. (2017). The nutraceutical benefits of subfractions of Abelmoschus esculentus in treating type 2 diabetes mellitus. *PLoS One.* 7; 12(12): e0189065.

Ibraheim, Z. Z.; Ahmed, A. S. & Gouda, Y. G. (2011). Phytochemical and biological studies of Adiantum capillus-veneris L. Saudi. *Pharmaceut. J.*, 19(2): 65-74.

Jane M, McKay J, Pal S. (2018). Effects of daily consumption of psyllium, oat bran and polyGlycopleX on obesity-related disease risk factors: A critical review. Nutrition. 2019 Jan;57:84-91. doi: 10.1016/j.nut.2018.05.036. Epub 2018 Jul 12.

Jaiswal, N.; Bhatia, V.; Srivastava, S. P.; Srivastava, A. K. & Tamrakar, A. K. (2012). Antidiabetic effect of Eclipta alba associated with the inhibition of alpha-glucosidase and aldose reductase. *Natural. Product. Res.*, 26(24): 2363-2367.

Jiang W, Si L, Li P, Bai B, Qu J, Hou B, Zou H, Fan X, Liu Z, Liu Z, Gao L. (2018). Serum metabonomics study on antidiabetic effects of fenugreek flavonoids in streptozotocin-induced rats. J Chromatogr B Analyt *Technol Biomed Life Sci.* 15; 1092:466-472.

Jianfang Fu, Jufang Fu, Jun Yuan, Nanyan Zhang, Bin Gao, Guoqiang Fu, Yanyang Tu, Yongsheng Zhang. (2012). Anti-diabetic activities of *Acanthopanax senticosus* polysaccharide (ASP) in combination with metformin. *International Journal of Biological Macromolecule*s Vol. 50, Issue 3, 1: 619–623.

Jianfang Fu, Jun Yuan, YanYang Tu, Jufang Fu, Nanyan Zhang, Bin Gao, Guoqiang Fu,Yongsheng Zhang. (2012). A polysaccharide from *Acanthopanax senticosus* improves the antioxidant status in alloxan-induced diabetic mice. *Carbohydrate Polymers*Volume 88, Issue 2, 517–521.

Joshi BN, Munot H, Hardikar M, Kulkarni AA. (2013). Orally active hypoglycemic protein from Costus igneus N. E. Br.: an *in vitro* and *in vivo* study *Biochem Biophys Res Commun.* 436(2):278-82.

Kang, M. H., Lee, M. S., Choi, M. K., Min, K. S., & Shibamoto, T. (2012). Hypoglycemic activity of Gymnema sylvestre extracts on oxidative stress and antioxidant status in diabetic rats. *J. Agric. Food Chem.*, 60(10): 2517-2524.

Kamalakkannan N and Prince PS. (2005). The effect of Aegle marmelos fruit extract in streptozotocin diabetes: a histopathological study. *J Herb Pharmacother.* 5(3):87-96.

Kasabri, V.; Flatt, P. R.; & Abdel-Wahab, Y. H. (2010). Terminalia bellirica stimulates the secretion and action of insulin and inhibits starch digestion and protein glycation *in vitro. Brit. J. Nut.*, 103(02): 212-217.

Kavitha, K. N. and Dattatri, A. N. (2013). Experimental Evaluation of antidiabetic activity of Swertia Chirata – Aqueous Extract. *J. Pub. Health. Med. Res.* 1(2): 71-75

Keller, A. C.; Ma, J.; Kavalier, A.; He, K.; Brillantes, A. M. B. & Kennelly, E. J. (2011). Saponins from the traditional medicinal plant Momordica charantia stimulate insulin secretion *in vitro. Phytomed.*, 19(1): 32-37.

Khan, K. Y.; Khan, M. A.; Ahmad, M.; Hussain, I.; Mazari, P.; Fazal, H.; Ali, B. and Khan, I. Z. Hypoglycemic potential of genus Ficus L.: A review of ten years of Plant Based Medicine used to cure Diabetes. *J. Appl. Pharma. Sci.* 1(06): 223-227.

Kim, K.; Kim, H.; Kwon, J.; Lee, S.; Kong, H.; Im, S. A. & Kim, K. (2009). Hypoglycemic and hypolipidemic effects of processed Aloe vera gel in a mouse model of non-insulin-dependent diabetes mellitus. *Phytomed.*, 16(9): 856-863.

Kim HJ, Jung SW, Kim SY, Cho IH, Kim HC, Rhim H, Kim M, Nah SY. (2018) *Panax* ginseng as an adjuvant treatment for Alzheimer's disease. *J Ginseng Res.* Oct;42(4):401-411. doi: 10.1016/j.jgr.2017.12.008. Epub 2018 Jan 12.

Ko JH, Arfuso F, Sethi G, Ahn KS. (2018). Pharmacological Utilization of Bergamottin, Derived from Grapefruits, in Cancer Prevention and Therapy. *Int J Mol Sci.* 14;19(12). pii: E4048.

Kondeti VK, Badri KR, Maddirala DR, Thur SK, Fatima SS, Kasetti RB, Rao CA. (2010). Effect of Pterocarpus santalinus bark, on blood glucose, serum lipids, plasma insulin and hepatic carbohydrate metabolic enzymes in streptozotocin-induced diabetic rats. *Food Chem Toxicol.* 48(5):1281-7.

Kulczyński B, Gramza-Michałowska A. (2016). The importance of selected spices in cardiovascular diseases. *Postepy Hig Med Dosw* (Online). Nov 14;70(0):1131-1141.

Kumar, M.; Gautam, M. K.; Singh, A. & Goel, R. K. (2013). Healing effects of Musa sapientum var. paradisiaca in diabetic rats with co-occurring gastric ulcer: Cytokines and growth factor by PCR amplification. BMC. Complement *Altern Med.*, 13(1): 305.

Kumar, S.; Kumar, V. & Prakash, O. (2011). Antidiabetic, hypolipidemic and histopathological analysis of Dillenia indica (L.) leaves extract on alloxan induced diabetic rats. Asian. Pacific. *J. Tropical. Med.*, 4(5): 347-352.

Kumar, V., Ahmed, D., Verma, A., Anwar, F., Ali, M., & Mujeeb, M. (2013). Umbelliferone beta-D-galactopyranoside from Aegle marmelos (L.) corr. an ethnomedicinal plant with antidiabetic, antihyperlipidemic and antioxidative activity. BMC Complement. *Altern. Med.*, 3(1): 273.

Kumar V, Ahmed D, Gupta PS, Anwar F, Mujeeb M. (2013). Anti-diabetic, anti-oxidant and anti-hyperlipidemic activities of Melastoma malabathricum Linn. leaves in streptozotocin induced diabetic rats. BMC Complement *Altern Med.* 9; 13:222.

Kumari K and Augusti KT. (2002) Antidiabetic and antioxidative effects of S-methyl cysteine sulfoxide isolated from onions(*Allium cepa* Linn) as compared to standard drugs in alloxan diabetic rats. *Indian J Exp Biol.* 40:1005–1009

Kumar A, D'silva M, Dholakia K, Levenson AS. (2018). *In Vitro* Anticancer Properties of Table Grape Powder Extract (GPE) in Prostate Cancer. *Nutrients.*10(11). pii: E1804. doi: 10.3390/nu10111804.

Latha, R. & Daisy, P. (2011). Insulin-secretagogue, antihyperlipidemic and other protective effects of gallic acid isolated from Terminalia bellerica Roxb. in streptozotocin-induced diabetic rats. *Chemico-Biol. Interactions.*, 189(1): 112-118.

Lee, J., & Pilch, P. F. (1994). The insulin receptor: structure, function, and signaling. *Am. J. Physiol. Cell. Physiol.*, 266(2): 319-334.

Lima, C. R.; Vasconcelos, C. F.; Costa-Silva, J. H.; Maranhão, C. A.; Costa, J.; Batista, T. M.; Carneiro, E. M.; Soares, L. A.; Ferreira, F.; Wanderley, A. C. (2012). Anti-diabetic activity of extract from *Persea americana* Mill. leaf via the activation of protein kinase B (PKB/Akt) in streptozotocin-induced diabetic rats. *J. Ethnopharmacol.*, 141: 517–525.

Liu IM, Liou SS, Lan TW, Hsu FL, Cheng JT. (2005). Myricetin as the active principle of Abelmoschus moschatus to lower plasma glucose in streptozotocin-induceddiabetic rats. *Planta Med.* 71(7):617-21.

Liu IM, Tzeng TF, Liou SS, Lan TW. (2007). Improvement of insulin sensitivity in obese Zucker rats by myricetin extracted from Abelmoschus moschatus. *Planta Med.* 73(10):1054-60.

Liu IM, Tzeng TF, Liou SS, Lan TW. (2007). Myricetin, a naturally occurring flavonol, ameliorates insulin resistance induced by a high-fructose diet in rats. *Life Sci.* 10;81(21-22):1479-88.

Liu IM, Tzeng TF, Liou SS. (2010). Abelmoschus moschatus (Malvaceae), an aromatic plant, suitable for medical or food uses to improve insulin sensitivity. *Phytother Res.* 24(2):233-9.

Magrone T, Perez de Heredia F, Jirillo E, Morabito G, Marcos A, Serafini M. (2013). Functional foods and nutraceuticals as therapeutic tools for the treatment of diet-related diseases. *Can J Physiol Pharmacol.* 91(6):387-96.

Mahajan S., P. Chauhan, M. Mishra, D. Yadav, M. Debnath and GBKS Prasad. (2018). Antidiabetic Potential of Eugenia jambolana Ethanolic Seed Extract: Effect on Antihyperlipidemic and Antioxidant in Experimental Streptozotocin-Induced Diabetic Rats. *Advances in Complementary & Alternative Medicine.* 2(3), 1-9.

Majewska-Wierzbicka M, Czeczot H. (2012). Flavonoids in the prevention and treatment of cardiovascular diseases. *Pol Merkur Lekarski.* 32(187):50-4.

Maleki SJ, Crespo JF, Cabanillas B. (2019). Anti-inflammatory effects of flavonoids. *Food Chem.* 30;299:125124.

Mariadoss AVA, Vinyagam R, Rajamanickam V, Senthilkumar V, Venkatesan S, David E. (2019). Pharmacological aspects and potential use of phloretin: a systemic review. *Mini Rev Med Chem.* doi: 10.2174/1389557519666190311154425.

Manikandan, R., Anand, A. V. and Muthumani, G. D. (2013). Phytochemical and in vitro anti-diabetic activity of methanolic extract of Psidium guajava leaves. *Int. J. Curr. Microbiol. App. Sci.* 2(2): 15-19.

Mao QQ, Xu XY, Cao SY, Gan RY, Corke H, Beta T, Li HB. (2019). Bioactive Compounds and Bioactivities of Ginger (*Zingiber officinale* Roscoe). *Foods.* 30;8(6). pii: E185. doi: 10.3390/foods8060185.

Manikandan, R.; Thiagarajan, R.; Beulaja, S.; Sivakumar, M. R.; Meiyalagan, V.; Sundaram, R.; & Arumugam, M. (2011). 1, 2 di-substituted idopyranose from Vitex negundo l. Protects against streptozotocin-induced diabetes by inhibiting nuclear factor-kappa B and inducible nitric oxide synthase expression. *Microscopy. Res. Technique.*, 74(4): 301-307.

Micheli L, Mattoli L, Maidecchi A, Pacini A, Ghelardini C, Di Cesare Mannelli L. Effect of **Vitis** vinifera hydroalcoholic extract against oxaliplatin neurotoxicity: *in vitro* and in vivo evidence. Sci Rep. 2018 Sep 25;8(1):14364. doi: 10.1038/s41598-018-32691-w.

Mishra N, D Kumar, SI Rizvi. Protective effect of *Abelmoschus esculentus* against alloxan-induced diabetes in Wistar strain rats. *Journal of dietary supplements*, 2016, 13 (6), 634-646

Misra, P. & Chakrabarti, R. (2007). The role of AMP kinase is diabetes. *Indian. J. Med. Res.* 289-398.

Mohammadi, S., Kouhsari, S. M. and Feshani, A. M. (2010). Antidiabetic properties of the ethanolic extract of Rhus coriaria fruits in rats. *J. Faculty Pharmacy.*, 18(4): 270-275.

Mohankumar, S. K.; O'Shea, T. & McFarlane, J. R. (2012). Insulinotrophic and insulin-like effects of a high molecular weight aqueous extract of Pterocarpus marsupium Roxb hardwood. *J. Ethnopharmacol.*, 141(1): 72-79.

Mooventhan A. and L. Nivethitha. A Narrative Review on Evidence-based Antidiabetic Effect of Fenugreek (*Trigonella Foenum-Graecum*). *Intl. J. Nutr. Pharmacol. And Neurological Disorders.* 2017, 7 (4), 84-87.

Mootoosamy, A. and Mahomoodally, F. M. (2014). Ethnomedicinal application of native remedies used against diabetes and related complications in Mauritius. *J. Ethnopharmacol.* 151(1): 413-44.

Morshed, M. A.; Haque, A.; Rokeya, B.; & Ali, L. (2011). Anti-hyperglycemic and lipid lowering effect of Terminalia arjuna bark extract on Streptozotocin induced Type-2 Diabetic Model Rats. *Intl. J. Pharma. Pharmaceut. Sci.*, 3(4): 449-453.

Muhammad A, Ibrahim MA, Erukainure OL, Malami I, Adamu A. Spices with Breast Cancer Chemopreventive and Therapeutic Potentials: A Functional Foods Based-Review. *Anticancer Agents Med Chem.* 2018;18(2):182-194.

Mukherjee, S.; Mitra, A.; Dey, S. & Thakur, G. (2010). Alpha-amylase activity of tannin isolated from Terminalia chebula. *Intl Conf on Med and Biol.*, 443-445.

Mukhtar M. H., W. H. Almalki, A. Azmat, M. R. Abdalla and M. Ahmed. Evaluation of anti-diabetic activity of Acacia tortilis (Forssk.) Hayne leaf extract in streptozotocin-induced diabetic rats. *Intl. J. Pharmacology.* 13 (5): 438-447, 2017.

Mwangi J Mukundi, Ngugi M Piero, Njagi EN Mwaniki, Njagi J Murugi, Agyirifo S Daniel, Gathumbi K Peterand Muchugi N Alice. (2015). Antidiabetic effects of aqueous leaf extracts of Acacia nilotica in Alloxan induced diabetic mice. *Journal of Diabetes & Metabolism.* 6:7, 1000568.

Munasinghe, M. A. A. K.; Abeysena, C.; Yaddehige, I. S.; Vidanapathirana, T.; & Piyumal, K. P. B. (2011). Blood sugar lowering effect of Coccinia grandis (L.) J. Voigt: path for a new drug for diabetes mellitus. Exptl. Diab. Res., Article ID 978762, 4 pages

Nayak, B. S.; Marshall, J. R.; Isitor, G.; & Adogwa, A. (2010). Hypoglycemic and hepatoprotective activity of fermented fruit juice of *Morinda citrifolia* (Noni) in diabetic rats. Evid. Based. Complement. *Alternat. Med. Article* ID 875293, 5 pages

Neeraj, O.; Maheshwari. and Khan, A. And Chopade, A. B. (2013). Rediscovering the medicinal properties of Datura sp.: *A review. J. Med. Plants Res.*, 7(39): 2885-2897.

Nim, D. K.; Shankar, P.; Chaurasia, R., Goel, B., Kumar, N., & Dixit, R. K. (2013). Clinical evaluation of anti-hyperglycemic activity of Boerhaavia diffusa and Ocimum sanctum extracts in streptozocin induced T2DM rat models. *Int. J. Pharm.*, 4(1): 30-34.

Nobutomo Ikarashi, Takahiro Toda, Takehiro Okaniwa, Kiyomi Ito, Wataru Ochiai, and Kiyoshi Sugiyama. (2011). Anti-obesity and anti diabetic effects of Acacia polyphenol in obese diabetic KKAy mice fed high-fat diet. *Evid Based Complement Alternat Med.* 2011: 952031. PMCID : PMC3137845.

Ozougwu, Jevas C. Anti-diabetic effects of Allium cepa (onions) aqueous extracts on alloxan-induced diabetic Rattus novergicus. *Journal of Medicinal Plants Research* 5(7), 1134-1139, 2011.

Panda, A. K.; Das, M. C.; Panda, P. K.; Kumar, S. & Pani, S. R. (2013). In-vivo, anti-hyperglycemic and anti-hyperlipidemic activity of annona squamosa (linn.) leaves, collected from southern odisha. World. *J. Pharma. and Pharmaceutic. Sci.* 2(5): 3347-3359.

Pandit, R.; Phadke, A. and Jagtap, A (2010). Antidiabetic effect of Ficus religiosa extract in streptozotocin-induced diabetic rats. *J. Ethnopharmacol.*, 128(2): 462-466.

Park J Y, Ji H-D, Lee W-M, Park EY, Jeong K-S, Kim H-K, Cho J-H, Baik S-O and Rhee MH. (2013) Antidiabetic effects of fermented Acanthopanax senticosus on rats with streptozotocin-induced type 1 diabetic mellitus. *J. Med. Plant Res.*, 7(27), 1994-2000.

Parvin, A; Alam, M; Haque, M; Bhowmik, A; Ali, L. & Rokeya, B. (2013). Study of the Hypoglycemic Effect of *Tamarindus* indica Linn. Seeds on Non-Diabetic and Diabetic Model Rats. *Brit. J. Pharmaceut. Res.*, 3(4): 1094-1105.

Patil, R.; Patil, R.; Ahirwar, B.; & Ahirwar, D (2011). Isolation and characterization of anti-diabetic component (bioactivity—guided fractionation) from *Ocimum sanctum* L.(Lamiaceae) aerial part. *Asian. Pacific J. Tropical. Med.*, 4(4): 278-282.

Peng CH, Lin HC, Lin CL, Wang CJ, Huang CN. (2019). Abelmoschus esculentus subfractions improved nephropathy with regulating dipeptidyl peptidase-4 and type 1 glucagon-like peptide receptor in type 2 diabetic rats. *J Food Drug Anal.* (1):135-144.

Perez-Pardo P, Kliest T, Dodiya HB, Broersen LM, Garssen J, Keshavarzian A, Kraneveld AD. (2017). The gut-brain axis in Parkinson's disease: Possibilities for food-based therapies. *Eur J Pharmaco*l. 15;817:86-95.

Phoboo, S.; Pinto, S. M.; Barbosa, A. C.; Sarkar, D.; Bhowmik, P. C.; Jha, P. K.; Shetty, K. (2013). Phenolic-linked biochemical rationale for the anti-diabetic properties of Swertia chirayita (Roxb. ex Flem.) Karst. *Phytother. Res.*, 27(2): 227-235.

Poonam, T.; Prakash, G. P.; Kumar, L. V. (2011). Influence of Allium sativum extract on the hypoglycemic activity of glibenclamide: an approach to possible herb-drug interaction. *J. Ethnopharmacol.*, 138(2): 345-50.

Puranik N., K. F. Kammar , S. Devi. (2010). Anti-diabetic activity of *Tinospora cordifolia* (Willd.) in streptozotocin diabetic rats; does it act like sulfonylureas? *Turk J Med Sci* 40 (2): 265-270.

Rahimlou M, Yari Z, Rayyani E, Keshavarz SA, Hosseini S, Morshedzadeh N, Hekmatdoost A. (2019). Effects of ginger supplementation on anthropometric, glycemic and metabolic parameters in subjects with metabolic syndrome: A randomized, double-blind, placebo-controlled study. *J Diabetes Metab Disor*d. 18(1):119-125.

Rahmatullah, M.; Sultan, S.; Toma, T.; Lucky, S.; Chowdhury, M.; Haque, W. & Jahan, R. (2010). Effect of Cuscuta reflexa stem and Calotropis procera leaf extracts on glucose tolerance in glucose-induced hyperglycemic rats and mice. *Afr. J. Tradit. Complement. Altern. Med.*, 7(2): 109-112

Rai, P. K.; Jaiswal, D.; Rai, D. K.; Sharma, B.; & Watal, G. (2010). Antioxidant potential of oral feeding of cynodon dactylon extract on diabetes-induced oxidative stress. *J. Food. Biochem.*, 34(1): 78-92.

Ramachandran, V., & Saravanan, R. (2013). Efficacy of Asiatic acid, a pentacyclic triterpene on attenuating the key enzymes activities of carbohydrate metabolism in streptozotocin-induced diabetic rats. *Phytomed.*, 20(3-4): 230-6.

Ramakrishnan. M.; Bhuvaneshwari, R.; Duraipandyan, V. and Dhandapani, R. (2011). Hypoglycaemic activity of *Coccinia indica* Wight & Arn. fruits in Alloxan-induced diabetic rats. Ind. *J. Natural. Products. Resources.*, 2(3): 350-353.

Rawat Mukesh and Parmar Namita Medicinal Plants with Antidiabetic Potential - A Review. *American-Eurasian J. Agric. & Environ. Sci.*, 13 (1): 81-94, 2013.

Rissanen TH, Voutilainen S, Virtanen JK, Venho B, Vanharanta M, Mursu J and Salonen JT. Low Intake of Fruits, Berries and Vegetables Is Associated with Excess Mortality in Men: the Kuopio Ischaemic Heart Disease Risk Factor (KIHD) Study. *J Nutr.* 2003; 133: 199-204.

Roozbeh Nasibeh, Leili Darvish and Fatemeh Abdi. Hypoglycemic effects of Acacia nilotica in type II diabetes: a research proposal.*BMC Res Notes.* 2017; 10: 331.

Sharma P, McClees SF, Afaq F. (2017). Pomegranate for Prevention and Treatment of Cancer: *An Update. Molecules.* 24; 22(1).

Salehi B, Fokou PVT, Yamthe LRT, Tali BT, Adetunji CO, Rahavian A, Mudau FN, Martorell M, Setzer WN, Rodrigues CF, Martins N, Cho WC, Sharifi-Rad J. Phytochemicals in Prostate Cancer: From Bioactive Molecules to Upcoming Therapeutic Agents. Nutrients. 2019 Jun 29;11(7). pii: E1483. doi: 10.3390/nu11071483.

Salehi B, Berkay Yılmaz Y, Antika G, Boyunegmez Tumer T, Fawzi Mahomoodally M, Lobine D, Akram M, Riaz M, Capanoglu E, Sharopov F, Martins N, Cho WC, Sharifi-Rad J. (2019). Insights on the Use of α-Lipoic Acid for Therapeutic Purposes. *Biomolecules.* 9;9(8). pii: E356. doi: 10.3390/biom9080356. Review

Sabitha V, Ramachandran Subramaniam, K R Naveen and K Panneerselvam. (2011). Antidiabetic and antihyperlipidemic potential of *Abelmoschus esculentus* (L.) Moench. in streptozotocin-induced diabetic rats. *J.Pharm Bioallied Sci* 3(3): 397-402.

Sabu MC, Kuttan R. (2004). Antidiabetic activity of Aegle marmelos and its relationship with its antioxidant properties. *Indian J. Physiol Pharmacol.* 48(1):81-8.

Sankar, P.; Subhashree, S. and Sudharani, S. (2012). Effect of Trigonella foenum-graecum seed powder on the antioxidant levels of high fat diet and low dose streptozotocin induced type II diabetic rats. *Eur. Rev. Med. Pharmacol. Sci.* 16(3): 10-7.

Sankeshi, V.; Kumar, P. A.; Naik, R. R.; Sridhar, G.; Kumar, M. P.; Gopal, V. V.; Raju, T. N. (2013). Inhibition of aldose reductase by Aegle marmelos and its protective role in diabetic cataract. *J. Ethnopharmacol.*, 149(1): 215-221.

Saxena, A. and Vikram, N. K. (2004). Role of Selected Indian Plants in Management of Type 2 Diabetes: A Review. *J. Alternative. Complementary. Med.*, 10(2): 369-378.

Shareef, S. M.; Sridhar, I.; Mishra, S. S.; & Venkata Rao, Y. (2013). Evaluation of hypoglycemic effect of Lagerstroemia speciosa (Banaba) leaf extract in alloxan induced diabetic rabbits. *Intl. J. Med. Res. Health Sci.*, 2(2): 217-222.

Sharma, B.; Salunke, R.; Balomajumder, C.; Daniel, S.; & Roy, P. (2010). Anti-diabetic potential of alkaloid rich fraction from Capparis deciduas on diabetic mice. *J.Ethnopharmacol.*, 127(2): 457-462.

Shibib, B. A.; Amin, M. A.; Hasan, A. K.; Rahman, R. (2012). A creeper, Coccinia indica, has anti-hyperglycaemic and anti-ureogenic effects in diabetic rats. *J. Pak. Med. Assoc.* 62(11): 1145-8.

Shih CC, Shlau MT, Lin CH, Wu JB. (2014). Momordica charantia ameliorates insulin resistance and dyslipidemia with altered hepatic glucose production and fatty acid synthesis and AMPK phosphorylation in high-fat-fed mice. *Phytother Res.* 28(3):363-71.

Shirazi-Beechey, S. P., Moran, A. W., Batchelor, D. J., Daly, K., Al-Rammahi, M. (2011). Glucose sensing and signalling; regulation of intestinal glucose transport. *Proc. Nutr. Soc.* 70 (2): 185-93.

Singh, S. K.; Rai, P. K.; Mehta, S.; Singh, R. K.; & Watal, G. (2009). Curative effect of Cynodon dactylon against STZ induced hepatic injury in diabetic rats. *Ind. J. Clin. Biochem.*, 24(4): 410-413.

Sole, S. S and Srinivasan, B. P. (2012). Aqueous extract of tamarind seeds selectively increases glucose transporter-2, glucose transporter-4, and islets' intracellular calcium levels and stimulates β-cell proliferation resulting in improved glucose homeostasis in rats with streptozotocin-induced diabetes mellitus. *Nutr. Res.* 32(8): 626-36.

Solfrizzi V, Custodero C, Lozupone M, Imbimbo BP, Valiani V, Agosti P, Schilardi A, D'Introno A, La Montagna M, Calvani M, Guerra V, Sardone R, Abbrescia DI, Bellomo A, Greco A, Daniele A, Seripa D, Logroscino G, Sabbá C, Panza F. (2017). Relationships of Dietary Patterns, Foods, and Micro- and Macronutrients with Alzheimer's Disease and Late-Life Cognitive Disorders: A Systematic Review. *J Alzheimers Dis.* 59(3):815-849. doi: 10.3233/JAD-170248.

Simunkova M, Alwasel SH, Alhazza IM, Jomova K, Kollar V, Rusko M, Valko M. (2019). Management of oxidative stress and other pathologies in Alzheimer's disease. *Arch Toxicol.* doi: 10.1007/s00204-019-02538-y.

Somani, R., Singhai, A. K., Shivgunde, P., & Jain, D. (2012). Asparagus racemosus Willd (Liliaceae) ameliorates early diabetic nephropathy in STZ induced diabetic rats. *Ind. J. Exptl. Biol.* 50 (7): 469-475.

Soni, R.; Mehta, N. M.; Srivastava, D. N. (2013). Effect of ethanolic extract of cinnamomum tamala leaves on wound healing in stz induced diabetes in rats. *Asian. J. Pharm. Clin. Res.*, 6(4): 39-42.

Sosuke Ogawa, Tomoyuki Matsumae, Takeshi Kataoka, Yoshikazu Yazaki, Hideyo Yamaguchi. Effect of acacia polyphenol on glucose homeostasis in subjects with impaired glucose tolerance: A randomized multicenter feeding trial. *Experimental and Therapeutic Medicine* 06/2013; 5(6):1566-1572.

Subramoniam A., P. Pushpangadan, S. Rajasekharan, D.A. Evans, P.G. Latha and R.Valsaraj. Effects of Artemisia pallens Wall. on blood glucose levels in normal and alloxan-induced diabetic rats. *Journal of Ethnopharmacol.* Vol. 50 (1), 1996, 13-17

Sundaram, R.; Naresh, R.; Shanthi, P. & Sachdanandam, P. (2012). Antihyperglycemic effect of iridoid glucoside, isolated from the leaves of Vitex negundo in streptozotocin-induced diabetic rats with special reference to glycoprotein components. *Phytomed.*, 19(3): 211-216.

Sutar, N. G.; Sutar, U. N.; & Behera, B. C (2009). Antidiabetic activity of the leaves of Mimosa pudica Linn. in albino rats. *J. Herbal. Med. Toxicol.*, 3(1): 123-126.

Tag, H.; Kalita, P.; Dwivedi, P.; Das, A. K. Namsa, N. D (2012). Herbal medicines used in the treatment of diabetes mellitus in Arunachal Himalaya, northeast, India. *J. Ethnopharmacol.*, 141(3): 786-95.

Talukder FZ, Khan KA, Uddin R, Jahan N, Alam MA. In vitro free radical scavenging and anti-hyperglycemic activities of Achyranthes aspera extract in alloxan-induceddiabetic mice. *Drug Discov Ther.* 2012;6(6):298-305.

Tanaka T, Iuchi A, Harada H, Hashimoto S. (2019)Potential Beneficial Effects of Wine Flavonoids on Allergic Diseases. Diseases. Jan 15;7(1). pii: E8. doi: 10.3390/diseases7010008

Teja, B. B., Subbarao, M., Basha, D. P. & Kumar, K. P. (2011). Antidiabetic activity on extracts of Mangifera indica in Alloxan monohydrate induced diabetic rats. *Drug Invention Today.*, 3(7): 165-168.

Temple WJ and Gladwin KK. Fruits, vegetables, and the. Prevention of cancer: Research challenges. *Nutrition.* 2003; 19: 467-470.

Tolhurst, G.; Reimann, F.; Gribble, F. M. (2012). Intestinal sensing of nutrients. Handbook. *Exp. Pharmacol.* 209: 309-35.

Tomar, R. S. and Sisodia S. S. (2012). Antidiabetic Activity Of Annona squamosa L In Experimental Induced Diabetic Rats. *Intl. J. Pharmaceut. Biol. Archive.*, 3(6): 1492-1495.

Tsui PF, Lin CS, Ho LJ, Lai JH. (2018). Spices and Atherosclerosis. *Nutrients.* 10:10(11). pii: E1724. doi: 10.3390/nu10111724. Review.

Udayakumar, R., Kasthurirengan, S., Mariashibu, T. S., Rajesh, M., Anbazhagan, V. R., Kim, S. C. & Choi, C. W. (2009). Hypoglycaemic and hypolipidaemic effects of Withania somnifera root and leaf extracts on alloxan-induced diabetic rats. *Intl. J. Mol. Sci.*, 10(5): 2367-2382.

Udayakumar, R., Kasthurirengan, S., Vasudevan, A., Mariashibu, T. S., Rayan, J. J. S., Choi, C. W., & Kim, S. C. (2010). Antioxidant effect of dietary supplement Withania somnifera L. reduce blood glucose levels in alloxan-induced diabetic rats. *Plant. Foods. Hum. Nut.*, 65(2): 91-98.

Ugochukwu, N.H., Babady, N.E. Antihyperglycemic effect of aqueous and ethanolic extracts of Gongronema latifolium leaves on glucose and glycogen metabolism in livers of normal and streptozotocin-induced diabetic rats. *Life Sci.* 2003 Aug 29;73(15):1925-38.

Veerapur, V. P., Prabhakar, K. R., Thippeswamy, B. S., Bansal, P., Srinivasan, K. K. & Unnikrishnan, M. K. (2012). Antidiabetic effect of Ficus racemosa Linn. stem bark in high-fat diet and low-dose streptozotocin-induced type 2 diabetic rats: A mechanistic study. *Food. Chem.*, 132(1): 186-193.

Verma, V. K., Sarwa, K. K. and Zaman, K. M. D. (2013). Antihyperglycemic activity of Swertia chirayita and andrographis paniculata plant extracts in streptozotocin-induced diabetic rats. *Int. J. Pharm. Pharm. Sci.*, 5(3): 305-311.

Verma R. K., Mishra G., Singh P., K. K. Jha, and R. L. Khosa. (2015)Anti-diabetic activity of methanolic extract of *Alpinia galanga* Linn. aerial parts in streptozotocin induced diabetic rats. Ayu Jan-Mar; 36(1): 91–95.

Vidhya, R., R. Gandhi, G., Jothi, G., Radhika, J., Brindha, P. (2012). Evaluation of antidiabetic potential of Achyranthes aspera linn. On alloxan induced diabetic animals. *International Journal of Pharmacy & Pharmaceutical Sciences*; Supplement 5, Vol. 4, p577.

Wadood, A., Wadood, N., Shah, S.A. (1989). "Effects of Acacia arabica and Caralluma edulis on blood glucose levels of normal and alloxan diabetic rabbits". *J. Pak. Med. Assoc.* 39(8): 208-212.

Wang Y, Wang S, Song R, Cai J, Xu J, Tang X, Li N. (2018) Ginger polysaccharides induced cell cycle arrest and apoptosis in human hepatocellular carcinoma HepG2 cells. *Int J Biol Macromol.* 2019 Feb 15;123:81-90. doi: 10.1016/j.ijbiomac..10.169.

Wang, J., Song, Y., Gao, M., Bai, X., Chen, Z. (2016) Neuroprotective Effect of Several Phytochemicals and Its Potential Application in the Prevention of Neurodegenerative Diseases. Geriatrics (Basel). Nov 12;1(4). pii: E29. doi: 10.3390/geriatrics1040029.

Waheeb Dakhelallah, Mohammad Alharbi and Aisha Azmat. (2011) Hypoglycemic and hypocholesterolemic effects of Acacia tortilis (Fabaceae) Growing In Makkah. *Pakistan Journal of Pharmacology*. Vol.28, No.1, January pp.1-8.

Watson, R.T., Pessin, J.E. (2001)Intracellular organization of insulin signaling and GLUT4 translocation. *Recent Prog Horm Res.* ; 56:175-93.

Weicheng Hu Jin-Hee Yeo, Yunyao Jiang, Seong-Il Heo, and Myeong-Hyeon Wang. The antidiabetic effects of an herbal formula composed of *Alnus hirsuta, Rosa davurica, Acanthopanax senticosus* and *Panax schinseng* in the streptozotocin-induced diabetic rats *Nutr Res Pract.* Apr 2013; 7(2): 103–108.

Wesselman, LMP, Doorduijn, A.S., de Leeuw, F.A., Verfaillie, S.C.J., van Leeuwenstijn-Koopman M, Slot RER, Kester MI, Prins ND, van de Rest O, de van der Schueren MAE, Scheltens P, Sikkes SAM, van der Flier WM. (2019) Dietary Patterns Are Related to Clinical Characteristics in Memory Clinic Patients with Subjective Cognitive Decline: *The Science Project. Nutrients.* May 11;11(5). pii: E1057. doi: 10.3390/nu11051057.

Xu, J., Li, Z., Cao, M., Zhang, H., Sun, J., Zhao, J. & Yang, L. (2012). Synergetic effect of Andrographis paniculata polysaccharide on diabetic nephropathy with andrographolide. *Int. J. Biol. Macromol.*, 51(5):738-42.

Xi, M., Hai, C., Tang, H., Chen, M., Fang, K., Liang, X. (2008). Antioxidant and antiglycation properties of total saponins extracted from traditional Chinese medicine used to treat diabetes mellitus. *Phytother Res.* ; 22(2):228-37.

Yagi, A., Hegazy, S., Kabbash, A. & Wahab, E. A. E. (2009). Possible hypoglycemic effect of Aloe vera L. high molecular weight fractions on type 2 diabetic patients. Saudi. *Pharmaceut. J.*, 17(3): 209-215.

Yang, C., Li, L., Yang, LL.H., Wang, S., Sun, G. (2018). Anti-obesity and Hypolipidemic effects of garlic oil and onion oil in rats fed a high-fat diet. Nutr Metab (Lond). 2018 Jun 20;15:43. doi: 10.1186/s12986-018-0275-x. eCollection.

Yasir Mohammad, Prateek Jain, Debajyoti Debajyoti, M.D. Kharya. (2010) Hypoglycemic and antihyperglycemic effect of different extracts of *Acacia arabica* lamk bark in normal and alloxan induced diabetic rats. *Intl. J. Phytomedicine* 2, 133-138

Zakia, S., Jami, S. I., Ali, E., Begum, M., Haque, M. (2014). Investigation of Antidiabetic Effect of Ethanolic Extract of *Phyllanthus emblica* Linn. Fruits. in experimental animal models. Pharmacology. *Pharmacy.*, 5(1): 11-18.

Zambare, M.R., Bhosale, U.A., Somani, R.S., Yegnanarayan, R., & Talpate, K.A. (2011). *Achyranthes aspera* (Agadha): Herb that improves pancreatic function in alloxan induced diabetic rats. *Asian. J. Pharm. Biol. Res.*, 1(2): 99-104.

Zambare, M., et al. "Effect of Treatment with *Achyranthes aspera* (Agadha) Ethanol Extract on Various Hematological and Biochemical Parameters in Alloxan Induced Diabetic Rats." *Intl. J. Pharma. Frontier. Res.*, 1(1): 42-52.

Zambare, M.R., U.A., Bhosale, R.S., Somani, R. Yegnanarayan, K.A. Talpate. (2011)*Achyranthes aspera* (Agadha): Herb that improves Pancreatic function in Alloxan Induced Diabetic Rats. *Asian J Phar Biol Res.* 1(2): 99-104

Zia-Ul-Haq, M., Ćavar, S., Qayum, M., Imran, I. & Feo, V. D. (2011). Compositional studies: antioxidant and antidiabetic activities of *Capparis decidua* (Forsk.) Edgew. *Intl. J. Mol. Sci.*, 12(12): 8846-8861.

Zhang, Y., Yang, X., Wang, S., Song, S. (2019). Ginsenoside Rg3 prevents cognitive impairment by improving mitochondrial dysfunction in the rat model of Alzheimer's disease. *J. Agric Food Chem*. Aug 18. doi: 10.1021/acs.jafc.9b03793.

Zhao, X., Zhang, M., Li, C., Jiang, X., Su, Y., Zhang, Y. (2019). Benefits of Vitamins in the Treatment of Parkinson's Disease. Oxid Med Cell Longev. Feb 20;2019:9426867.

3

Current Expansions in Microbiology for Food Preservation

K. Ranjitha

Scientist (Microbiology), Division of Post Harvest Technology & Agricultural Engineering, ICAR- Indian Institute of Horticultural Research, Hessaraghatta Lake (PO), Bengaluru- 560 089, Karnataka, India

Introduction

Advances in food preservation methods have supported the journey of food science from that of a cookery science to a fast growing industrial technology. Food preservation knowledge deals with the way to prevent spoilage and maintenance of quality in foods. Spoilage is the deterioration in quality brought about by inherent biochemical and chemical changes, as well as microbial metabolism during food storage. Basic principles of food preservation (Frazier et al., 2008) remains the same, but newer methods for achieving the goal is added through research fuelled by a consumer demand for fresh-like foods with convenience and safety. Scientific information on novel processing methods are beautifully compiled (Cullen et al., 2011; Da-Wen Sun, 2014; Tokuþoðlu and Swanson 2015; Doona, 2018).

In-depth knowledge of physiological activities of micro-organisms in presence of preservation agents is essential for preparation of safe foods with prolonged shelf life. Besides this, it is necessary to understand the effect of food constituents on microbial resistance against inactivation agents, to identify critical dose requirement. This profound knowledge would also help to understand kinetics of microbial inactivation and to develop mathematical models suitable for a realistic design of process parameters. For better appreciation of the following sections, readers are encouraged to have fundamental knowledge in microbiology and food processing. The sections in this chapter cover advanced microbiological knowledge pertaining to the preservation aspects of food.

Food spoilage microorganisms

Microbes use food as a source of nutrients for cell growth and energy. In the process, the food texture, odour, appearance and quality changes. Multiple types of microbial species are often found in spoiled food but specific spoilage organisms (SSO) are spoilage initiators responsible for deteriorative; changes making food unfit for consumption. Common specific spoilage organisms in foods are summarized in Table 1. With newer development of molecular typing methods, scientific names of certain microorganisms have changed in recent years, and some older names are no longer in use.

Table 1: Common microorganisms associated with food spoilage

Bacteria	Yeasts	Molds
Acetobacter	*Candida*	*Aspergillus*
Alicyclobacillus	*Dekkera*	*Botrytis*
Alcaligenes	*Rhodotorula*	*Ceratocystis*
Bacillus	*Saccharomyces*	*Colletotrichum*
Clostridium	*Zygosachharomyces*	*Diploidia*
Desulfotomaculum		*Fusarium*
Erwinia		*Monilinia*
Flavobacterium		*Mucor*
Lactobacillus		*Penicillium*
Leuconostoc		*Phomopsis*
Moraxella		*Rhizopus*
Pediococcus		
Photobacterium		
Proteus		
Pseudomonas		
Serratia		
Streptococcus		
Weissella		

Metabolic properties of food spoilers

Correlation of food spoilage characteristics with the functional properties of spoiler organisms is a difficult task. This is due to the diversity in the food composition and metabolic characteristics of microbes. Spoilage microorganisms are responsible for visual, texture and flavor defects, but these sensory descriptors are not directly correlated with enzymatic functions and metabolic pathways. Some studies reported the detection and measurement of volatile molecules, amines, organic acids etc. in food spoiled by known microorganisms (Table 2). However, correlating the production of these molecules to the metabolic functions of spoilage microorganisms is not always possible; the reason being complexity of reactions in spoilage, serial changes in the first formed degradation products (Eg. meat discolouration), enzymatic reactions originating from both the spoilage organisms and the food matrix (Eg. lytic enzymes of muscle cells) etc. Some of the spoilage enzymes are not well characterized. Usually, the end product molecule in a biological system is formed through multiple pathways and enzymes and identifying the entire enzymatic chain in the context of insufficient knowledge on metabolic properties of the microbes is a challenge (Remenant et al., 2015).

Table 2: Examples for spoilage compounds formed in food due to metabolism of specific spoilage organisms.

Sensory defect	Spoilage product	Spoilage substrate	Food product	Specific spoilage organism
Slime	Dextran	Sugars	Fermented vegetables	*Leuconostoc*
Fishy off odour	Trimethyl amine	Trimethyl amine oxide	Fish	*Shewanella putrefaciens, Photoobacterium phosphoreum,*
Medicinal flavor	2-methoxy phenol	Sugars	Juice	*Alicyclobacillus acidoterrestris*
Cheesy/buttery off flavor	Acetoin	Sugars	Meat	*Enterobacteriaceae*
Fruity smell	Esters		Fish	*Pseudomonas fragii*
Sweet curdling	Proteinaceous fat particles	Phospholipids	Milk	*Bacillus cereus*

Stress adaptation of food microorganisms

Microbes in food are able to outgrow the preservation factors due to the gradual fine tuning of their metabolisms to adapt the changes and thus to maintain cellular homeostasis. The change from a normal physiological situation to a situation of stress is a gradual one. Processing steps present multiple stresses to the organisms by changing the water availability, heating, pH changes, antimicrobial additives etc. Spoilage of processed foods is caused due to a set of organisms capable of overcoming these stresses. Most often, the stress adaptation is achieved through modifications of the cell wall and membrane, which represents 'outer and inner barriers' for the action of antimicrobial agents. Cell membrane adaptations in *Clostridium botulinum* cells resistant to heat and the bacteriocin nisin have been reported (Mazzotta and Montville, 1999). Acid-adaptation of *Listeria monocytogenes* resulted in enhanced tolerance against the biological preservatives nisin and lacticin 3147; and this observation correlated with changes in fatty acid profile of cell membrane (van Schaik et al., 1999). Induced changes in bacterial cell wall have also been reported as a response to the presence of antibacterial compounds. These studies on these aspects were focused on food borne pathogens, rather than on the spoilers, but it is logical to assume that similar mechanisms exist across similar bacterial genera.

Yeast cells adapt against membrane active peptides by significantly increasing the chitin and cell wall protein levels, specifically of cell wall mannoproteins. Role of proton-pumping ATPase, in resistance development against weak organic acids in yeasts are well established. Brul and Coote (1999) have shown that in adapted organisms, transportation of preservatives, sorbic acid, benzoic acid, acetic acid from the cytosol to the extracellular environment occurs. Microarray and proteomics analysis indicated mechanisms such as the activation of heat-shock proteins and indications for the activation of the cell integrity pathway as adaptive responses in weak organic acid resistant cells (Smelt and Brul., 2014).

The phenomenon of adaptation to one type of stress improving resistance of other unrelated stresses is called cross adaptation or cross protection. For example, acid adaptation and starvation induces thermotolerance in *Listeria monocytogenes, Escherichia coli* and *Salmonella*. Thus, initial adaptation of the microbes to one type of stress helps in resistance development and survival of the microorganisms in food. A study by Gahan et al. (1996) revealed that acid adaptation improved the survival of *L.monocytogenes* in low-pH dairy products, including cottage cheese, yogurt, and whole-fat cheddar cheese, and low pH foods such as orange juice and salad dressing. This is because, many stresses can have similar effects on bacterial cells, and bacteria have evolved to build up an all-purpose general stress response. The best-known general stress responses are controlled at the gene expression level by controlling transcription by alternative transcription initiator proteins (alternative sigma factors). In bacteria, a constitutive sigma factor controls housekeeping genes under non-stress conditions, but an alternative sigma factor binds under stress conditions, and this changes specificity of RNA polymerase to achieve the transcription a different set of genes that protect the cell against adverse conditions. This coordinated regulation allows the simultaneous expression of numerous genes in response to a single stimulus in the environment. These regulators are key players in raising a coordinated stress response involved in tolerance of stresses such as acids, low water activity, temperature, bacteriocins, antibiotics, ethanol, and starvation. Most of the proteins formed as a result of stress adaptation responses have role in biofilm formation and sporulation. (Begley and Hill, 2016).

Microbiology of food preservation methods

Thermal inactivation of microorganisms

Application of heat treatment to destroy pathogenic and spoilage microorganisms is one of the most successful food preservation method used since centuries, and is widely used even today. Based on the extent of heat treatment, the processes are classified into major groups-

1. **Low heat processing or pasteurization**: The purpose is to destroy the vegetative growth of all spoilage microbes and the time temperature combination required for this treatment should be sufficient to destroy the most heat resistant pathogens viz., *Mycobacterium tuberculosis* and *Coxiella burnetti*. The shelf life of pasteurized foods is further maintained through refrigeration.
2. **High heat processing or sterilization**: This heat processing is meant for achieving the complete destruction of microorganisms. *Bacillus stearothermophilus* spores, being extremely heat resistant are used as evaluator for efficiency of sterilization. But the old type 'canning' sterilization is replaced now by ultra high temperature (UHT) processing, which combines continuous flow thermal processing with aseptic packaging. Some times to denote this combination, the term 'aseptic processing' is used. By UHT processing, quality of the products can be retained due to the low heating time and fast cooling time. UHT process is followed in milk industry, beverage industries etc.

Process design suitable for thermal inactivation of microorganisms in foods is achieved through insights gained so far on the inactivation pattern through heating. This pattern help to decide the temperature-time combination required for the target organisms in heat processing.

When a microbial population is heated at a specific temperature, the cells die at a constant rate (first order reaction). To describe the thermal inactivation kinetics, two parameters viz., D value and Z value is vital. D value is the time at particular temperature necessary to destroy 90% of the population of a specific organism. Z value is the change in temperature needed to change the D value by 90%. The thermal deactivation kinetics with respect to foods is well worked out (Stumbo, 1973; Singh & Heldman, 2014). Although the results of D value experiments in buffer are not always the same as in real food, the data are comparable taking into account the simulation of pH and water activity in the studies. For most vegetative cells, the reported *z* values are generally around 5 to 6 with a range of 4 - 9. For *Listeria monocytogenes*, D_{55} values of 3.2 minutes are reported for vacuum-packed minced meat to 47 minutes in ground pork with *z* values between 4 and 5. For *E. coli* O157:H7 and *Salmonella*, similar values have been reported ranging about six minutes in minced meat and in ground beef or ground pork. *Yersinia enterocolitica* in ground beef and milk or in liquid egg (Toora et al., 1992., Bolton et al., 2000, Hunag and Juneja., 2001; Favier, 2008.).

Variation occurs from the standard thermal inactivation model in real situations. The variation occurs as appearance of a shoulder and tailing in the log survivor Vs time graph. (Fig. 1). Shoulder is the lag period contributed by poor heat transfer of the heating medium (food) and due to the increase in temperature required to bring out sufficient injury before the first order inactivation kinetics in the log number of survivors with time. The tailing effect may be due to either the stress adaptation response of the cells or due to the heterogeneous flora with difference in their heat resistance levels. Heat shock also activate physiological responses leading to the synthesis of an array of proteins known as heat shock proteins (HSPs), which provide resistance to further heat treatment. In general, heat shocked cells have to be heated twice than that of non-heat-shocked cells to achieve the same extent of lethality. Increased heat resistance due to heat shock proteins have been reported in a variety of *Enterobacteriaceae* members, spore formers like *Clostridium spp., Bacillus stearothermophilus* etc., Most of the HSPs acts as molecular chaperones to helping in proper folding of proteins denatured by heat.

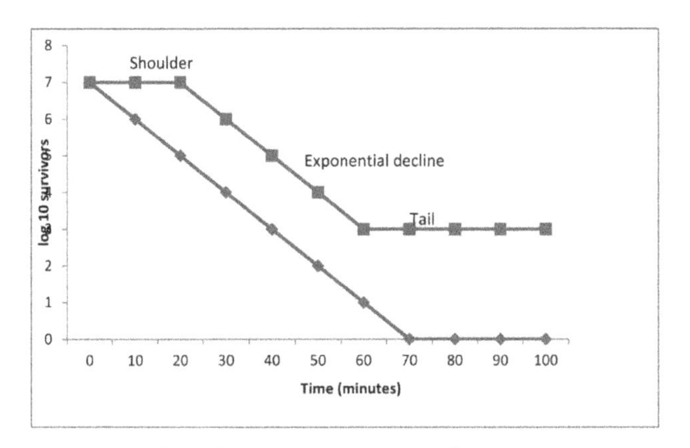

Fig 1. Thermal inactivation pattern of microorganisms

Bacterial endospores are heat resistant. Heat resistance of bacterial spores have positive correlation with DNA content, dipicolinic acid and calcium level, mineralization and dehydration, while the vegetative cells heat resistance is associated with high level of saturated fatty acids in the cell membrane. Apart from these inherent heat resistance, stage of growth, growth temperature and growth medium affects the heat resistance of organisms in food. pH of the medium is recognized as one of the most important factors affecting the heat resistance of bacteria, typified by reduction in heat resistance at acidic pH. Logarithm of heat- destruction rate shows first order reaction curve at acid to alkaline range with a minimum rate at the optimum pH for growth.

Fatty materials in foods prevent the faster heat penetration and thus provide protective effect to the microorganisms against heat treatment. Soluble solids in the medium also affect the heat resistance of microbes. For example, dissolved sodium chloride increases heat resistance by stabilizing membrane proteins, increase melting temperature of membrane phospholipids. By doing so, cell membrane damage and cell leakage is retarded. The heat resistance is also influenced by oxygen concentration. In certain cases, eight fold increase in heat resistance was recorded when the organisms are heated at low redox potential conditions. Stress adaptation of *Listeria* cells to starvation, ethanol, hydrogen peroxide and acid significantly increase the resistance to heat through cross protection mechanisms (Juneja and Sofos, 2001).

Novel thermal methodologies for inactivation of microbes in foods

Food technologists have adopted the concepts of alternative methods for heat generation such as radiofrequency (RF), microwave (MW), infrared (IR) and ohmic heating (OH) methods, for faster heating of foods and thus to retain the better sensory qualities along with microbial inactivation. Novel thermal food-preservation technologies, causes self heating of the product, helping to reduce the heating time. Lower heating time reduce the loss of sensory attributes, making these method as potential next generation food processing methods.

Ohmic heating

Ohmic or electrical resistance heating involves an application of a low-voltage alternating current to a continuously flowing food product. Heat is generated within the food product during this process due to electrical conductivity of food. This method has potential applications in inactivation of microbes in highly viscous products and liquid-particle mixtures. Earlier studies on microbial inactivation kinetics have shown that D values due to ohmic heating is usually less than conventional heating, showing additional mechanisms of microbial inactivation operating synergistically with heat generation. The mechanisms have been reported as thermal, chemical and mechanical effects. The thermal effect is due to self heating of foods, while chemical effects are due to generation of highly oxidizing compounds such as hydrogen peroxide, chlorine etc. Extent of chemical damage depends on type of electrode, current density, frequency and the medium used. More recent studies suggest that the electricity flow results in pores in cell membrane (electrical permeation mechanism or electroporation) causing leakage of cellular constituents and cell death (Cunha et al., 2017; Cappato et al., 2017).

Microwave heating

Microwave heating is a very popular, extensively studied technology in food processing worldwide. This process has advantage of high heating rates, which result in reduced processing time, easy operation, cleaner work environment, lower maintenance requirements, and reduced use of plant space. Microwave pasteurization or sterilization can be explained by various mechanisms, such as selective heating, electroporation, cell membrane rupture and magnetic field coupling to cause reactivity of biomolecules. For example, due to the microwave selective heating, microbial bodies can reach a higher temperature than the surrounding fluid, leading to faster microbial destruction (Tang, 2015).

Nonthermal physical methods of microbial inactivation

Increased consumer demand for fresh-like foods has promoted the development of nonthermal methods for microbial inactivation for food preservation. These technologies are named according to the main processing parameter leading to cell inactivation. The alternative physical technologies food processing with high microbial inactivation properties are ultrasound, high hydrostatic pressure (HHP), pulsed electric field (PEF), and electromagnetic radiations such as gamma rays, electron beam & UV radiations.

Ultrasound

Sound waves with frequencies above the threshold for human hearing (>16 kHz) are called ultra sound. High-power ultrasound (>1 W/cm^2 and & 20 - 500 kHz) is a green substitute to the traditional methods of food preservation. The cavitation force produced due to ultrasound device in the liquid food medium causes lethality to microbes. The antimicrobial efficiency of high-intensity ultrasound in foods is determined by contact time with the microorganism, microorganism type, food quantity, composition, and treatment temperature. Intracellular cavitation ruptures cell membranes, and produces free radicals and H_2O, which also assist in microbial inactivation. Greater surface area of the cell, more is the lethality Thus in general rod shaped bacteria are more vulnerable to death by sonication than cocci. Bacteria that form spores of *Clostridium* and *Bacillus* species have shown greater resistance to ultrasonic treatment than vegetative cells, these bacteria are more resistant to thermal treatment as well as to ultrasound. Microbial lethality improves when sonication is combined with heat (thermosonication) or pressure (manosonication). Unlike many other types of cross adaptations, the stress adaptation of microorganism to heat does not affect their sensitivity to ultrasound. This characteristic makes manosonication and manothermosonication as a potential tool in processing. The other advantage is that damage to cells caused by heat can be reversible, in contrast to this, damage caused by mano sonication is irreversible (Piyasena et al., 2003) .

Irradiation

Gamma irradiation for inactivation of insect pests, pathogens and sprouting etc. is already a technology, even though the public acceptances is slightly less. Medium doses (0.75- 2.5 kGy) impart a pasteurising effect in food (radurization). High doses (30- 40kGy) can be applied to commercial sterilization for purposes similar to canning (radappertization).

Mechanism of action of gamma rays occurs through both direct and indirect effects of irradiation on cellular components. Irradiation results in radiolysis of water resulting in hydrogen and hydroxyl free radicals generation. In the direct action, rays cause damage to the DNA of of organisms (Lacrox, 2005).

Anaerobic conditions reduce the lethality due to irradiation. This is due to the lowered rate of oxidizing reactions which produce free radicals and toxic oxygen derivatives. Prokaryotes are more resistant to irradiation damage than eukaryotic microorganisms. Viruses exhibit higher resistance to irradiation treatments than bacterial spores, which are more resistant than vegetative cells. Vegetative cells of bacteria are more resistant than yeasts and molds. Among bacteria, gram negative bacteria are more sensitive to irradiation than gram positives. Researchers have demonstrated that irradiation doses of at least 1.0 kGy, which could virtually eliminate gram negative bacteria in foods, have a much little effect on gram positive bacteria. Similar to heat resistance pattern, spore formers and stationary phase cultures offer more resistance to irradiation. A radiation dose of 1.5kGy to 3.0kGy is required to inactivate spoilage fungi such as *Aspergillus, Penicillium* etc. in commodities like pulses.

Proteins, antioxidants and low water activity of food matrices protect the microbes from the lethality due to irradiation, while fat content does not affect the sensitivity of microbes in foods. The vast majority of scientific literature on effect of temperature on ionizing radiations indicates that microbial sensitivity to irradiation is higher at ambient temperatures than at subfreezing temperatures. For example, 8.5 fold higher irradiation doses were required for a 90% reduction in *Pseudomonas spp.* at subfreezing temperatures than an ambient temperature. The D value for E coli O157:H7 was 0.28kGy at 5 °C and 0.44 kGy at -5°C. This is due to low water activity subfreezing temperatures, and thus low level of radioloysis, resulting in less production and movement of free radicals. Free radicals such as hydroxyl radicals and hydrogen radicals accounts for 85% of lethality factors linked to irradiated organism when *E coli*challenge tests were conducted. The atmospheric composition in contact with microbes influences their inactivation by irradiations. For example, Hastings et al. (1986) observed that *Lactobacillus sake, Lactobacillus alimentarius, Lactobacillus curvatus* were more sensitive to gamma irradiation in ground meat packaged under 100% carbon dioxide than under nitrogen.

Electron beam Irradiation(EBI) is another kind of irradiation used for inhibit microbial growth in food. The energy transfer within the body is due to EBI destruction of chemical and molecular bonds of the cells. Ultra violet radiations of the wavelength ranging between 200- 280 nm (UV C) are also germicidal, and finds application in very popularly water purification systems. UV light exposure causes the formation of covalent links between adjacent bases, resulting in thymine dimer ormation. These thymine dimers inhibit DNA functions and cell growth stopped. Ultra violet exposure reduces the heat resistance in microbes. Therefore UV light in combination with thermal treatment might be an alternative to thermal sterilization (Gayan et al., 2013)

High hydrostatic pressure (HHP) processing

HHP processing, the application of high pressures (100–800 MPa) to foods, is a well-developed non-thermal treatment, to prolong shelf life of solid or liquid foods. The pressure is transmitted uniformly within the product through a liquid medium, usually water (Isostatic Principle). As the pressure build up in the medium, adiabatic heating occurs and water temperature increases around 3 °C every 100 MPa. Therefore, HHP should be applied in high-moisture foods; while porous and dry foods are not suitable for this process. Along the years, successful results on microorganism inactivation by HHP in plant- and animal based food products have been reported.

HHP acts a microbial agent by inducing the structural changes in the cell membrane and enzymes. For example, 50 MPa of pressure inhibit protein synthesis in microorganisms and reduce the number of ribosomes. A pressure of 100 MPa can induce partial protein denaturation, and 200 MPa causes damage to the cell membrane and internal cell structure. Increasing the pressure to 300 MPa or more induces irreversible denaturation (Srinivas et al., 2018). Fungi are the most sensitive microorganisms to HHP, while some fungal spores are highly resistant. Vegetative cells are killed at 400-600 MPa, spores can survive upto 1000 MPa. Temperature of sporulation determines pressure resistance, as they are most resistant when sporulated at lower temperatures. Initiation of pressure induced germination is also affected by sporulation at lower temperatures (Barba et al., 2015).

Eukaryotic microorganisms are more pressure sensitive than prokaryotes, and consequently fungi are easy to be killed than bacteria using HPP. However, heat resistant fungus Byssochlamys produce pressure resistant spores. Since fungi are easily killed, HPP is commercially used in food products such as fruit juices, where bacterial spoilage is lesser concern.

Besides intrinsic parameters of foods, pressure, temperature, and treatment time are the most important parameters to be considered for processing optimization using HPP. HHP has efficiency similar to conventional pasteurization in microbial inactivation, with the advantage of not modifying nutritional and sensory properties of the product.

The intrinsic parameters refer to the type of substrate and composition of the food. All major nutrients like carbohydrates, proteins, lipids confer a protective effect to the microbes. The higher protective effect due to the presence of vitamins is attributed to the faster recovery of organisms from sub lethal injury with the help of these nutrients (Black et al., 2007). Baroprotective effect of ions like calcium, magnesium and phosphates etc. is also known. Black et al. (2007) reported the increased survival of *Listeria innocua* in simulated milk ultra filtrate with added calcium, magnesium, citrate and phosphate. The authors concluded that magnesium stabilises ribosomes, calcium stabilizes cell membranes. Sucrose, a very common ingredient many of the fruit based and cereal based products, can impart protection to bacterial cells from the damaging effects of HPP stabilizing membrane protein functionality (Manas & Pag ˜ an, 2005). Foods with low water activity (aw) exerts a protective effect to microbial cells against HPP. On the other hand, microbes injured through HPP are more sensitive to death by lowering the water activity as a strategy for further preservation. It is not only the osmolarity and water availability that has an impact on resistance to HPP but also the nature of the solute. Application of high pressure brings down the pH of the food, which can also inhibit the proton pumping force and energy generation. Oscillatory high pressure treatments (several cycles of high pressure) are more effective than equivalent continuous pressure of comparable holding time. For example, Hayakawa et al (1994) compared continuous and oscillatory (cycles of 5 min each) pressurizations in Bacillus stearothermophilus spore suspensions. Treatments at 800MPa for 60 minutes at 60 °C and 70 °C decreased spore counts by 3.5 and 4.5 log CFU/mL., Oscillatory pressurization at 70 °C could achieve 6 log reductions (Hayakawa et al., 1994).

Numerous HHP treated products are available in the market. Important challenge in HHP processing is the development of basic research and in the equipment design of high temperature and high pressure (HTHP) processing to achieve food sterilization levels (Considine et al., 2004).

Pulsed Electric Fields (PEF)

PEF process involves the application of voltage pulses of the range 15–80 kV/cm for a short time to pumpable foods passing between two electrodes. Therefore, only homogeneous liquids can be pasteurized by PEF treatment. Unlike other non-thermal methods, PEF has several critical parameters to be considered for processing optimization. This includes electric field strength (E), treatment time, pulse-shape, pulse-width, pulse-frequency, pulse-polarity, and temperature. Electric field strength is considered the most influencing factor on producing cell damages and thus microbial death. By increasing E and treatment time, higher microbial inactivation is expected. Likewise, PEF effectiveness is highly impacted by the type of microorganism and medium characteristics. PEF is not suitable for food sterilization, but it can be a potential option for acid products pasteurization, like fruit juices.

Inactivation rates of microorganisms using PEF increase with decreasing conductivity of the treatment medium. Several investigators have reported protective effect of food constituents such as xanthan gum, proteins and fats. The presence of nonconductive large solid particles and air bubbles also reduce the efficacy of PEF mediated microbial inactivation. The main microorganism inactivation mechanism is the electroporation, which is the formation of pores on the cellular membrane. Nonetheless, structural arrangement of microbial enzymes, electromechanical compression, and osmotic imbalance also could explain microbial death due to PEF. (Vega-Mercado et al., 1997).

Sublethal effects by PEF with subsequent high resistance of microbial cells to PEF have been reported. For example, exposure of the yeast *S. cerevisiae* to PEF induced oxidation stress responses which activated glutathione synthesis enzymes. Such a response results in protection against the injury and the recovery of the cells. Subsequently, it was shown that the glutathione-dependent biochemical defense system against oxidation stress in sublethally injured cells is closely related to their level of resistance toward PEF stress and PEF can be used for improving this antioxidant in foods (Wang etal., 2014).

Non-thermal Plasma (NTP)

Plasma, fourth state of matter where increases in the material's energy levels converts matter to an ionised state. Cold plasma (CP) is comprised of several excited atomic, molecular, ionic, and radical species, co-existing with numerous reactive species, including electrons, positive and negative ions, free radicals, gas atoms, molecules in the ground or excited state and quanta of electromagnetic radiation (UV photons and visible light). In recent years, NTP has emerged as a potential alternative zero-residue, nonthermal processing technology for decontamination of foods, which does not leave any toxic residuals on processed foods (Misra et al., 2011). Plasma can be generated by different methods, like gas discharge, photo ionization, heat radiation, radio frequencies etc. Among these methods, gas discharge is the most common way to create NTP.

Since plasma is a complex mixture, several mechanisms applicable to each individual component is attributed to contribute in antimicrobial action. Some researchers hypothesize that UV-C radiation present in plasma plays an important inactivating role result; while various other researchers identifies the mechanical or oxidative damage caused to cellular envelopes as the main cause of death. NTP cause an uncontrolled process of lipid peroxidation resulting in cell membrane rupture. Reactive species, such as singlet oxygen, hydrogen peroxide, nitric oxide and excited atoms and molecules, can rapidly and easily diffuse into the cells, even when the membrane is intact, and oxidize many macromolecules. The oxidation of DNA bases leads to substitution of a purine for a pyrimidine or *vice versa*, leading to cell death. The reactive species are able to break peptide bonds, oxidize amino acid side chains, produce protein cross links to produce protein aggregate. All these effects result in modifications in the conformation and three-dimensional structure of proteins and enzymes. In addition, the enzymes are inhibited also by the oxidation of their cofactors (Lopez et al., 2019).

Environmental factors, such as pH, food matrix, and relative humidity, have significant influences on the sterilization efficacy of NTP. Kayes et al. (2007) found that the *Bacillus cereus* reduction at pH 5 was about 4.9 logs, while it was only 2.1 logs at pH 7 after 30 s plasma treatment. Muranyi et al. (2008) demonstrated that increasing relative humidity had improved the rate of NTP inactivation of *Aspergillus niger* because of the decomposition of additional water molecules into more hydroxyl radicals.

Preservation of foods through additives

One of the fastest growing areas of research on food science is pertaining to the antimicrobial food additives. Preservatives are preferred by the processors due to their wide availability, broad spectrum antimicrobial activity and low cost. Besides the conventional preservatives, novel molecules with antimicrobial activities are also explored and introduced for food preservation. The microbiological knowledge of preservatives is summarized below.

Organic acids and their derivatives

Organic acids and their derivatives are the conventional, vastly used preservatives. Acetic, propionic, lactic, sorbic, benzoic, citric, malic, fumaric, tartaric and adipic acids, esters of hydroxyl benzoic acids (parabens) and glucono-delta-lactone. Benzoate, propionate and sorbates are used in a wide variety of foods, where as lactate and acetates are used often in meat and meat products.

Among the organic acids, benzoic acids and parabens have good antimicrobial effect against bacteria, yeasts and molds, while propionic acid is more effective against yeasts and molds. Acetic acid and citric acids are more effective against bacteria than yeasts and molds. The mechanisms of antimicrobial activity of organic acids and their salts are mainly by the inhibition of ATP generation by affecting the proton motive force (pmf) of membrane.

Plant derived compounds

Apart from the above, traditionally used organic acids or their derivatives, long chain fatty acids, phenolic acids such as oleic, chlorogenic, hydroxyl cinnamic, caffeic, p-coumaric, ferulic and lauric acids are currently being investigated for their antimicrobial activities in foods. Medium and long chain unsaturated fatty acids are also antimicrobial compounds and they are more efficient against Gram positive, than against Gram negative bacteria (Desbois et al., 2010, Pisoschi et al., 2018).

During last century, the antimicrobial principles from numerous plants have been elucidated. These include essential oils of cinnamon, clove, thyme, sage, rosemary, oregano, allspice, nutmeg, coriander, lemon grass, cumin, black pepper etc. The essential oil components mainly include terpenes, phenols, aldehydes, esters and ketones. Plant based antimicrobials show a broad spectrum activity against gram positive and gram negative bacteria and fungi, although majority of the studies show better activity against gram positive than gram negative bacteria. Even though very good in vitro antimicrobial activity is shown by essential oils, the efficacy is weak in food sytsems, probably due to interference. Thus, appreciably high amount of the essential oils have to be added for inactivation of microbes in food. In majority of the cases, this will lead to unacceptable sensory effects.

Plant derived phenolic compounds acts by altering microbial cell membranes. This is accomplished through the interaction of phenolics with membrane proteins inducing their structure and function alteration, creating membrane instability. So Membrane functions like electron transport, nutrient uptake, synthesis of proteins and nucleic acids, and enzyme activity are affected. (Bajpai et al., 2008, Hayek et al., 2013). Alkyl substitution into the aromatic nucleus contributes the antibacterial activity of phenolics. The antimicrobial activity of phenolics depends on concentration at low concentration, phenols inhibit microbial enzyme activity, while high concentrations induce denaturation of proteins.

Presence of functional groups such as hydroxyl groups promote electron delocalization, and can act as proton exchangers, diminishing the pH gradient across the cytoplasmic membrane of bacterial cells. This causes the diminished proton motive force, energy depletion, and subsequent cell death. Isothiocyanates present in crucifers, garlic etc. another group of plant derived secondary metabolites with potential use as antimicrobials in foods. These inactivate extracellular enzymes through oxidative dissociation of -S-S- bonds.

Chitosan

Chitosan, the biopolymer obtained by deacetylation of chitin, is a molecule approved by several food regulatory agencies across the globe, is another prospective antimicrobial for food preservation. The net charge in the chitosan molecule (extent of acetylation) and pH are the main factors affecting the antimicrobial efficacy of chitosan. The charge density on the cell surface is the microbial characteristic affecting the sensitivity to chitosan (Kong et al., 2010, Zou et al., 2016). Several studies have shown that the biological activity of chitosan depends significantly on its molecular weight (MW) and degree of acetylation (DA). Both parameters affect the antimicrobial activity of chitosan independently, though it has been suggested that the influence of the MW on the antimicrobial activity is greater than the influence of the DA. An increase in the DA of the amino group in the chitosan molecule cause less activity. Minimum inhibitory concentration (MIC) of chitosan has been found to range from 0.05% to 0.1%, depending on the bacterial species and the molecular weight of chitosan. At present, chitosan has gained more importance as an edible coating with antimicrobial properties, rather than as a standalone preservative.

Bacteriocins

Bacteriocins are ribosomally-synthesized cationic and hydrophobic peptides with approximately 20-60 amino acids length, possesses anti-microbial peptides (AMP) and secreted by the producer-organism. Among these, nisin, an AMP produced by LAB such as *Lactococcus lactis*, has been granted "generally regarded as safe" (GRAS) status for certain applications by Food and Drug Administration (FDA), and is now permitted in more than 50 countries for food use. A multitude of other bacteriocins are also reported from lactic acid bacteria. Some other bacteriocins include piscicolin, subtilin, lichenicidin, divergicin, cinnamycin, actagardine, epidermin, lacticin, carnobacteriocin, mutacin, mesenterocin, enterocin, sakacin, leucocin, curvacin, lysostaphin,enterocin, duramycin, brevinin, mundticin, ruminococcin, curvaticin, and columbicin (Kumariya et al., 2019). Most of the food applications involving bacteriocins can be divided into three categories Partially purified bacteriocins (e.g., Nisaplin®, containing 2.5% nisin), dairy and other food-grade fermented products containing bacteriocins in the form of a crude fermentate (e.g., MicroGARD® series of bacteriocin-containing products and bacteriocin-producing protective cultures. One of the biggest challenge in bacteriocins use is their suceptability to proteases/peptidases and narrow spectrum of activity. Researchers are continuously working to genetically engineer the producer strains to overcome these limitations to obtain ideal bacteriocins (Chikindas et al., (2018).

The anionic molecules on bacterial cell surface are receptors for cationic bacteriocins. The anionic molecules include phospholipids, lipopolysaccharides, lipotechoic acid and cardiolipin. The bacteriocins align themselves on the membrane and positively charged groups interact electrostatically with the negatively charged bacterial cell surface, while the hydrophobic surfaces align toward the membrane to traverse the lipid bilayer. After piercing into the lipid bilayer, the peptides self-associate or polymerize to form complexes. These changes cause pore formation in bacterial membrane and finally cell death. Of late, there are scientific reports available on the resistance development by microbes against bacteriocins. Since charge and fluidity of membranes are the two bacterial properties exploited by bacteriocins during the attack, manipulation in these properties renders the bacteriocins ineffective, resulting in bacteriocins resistance. Bacteriocin combinations is a good way of tackling the bacteriocins resistance. Since bacteriocins are ribosomally-synthesized peptides, they can also be bioengineered at specific amino acid residues to make them more effective against the food-spoilage bacteria (Johnson et al., 2019).

Food preservation using microorganisms: Fermentation and protective cultures

Fermented foods are forming an important part of diet globally. Preservation of foods using fermentation by various microorganisms resulted in the vast expansion of dairy science, enology, brewing and lactic fermentation science. Primarily, lactic acid bacteria, yeasts and certain molds are used for food fermentations. Qualities of starter culture for different fermentations are well defined and scientists are trying to either isolate or to genetically modify the existing strains to materialize the 'ideal strain concept' for different fermentation. Extensive literature is available on microbiology of wine, beer and lactic fermented products. Delfini and Formica (2001), Ameen et al (2017), Bokulich and Bamforth (2017). All fermentations result in consumption of sugars present in raw food, induces flavour changes and either acid or alcohol production, thus resulting in formation of distinct category of food products. Apart from the role of preservation, fermentation is a means to improve the functionality of foods. For example, soybean fermentation produces different bioactive peptides with health benefits such as antihypertensive and antimicrobial activities. The type of bioactive peptide formed is largely dependent on the fermentation agent (Sanjukta and Rai, 2016).

Apart from the fermented product, the use of protective cultures for preservation of fresh and fresh-cut products are also gaining importance. These microbes bring in the control of pathogens through competition for nutrients and niches, antibiosis, lytic enzymes, volatile inhibitory metabolites, pH decrease, organic acid production, parasitism and induction of defense responses from the harvested plant product and others (Montesinos et al., 2015). These antagonistic organisms used for preservation of fresh and fresh-like products. Lactic acid bacteria of genera *Lactococcus, Lactobacillus, Lactosphaera, Leuconostoc, Oenococcus, Pediococcus, Streptococcus, Vagococcus, and Weissella* are reported to have biopreservation properties.

In certain cases, yeasts are also used for biopreservation. Mode of action of antagonistic yeasts is different than lactic acid bacteria. Studies suggest that the action is related to production of antifungal hydrolases, siderophores (iron binding molecules), induction of defense related proteins, production of volatile organic compounds and induction of ROS. Among them, volatile organic compound production has been adjudged as the major mechanism of antagonism. A volatile-producing fungus of *Ceratocystis fimbriata* was used to control two postharvest diseases caused by *M. fructicola* and *P. digitatum* on peaches and citrus respectively. The exposure to VOCs significantly inhibited the test fungi *in vitro* and *in vivo* trials. The most abundant VOCs that accounted for 97% of the total volatile compound yield were butyl acetate, ethyl acetate and ethanol (Ribes et al., 2017)

Microbiological implications of food packaging methods

Advances in food packaging system have revolutionized methods of preservation. Quality maintenance by packaging is achieved by maintaining the gas, moisture, water activity and sunlight exposure at optimal level. Among this in-pack atmospheric modification is the most important factor deciding the microbial proliferation in foods. Low oxygen level and high CO_2 in the packs either formed as a result of tissue respiration as in case of fresh fruits and vegetables (passive), or by flushing with suitable gas mixtures (active) prevents the spoilage by aerobic microbes especially fungi. But, very high CO2 levels can support the proliferation of organisms such as *Clostridium*. Thus, apart from the sensory quality mainatance, the microbiological safety concern is an important parameter to be taken care in modified atmosphere packaging. Ample evidences are generated on this regard. (Zahra et al., 2016, Oliveira et al., 2015).

Another packaging strategy to control undesirable microorganisms is the incorporation of antimicrobial substances onto food packaging materials. Natural antimicrobial agents, silver, volatile compounds and bacteriocins are usually incroprporated for this purpose. The silver zeolite is laminated as a thin layer on the food contact surface of the laminate. Commercial examples of silver substituted zeolites include Zeomic®, Apacider®, AgIon, Bactekiller and Novaron etc. Volatile antimicrobials generating systems to release chlorine dioxide, sulphur dioxide, carbon dioxide and ethanol are also proved efficient in controlling microflora in packed foods. The packaging material used for the volatile antimicrobial system should possess high barrier properties to prevent the loss through permeation. Ethicap® and Antimold® sachets release ethanol absorbing moisture and releasing ethanol vapour. Ethanol vapour generators are commercially used for packaging high moisture bakery products, cheese and fish. Edible materials such as carbohydrates (Eg. methyl cellulose, starch), proteins (Eg. Gelatin, casein), lipids (Eg Bees wax, carnauba wax) incorporated with antimicrobial substances are directly applied on the food material to make edible films. It makes use of organic acids, salts of organic acids, bacteriocins, fungicides, enzymes, and compounds like silver zeolites etc. as antimicrobial for incorporating to films (Biji et al., 2015).

Enhancing the efficacy of individual preservation factors: Hurdle concept

Earlier discussions have shown that adaptation of pathogens to environmental and processing stresses can result in survival of the cells in the foods even after the processing is completed. Therefore, food industry considers multiple barriers (hurdles) to improve the microbial stability of foods. Application of a single factor alone can bring in stress adaptations, ultimately making the product unstable. Better shelf stability results from the synergism of combination of factors. The hurdle technology produces minimal sensory changes, which makes the products more acceptable than those obtained by conventional methods (Aguilera and Chirife, 1994). Hurdle technology represents the application of well known strategies in reduced "doses" to interact synergistically to prolong overall quality of foods. Some of the well known hurdles are pH control, water activity reduction, mild heat, natural antimicrobials, aseptic packaging, rapid cooling, maintenance of refrigerated storage, modified atmosphere packaging, active packaging etc. Therefore all antimicrobial agents discussed earlier in this chapter can be used as a tool for developing suitable hurdle technologies for specific food products. Vast number of combinations of hurdles such as lowering water activity, pH, mild heating (blanching), chemical preservatives at low dose etc., have been successful in food preservation conventionally. In fact, most of the processed products available today in the market use the principle of hurdle technology deliberately or inadvertently (Singh and Shalini, 2016). At present, hurdle combinations of physical non thermal processes and natural preservatives and packaging are receiving significant attention as hurdles with an objective to reduce the use of chemical preservatives and heat. For example, nisin has a synergistic effect with PEF treatment, and an additive effect with HPP treatment (Yuste et al., 2002; Black et al., 2005). This phenomenon is explained as facilitation of the entry of a nisin molecule through the membrane during HPP treatments, as the combination helps cause membrane pores rapidly in bacteria.

Indicator microorganisms for testing for efficacy of processing and preservation

Studies on identification and behavior of indicator organisms are very important for processing parameter optimization of food. The classical example is successful process design in canning due to the scientific understanding of *Clostridium botulinum as* an indicator organism to check the efficiency of canning process, and design of 12D (time required to reduce 12 log population of *C. botulinum*) process. Similarly, safe pasteurized milk is 'virtually free' of *Mycobacterium* sp. In heat inactivation itself, the indicator organism has to be different based on the food properties. For example, Silva and Gibbs (2004) suggested *Alicyclobacillus acidocaldarius* as the reference microorganism to design heat processes for food products. USDA (2012) has requirements E. coli O157:H7 as the indicator strain for reprocessing, an HPP process that achieves a 5-log E. coli O157:H7 reduction should be sufficient for product produced to ensure microbial safety (Sonaliben et al., 2017). *Deinococcus radiodurans, due to its high resistance to irradiation, has been used to design radiation processing.* It is known that some of the human pathogens in food enters to a metabolic inactive state called viable but non culturable state (VBNC). VBNC form is also a stress adaptation of the organisms. With suitable environmental stimuli, these organisms inititate their growth. *Micrococcus luteus,* a milk spoilage bacteria has become a model organism for studying this growth initiation process, often called resuscitation (Galina et al., 1998). Challenge and validation studies using food-borne pathogens in the food processing facility pose contamination risk, and are prohibited by law. In such situation, surrogate biological indicators have been proposed as alternatives to pathogens in these studies. *Clostridium sporogenes* spore is the surrogate to predict the heat inactivation of *C. botulinum* spores while *Listeria innocua* is a non-pathogenic strain that grows in environments similar to those suitable for *L. monocytogenes* and this surrogate is suitable for studying efficacy of milk pasteurization. Surrogate microorganisms chosen should possess a slightly higher resistance than the targeted pathogens, in order to parctically estimate the pathogens surviving in the treated food. The rapid detection of low levels of pathogens or their corresponding nonpathogenic surrogates may be facilitated with markers such as bioluminescence, resistance to antibiotics etc. In the wake of increased use of alternative thermal and nonthermal processing methods for microbial inactivation, suitable indicators and their surrogates are to be clearly defined.

Measurement of cell survival and physiological state of microorganisms

Another remarkable microbiological expansion which contributed towards the food preservation science is the advances in detection of viability and physiological state of food organisms post food processing. Several methods are available to directly assess the extent and nature of cell injury within a microbial population. The conventional 'gold standard' method is to culture the organisms in selective mediase include culturing on selective agar media. But nowadays modern microscopic, immunological and molecular biology tools are used for rapid detection of food microbes, which in turn will give real-time status of microbiological load and physiological state in foods and thus to assess the efficacy of preservation methods. For example, epifluorescence staining using propidium iodide can be used to detect cells that lost selective permeability after HPP. Surface hydrophobicity studies and ATPase activity help for understanding the mechanism of action through membrane damage (Huang et al., 2014). The combination of fluorescent dyes viz., ethidium bromide, bis-oxonol and propidium iodide can be used to quantify, the populations that are reproductively viable, metabolically active and/or with intact polarized membrane (Von-caron, 1998). Acridine orange staining differentiates viable and dead cells based on their relative proportion of DNA and RNA. Commercial kits such as *Live/Dead BacLight*TM® provide information on viable cells in a population. The molecular biology applications are not useful so far in spoilage microbiology and preservation *per se*; but have found high application in food safety microbiology and fermentation microbiology for studying the population changes of a targeted microbe and gene expression changes due to stresses in the microbial cell.

Modelling and predictive microbiology as a tool for designing microbial inactivation in food

Predictive Microbiology (PM) is the prediction of microbiological consequences in a system achieved through the mathematical modelling. This has high application in food technology for evaluation of microbiological shelf life, and prediction of microbiological hazards connected with foods in HACCP and Microbiological risk assessment programmes. The most important death models used in predictive modelling are negative Gompertz, log-linear, shoulder/tail, Weibull, Weibull+tail, re-parameterized Weibull, biphasic approach, etc. . The details of mathematical models in foods were reviewed by Bevilacqua et al. (2015). One constraint with these models is the use of same types of mathematical models for different microbes, although the parameters of the models are specific to species or even strains. For example, the common model organisms used are *Morganella* sp (for histamine production), *Listeria monocytogenes* (as psychrotrophic organism) and a few lactic acid bacteria, *E. coli* etc. Several modelling packages are already available towards this purpose. Common ones are ComBase (www.combase.cc), Food Spoilage and Safety Predictor (http://fssp.food.dtu.dk). ARS Pathogen Modelling Program (https://pmp.errc.ars.usda.gov) and Sympревious (https://symprevius.eu),

Future prospects

This chapter summarized the role of advancements in understanding the antimicrobial mechanisms by different preservation strategies, their practical uses and the methods to study the microbiological dynamics in food preservation. Advancement of every aspect of microbiology has implications due to the fact that the knowledge can be applied in monitoring and curbing the microbial proliferation in foods. Every new methods of food processing has to be checked for their implications on microflora. Microbial stress adaptation is a tricky issue for the scientific world, considering the diversity of foods and food borne microbes or their strains. Having elucidated the physiology of microbial stress adaptation, the ultimate role of food microbiologists is to incorporate this knowledge into concepts that can be used by the food processors to warrant supply of safe foods coupled with better shelf life. Lot more scientific information need to be generated for developing real time quality monitoring system and process design for better preservation and implementation of the same through a quality control system. Additionally, fermentation is an area of food preservation now gaining more importance due to their acceptance as functional foods. More knowledge on microbe mediated biotransformations is required to establish the potential of fermented foods as functional foods.

References

Aguilera, J.M. & Chirife, J. (1994). Combined methods for the preservation of foods in Latin America and CYTED-D project. *Journal of Food Engineering* 22, 433–444. doi.: 10.1016/0260-8774(94)90045-0

Ameen, M.S. & Caruso G. (2017) Lactic Acid in the Food Industry. Springer

Bajpai, V.K., Rahman, A., Dung, N.T., Huh, M.K. and Kang, S.C. (2008). In vitro inhibition of Microorganisms, *Critical Reviews in Food Science and Nutrition*, 54:10, 1371-1385.

Barba, F.J. (2015) Current applications and new opportunities for the use of pulsed electric fields in food science and industry, *Food Research International* , 77, 773-778. DOI/10.1016/j.foodres.2015.09.015

Begley M. & Colin Hill C. (2016). Stress Adaptation in Foodborne Pathogens. *Annual Reviews in Food Science and Technology* 6, 191–210 DOI 10.1146/annurev-food-030713-092350

Bevilacqua, A., Speranza, B., Sinigaglia, M. & Corbo, M.P. (2015). A Focus on the death kinetics in predictive Microbiology: benefits and limits of the most important models and some tools dealing with their application in Foods. *Foods*, 4, 565-580; DOI10.3390/foods4040565

Biji K. B., Ravishankar, C. N., Mohan C. O. and Gopal T KS. (2015). Smart packaging systems for food applications: a review. *Journal of Food Science and Technology* (October 2015) 52(10):6125–6135. DOI 10.1007/s13197-015-1766-7

Brul. S. & Coote, P. (1999). Preservative agents in foods. Mode of action and microbial resistance mechanisms.International *Journal of Food Microbiology*, 15, 1-17.

Black, E.P. , Kelly, A.L. and Fitzgerald. G.F. (2005) The combined effect of high pressure and nisin on inactivation of microorganisms in milk. *Innovative Food Science & Emerging Technologies*. 6, 286-292. DOI 10.1111/j.1365-2672.2007.03722

Bokulich NA. & Bamforth CW. 2017. Brewing Microbiology: Current Research, Omics and Microbial Ecology . Caister Academic Press

Bolton, D.J., McMahon, C.M., Doherty, A.M., Sheridan, J.J., McDowell, D.A.,Blair,L.S.and Harrington, D. (2000).Thermal inactivation of *Listeria monocytogenes* and Yersinia enterocolitica in minced beef under laboratory conditions and in sous-vide prepared minced and solid beef cooked in a commercial retort. *Journal of Applied Microbiology*, 88,626–632.

Cappato, L.P., Ferreira M.V.S. , Guimaraes, J.T.. Portela J.B ,. Costa, A.L.R Freitas M.Q. , Ceuppens, S., & Li, D. Uyttendaele, M., Renault, P., Ross,P., Ranst, MV., Cocolin, L., & Donaghy J. (2014) Molecular methods in food safety microbiology: interpretation and implications of nucleic acid detection. *Comprehensive Reviews in Food Science and Food Safety*, 13. DOI 10.1111/1541-4337.12072 551-577.

Chikindas, M.L., Weeks, R., Drider, D., Chistyakov, V.A. and Dicks, M.T.L. (2017) Functions and emerging applications of bacteriocins. *Current Opinion in Biotechnology*, 49:23–28. DOI 10.1016/j.copbio.2017.07.011.

Considine, K.M., Kelly, A.L., Gerald F. Fitzgerald G F., Hill, H., Roy D. and Sleato, R. D. (2008). High-pressure processing effects on microbial food safetyand food quality. FEMS *Microbiology Letters*, 281, 1–9 . DOI10.1111/j.1574-6968.2008.01084.x

Cullen, PJ., Tiwari, BK, Valdramidis V. (Eds). (2011). Novel thermal and non-thermal technologies for fluid foods. Academic Press

Cunha R.L., Oliveira C.A.F., Mercali, G.D.. Marzack L.D.F. & Cruz, A.G. (2017) .Ohmic heating in dairy processing: Relevant aspects for safety and quality. *Trends in Food Science and Technology* 62, 104-112. doi. /10.1016/j.tifs.2017.01.010

Da-Wen Sun (2014). Emerging Technologies for Food Processing. 2nd Edition Academic Press.

Delfini, C., Formica J.V. (2001). Wine Microbiology: Science and Technology. CRC Press.

Desbois, A.P. and Smith, V.J. (2010). Antibacterial free fatty acids: activities, mechanisms of action and biotechnological potential. *Applied Microbiology and Biotechnology*, 85, 1629-1642 DOI 10.1007/s00253-009-2355-3.

Doona C J., (2018). Case Studies in Novel Food Processing Technologies: Innovations in Processing, Packaging, and Predictive Modelling. Woodhead Publishing, P.600.

Favier, G. I., Escudero, M.E. and De Guzman, A.M.S. (2008). Thermal inactivation of Yersinia enterocolytica in liquid egg products. *J. Food Safety*, 28: 157–169. DOI10.1111 /j.1745-4565.2008.00103

Frazier. W.C., Westhoff, D.C. (2008). Food Microbiology 4E. McGraw-Hill Education (India) Pvt Limited.

Gahan CGM, O'Driscoll B, Hill C. (1996). Acid adaptation of Listeria monocytogenes can enhance survival in acid foods and during milk fermentation. *Applied and Environmental Microbiology* 62:3123–28

Galina V. Mukamolova G.V., Nataliya D. Yanopolskaya N.D., Douglas B. Kell DB., Kaprelyants A.S. (1998). On resuscitation from the dormant state of Micrococcus luteus, *Antonie* van *Leeuwenhoek*, 73, 237–243.

Gayan E., Alvarez, I., Condon (2013). Inactivation of bacterial spores by UV light. *Innovative Food Science and Emerging Technologies,* 19 , 140-145. DOI 10.1016/j.ifset.2013.04.007

Hastings, J.W., Holzapfel, W.H. and Niemand, J.G. (1986). Radiation resistance of lactobacilli isolated from radurized meat relative to growth and environment. *Applied and Environmental Microbiology*, 52, 898-901.

Hayakawa I., Kanno, T., Tomita, M. and Fujio, Y. (1994). Application of High Pressure for Spore Inactivation and Protein Denaturation. *Journal of Food Science*, 59: 159-163. doi:10.1111/j.1365-2621.1994.tb06923.x

Hayek, S.A. , Gyawali, S.A. Ibrahim (2013). Antimicrobial natural products, In: A. Mendez-Vilas (Ed.), Microbial Pathogens and Strategies for Combating Them: Science, Technology and Education.

Huang, H., Lung, H. Yang, B., Wang C. (2014). Responses of microorganisms to high hydrostatic pressure processing. *Food Control.* 40, 250-259. DOI 10.1016/j.foodcont.2013.12.007

Huang, L. H. & Juneja, V. K. (2001). A new kinetic model for thermal inactivation of microorganisms: Development and validation using *Escherichia coli* O157:H7 as a test organism. *Journal of Food Protection*, 64: 2078–2082.

Johnson, EM., Jung, Y., Dr. Jin, Y., Jayabalan, R., Yang SW & Suh JW (2018) Bacteriocins as food preservatives: Challenges and emerging horizons, *Critical Reviews in Food Science and Nutrition*, 58:16, 2743-2767, DOI 10.1080/10408398.2017.1340870

Juneja, V.K. and Sofos J.N. (2001). Control of Foodborne Microorganisms. CRC Press. P.552

Kayes, M.M., Critzer, F J., Kelly, K., Wintenberg,J.R., Roth,T.C. and Montie,D.A.Golden (2007) Inactivation of foodborne pathogens using a one atmosphere uniform glow discharge plasma. *Foodborne Pathogens and Disease*, 4, 50-59. DOI 10.1089/fpd.2006.62

Kong, M., Chen, X.G., Xing K. & Park, H.J. (2010). Antimicrobial properties of chitosan and mode of action: A state of the art review. *International Journal of Food Microbiology,* 144 (2010) 51–63.DOI 10.1016/j.ijfoodmicro.2010.09.012

Kumariya, R., Garg A.K., Rajput , Y.S., Akhtar, N. & Patel S. (2019). Bacteriocins: Classification, synthesis, mechanism of action and resistance development in food spoilage causing bacteria. *Microbial Pathogenesis*,128, 171-177. DOI 10.1016/j.micpath.2019.01.002.

Lacroix, M. (2005). Irradiation of foods. In: Emerging technologies for food processing. (Ed) Da-Wen Sun. Elsevier Academic Press. 353-378.

Lopez., Calvo, T., Prieto, M., Mugica-Vidal, R., Muro- Fraguas, I., Alba-Elias, F. & Alvarez-Ordonez, A. (2019). A Review on Non-thermal Atmospheric Plasma for Food Preservation: Mode of Action, Determinants of Effectiveness, and Applications. *Frontiers in Microbiology*., 10:622. DOI 10.3389/fmicb.2019.00622

Montesinos E., Francés J., Badosa E., Bonaterra A. (2015) Post Harvest Control. In: Lugtenberg B. (eds) *Principles of Plant-Microbe Interactions*. Springer, Cham

Moreno-Vileta, H.M.L. Hernández-Hernándeza, S.J. and Villanueva-Rodríguez (2018). Current status of emerging food processing technologies in Latin America: Novel thermal processing. *J. Innovative Food Science and Emerging Technologies*, 50. 196-206 DOI 10.1016/j.ifset.2018.06.013.

Mazzotta, A.S. & Montville, T.J. (1999). Characterization of fatty acid composition, spore germination, and thermal resistance in a nisin-resistant mutant of *Clostridium botulinum* 169B and in the wild-type strain. *Applied and Environmental Microbiology*, 65, 659-64.

Misra, N.N., Tiwari B.J., Raghavarao, K.S.M.S. *and* Cullen, P.J. (2011). Nonthermal Plasma Inactivation of Food-Borne Pathogens. *Food Engineering Reviews*, 3. 159-170. 10.1007/s12393-011-9041-9.

Muranyi, P., Wunderlich, J. & Heise, M. (2008), Influence of relative gas humidity on the inactivation efficiency of a low temperature gas plasma. *Journal of Applied Microbiology*, 104, 1659-1666. doi:10.1111/j.1365-2672.2007.03691.x

Pisoschi, A. M., Pop, A., Georgescu, C., Turcus, V., Olah, N.K. & Mathe E. (2018). An overview of natural antimicrobials role in food. *European Journal of Medicinal Chemistry,* 143, 922-935 DOI 10.1016/j.ejmech.2017.11.095.

Piyasena P., Mohareb E. & McKellar R.C. (2003). Inactivation of microbes using ultrasound: a review. *International Journal of Food Microbiology* 87, 207 – 216. DOI10.1016/S0168-1605(03)00075-8

Remenant B., Jaffrès E., Dousset X., Pilet M. & Zagorec M. 2015. Bacterial spoilers of food: Behavior, fitness and functional properties. *Food Microbiology* 45: 45-53. DOI 10.1016/j.fm.2014.03.009.

Oliveira, M., Abadias, J. Usall, R. Torres, N. Teixid, o & I. Vi~nas. (2015). Application of modified atmosphere packaging as a safety approach tofresh-cut fruits and vegetables. A review. *Trends in Food Science and Technology*, 46, 13-26. DOI 10.1016/j.tifs.2015.07.017

Ribes, S.; Fuentes, A.; Talens, P. And Barat, J.M. (2017). Prevention of fungal spoilage in food products using natural compounds: A review. *Critical Reviews in Food Science and Nutrition*, 58, 2002-2016. DOI 10.1080/10408398.2017.1295017

Smelt J. P. P. M. & Brul S. (2014) Thermal Inactivation of microorganisms. *Critical Reviews in Food Science and Nutrition*., 54, 1371-85. DOI 10.1080/10408398.2011.637645.

Sanjukta, S. and Rai, A. K. (2016). Production of bioactive peptides during soybean fermentation and their potential health benefits. *Trends in Food Science & Technology*, 50, 1-10. DOI 10.1016/j.tifs.2016.01.010

Silva, F.V.M. and Gibbs, P. (2004). Target selection in designing pasteurization processes for shelf-stable high-acid fruit products. *Critical Reviews in Food Science and Nutrition*, 44, 353–360 DOI 10.1080/10408690490489251

Singh, R. P. and Heldman, D. (2014). Introduction to Food Engineering. Academic Press.

Singh, S. And Shalini, R. (2016) Effect of Hurdle Technology in Food Preservation: A Review, *Critical Reviews in Food Science and Nutrition*, 56,4, 641-649, DOI 10.1080/10408398.2012.761594

Sonaliben, L. Parekh, K.D. Aparnathi & Sreeja, V. (2017). High Pressure Processing: A Potential Technology for Processing and Preservation of Dairy Foods. *International Journal of Current Microbiology and Applied Sciences*. 6, 3526-3535. DOI 10.20546/ijcmas.2017.612.410

Srinivas, M.S., Madhu, B., Srinivas,G. & Jain, S.K. (2018). High Pressure Processing of Foods: A Review. *The Andhra Agricultural Journal*, 65, 467-476.

Stumbo C. R. (2013). Thermobacteriology in Food Processing.. 2nd Edition Academic Press

Tang, J. (2015). Unlocking Potentials of Microwaves for Food Safety and Quality. *Journal of Food Science*, 80, E1776–E1793. DOI 10.1111/1750-3841.12959

Tokuþoðlu, O. and . Swanson, B.G. (2015). Improving Food Quality with Novel Food Processing Technologies. CRC Press

Toora, S., Buduamoako, E., Ablett, R. F. & Smith, J. 1992. Effect of high-temperature short-time pasteurization, freezing and thawing and constant freezing, on the survival of Yersinia enterocolitica in milk. *Journal of Food Protection*, 55, 803–805.

Tremarin, A., Canbaz, E.A., Brandãoa, T.R.S., Silvaa, CLM . 2019. Modelling Alicyclobacillus acidoterrestris inactivation in apple juice using thermosonication treatments. *LWT - Food Science and Technology* 102, 159–163. DOI 10.1016/j.lwt.2018.12.027

Van Schaik, W., Gahan C.G.M. & Hill, C. (1999). Acid-adapted Listeria monocytogenes displays enhanced tolerance against the lantibiotics nisin and lacticin 3147. *Journal of Food Protection*. 62:536–39.

Vega-Mercado, H., Martin-Belloso, O., Qin, B., Chang, M. F. J. Gbngora-Nieto, M., Barbosa-Gnovas G V. and Swanson BG. (1997). Non-thermal food preservation: Pulsed electric fields. *Trends in Food Science and Technology*, 8, 151-157, DOI 10.1016/S0924-2244(97)01016-9

Von-caron, H.N., Stephens, P.J. and Badley, R. A. (1998). Assessment of bacterial viability status by flow cytometry and single sorting. *Journal of Applied Microbiology*, 84,988-98. DOI 10.1046/j.1365-2672.1998.00436.

Wang, J. Wang, K., Wang, Y., Lin, S., Zhao, P. & Jones, G. (2014) A novel application of pulsed electric field (PEF) processing for improving glutathione (GSH) antioxidant activity. *Food Chemistry*, 161, 361-366, DOI 10.1016/j.foodchem.2014.04.027

Wang, H., Qian, J., & Ding, F.(2018). Emerging Chitosan-Based Films for Food Packaging Applications. *Journal of Agricultural and Food Chemistry*, 66 , 395–413DOI 10.1021/acs.jafc.7b04528

Yuste, J., Pla, R., Capellas, M. & MMor-Mur. 2002. Application of high-pressure processing and nisin to mechanically recovered poultry meat for microbial decontamination. *Food Control*, 13, 451-455. DOI 10.1016/S0956-7135(01)00071-8

Zahra A.M.,Butt, Y.N., Nasar, S., Akram, S. & Fatima, Jannat Ikram J. (2016). Food packaging in perspective of microbial activity: a review. *Journal of Microbiology, Biotechnology and Food Sciences* / Zahra et al. 2016 : 6 (2) 752-757. DOI 10.15414/jmbfs.2016.6.2.752-757

Zou P., Yang X, Wang J, Y.H. Yu, Yanxin Zhang & Guangyang Liu . (2016). Advances in characterisation and biological activities of chitosan and chitosan oligosaccharides. *Food Chemistry* 190, 1174–1181. DOI 10.1016/j.foodchem.2015.06.076

4

Technological Advances in Nano-science for Specific Food and Nutrition Delivery

K.R. Anilakumar

Food Quality Assurance Division, Defence Food Research Laboratory DRDO, Mysuru, Karnataka, India

Nano-technology refers to technology on a nano- meter scale, specifically less than 100 nano-meters. The novel nano-particles are tiny spheres (fullerenes) or tubes (nano-tubes) made from carbon atoms. It facilitates the development of future inventions transversely a vast array of fields. New nano-particles are being made of silicon, ceramic, polymers or even natural ingredients that break down in the body. The characteristics that make these particles functional in the food industry include optical properties, reactivity to temperature, and biodegradability. They can be used as probes or filled with flavorings or nutrients for delivery. Highlights on the categorisation of agriculture with nano- food technology and nutrition are given in Fig 1.

There are three key elements in nano-technology. They are: (1) substances at nano-scale exhibit novel physical and chemical phenomena with novel functions (2) allowing to measure, control and operate material at the nano-scale that transform them to functions. (3) allowing the union of nano scales of materials for exceptional properties.

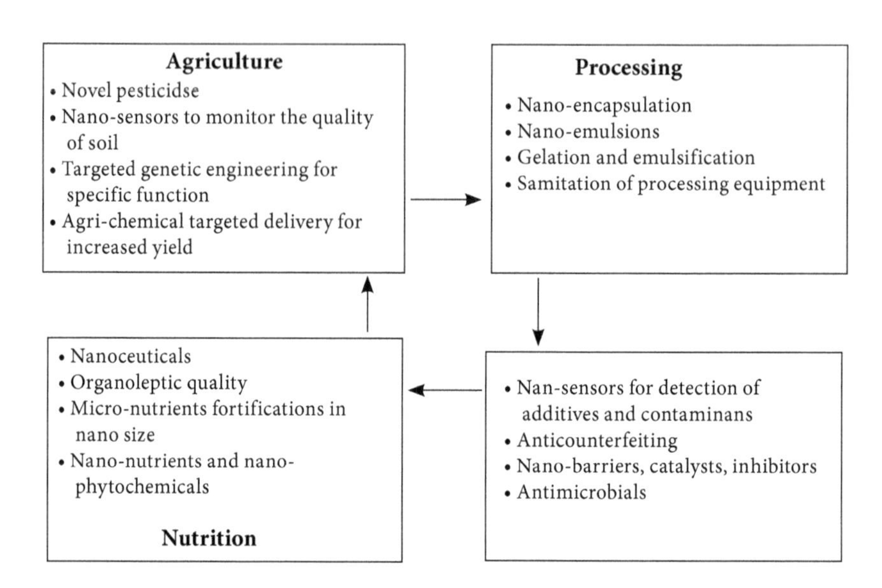

Fig. 1: Nano- food technology, agriculture and nutrition and its categorisation

Applications in food industry

Many foods contain nano- particles. Examples of foods that contain nano-particles include milk and meat. Milk contains caseins, a form of milk protein present at the nano-scale. Meat is made up of protein filaments that are much less than 100nm thin. The organisation and change to the structures of these affects the texture and properties of the milk or meat. Food packaging applications include plastic polymers containing or coated with nano-materials for improved mechanical, textural or functional properties. Nano-coatings on food contact surfaces for barrier or anti microbial properties, nano-sized agrochemicals (a chemical used in agriculture, such as a pesticide or a fertilizer) nano-sensors for food labelling, etc. are the other benefits.

At present studies are being carried out on nano-particles that are being used to deliver micro nutrients or macro nutrients in food and drinks without affecting the taste or appearance. These nano-particles encapsulate the nutrients and carry them through the stomach into the bloodstream. Nano-particle emulsions are being used in ice cream and various spreads to get better the texture and uniformity. New advancements in nano-science and nano-technology will allow more organize and have the potential of increased benefits. These include (1) healthier foods viz. lower fat, lower salt, etc. with desirable sensory properties (2) ingredients with improved properties (3) potential for removal of certain additives without loss of stability (4) and smart-processing aids to remove allergens such as peanut protein.

Researchers have produced smart packages that can inform consumers about the freshness of milk or meat. When oxidative damage occurs in the package, nano-particles provide indication that the colour is changed and the consumer can see if the product is fresh or not. Incorporation of nano-particles in packaging can improve the barrier to oxygen and decelerate degradation of food during ambient storage. Bottles made with nano-composites lessen the seepage of carbon dioxide out of the bottle. This increases the shelf life of fizzy and sparkling drinks without having to use heavier glass bottles or more pricey cans. Food storage waste bins have been embedded with silver nano-particles in the plastic. The silver nano-particles eradicate bacteria from any food formerly stored in the bins, minimising detrimental bacteria.

The use of nano-technology in the expansion of innovative packaging materials had a notable intensification in the recent years and is expected to have an important impact on the food market in the near future. This is due to the increasing information about nano-scopic technology uses in food packaging. This in turn was brought to the academia and industry new devices for the progress of new nano- size based technology with improved functions and possessions and advanced instrumentation. Concurrently, the impact on continuance and the interest on using innovative resources further added to reinforce intensification perspectives in this area (Cerqueira, et al., 2018).

The application of nano-technology and nano-particles in food are materializing at a fast pace. In today's competitive market technology is indispensable to keep headship in the food and food processing sectors. Today the consumers demand fresh, genuine, easy to use and flavour-rich foods. The potential of food industry belongs to newer products and newer processes, with the objective of enhancing the presentation of the product, which prolongs to the product preservative ability and originality, and recuperating the safety and superiority of food. Nano-technology/ nano-science is an enabling branch of technology that has the potential to revolutionize the food manufacturing. Nano-technology can be applied to widen nano-scale materials, controlled delivery systems, pollutant/adulterant detection and to generate nano-devices for molecular and cellular biology.

Nano-technology involves creating and manipulating organic and inorganic materials at the nano-scale. The present molecular biology techniques of genetic modification of crops are already kinds of what has been termed nano-technology. This innovative technology can provide for the future development of far more specific and effective methods of, and other forms of, planning of food polymers and polymeric assembly to provide custom-made improvements to thefood quality and food safety. Nano-technology agrees not only the creation of novel and precisely defined material properties; it also promises that these substances will have self-assembling, self-healing and maintaining properties.

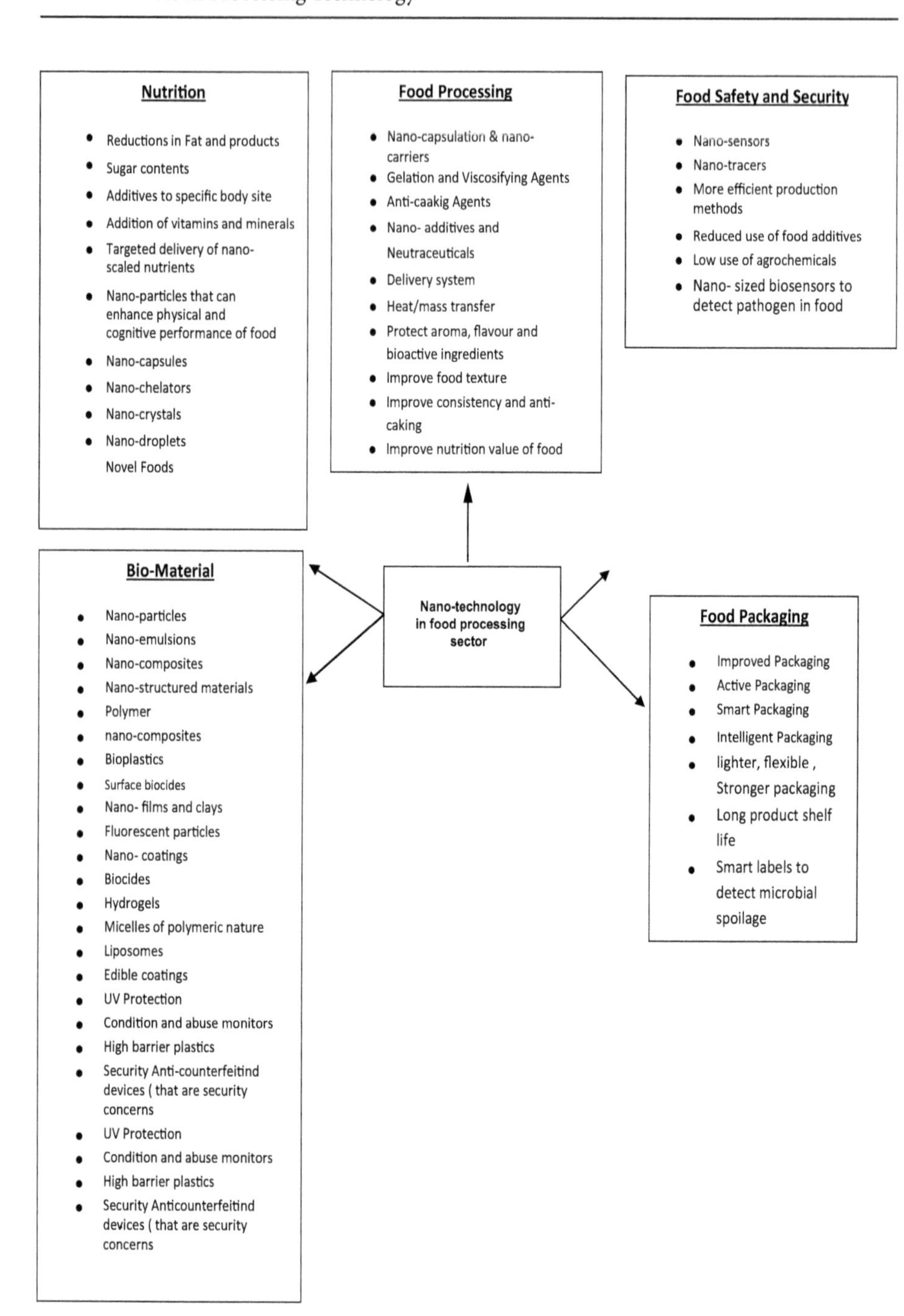

Fig. 2: Relevance of nano-technology in food processing industries

Nano-technology also promises to provide a means of altering and manipulating food products to more efficiently and resourcefully deliver nutrients, proteins and antioxidants to precisely target nutritional and health benefits to a specific site in the human body or to specific cells to augment their efficacy and bioavailability. Several researchers are studying on nano- encapsulation of certain nutrients, flavours and colours and release them upon need or over an extended period of time. This would also benefit functional food development from this new technology, followed by common food, nutraceuticals, health supplements and others.

Nano-technology is having an impact on numerous characteristics of the food industry, right growing the food to how it is packed (Krishnan et al., 2018). Nano-technology is envisaged for use in food manufacture, processing, preservation, flavouring and coloration, hygiene and safety and packaging. Relevance of nano-technology in agricultural field in particular, supporting farmer's problems is summarized in Fig. 2.

Microsieves for separation and fractionation which can also improve emulsification processes and can result in new products like low-fat mayonnaise. Control of materials at the nano- size will facilitate fine tuning of specific food characteristics viz. Texture, flavour, appearance, etc. to the demands of specific target groups.

Food safety / food microbiology

Nano- sensors

Detection of very tiny amounts of a chemical contaminant, virus or bacteria in food systems is another potential claim of nano-technology. The exhilarating possibility of combining biology and nano-scale technology into sensors holds the potential of increased sensitivity and therefore a significantly diminished response-time to understand potential problems. Nano-sensors make use of nano-particles, which can either be tailor-made to fluoresce different colours or, alternatively, be constructed out of magnetic materials. These nano-scale particles can then selectively connect themselves to any number of food pathogens. Microbiologists, who presently use either infrared light or magnetic materials, could then note the occurrence of even minuscule traces of harmful pathogens. The advantage of such a nano- sensors enabled system is that hundreds and potentially thousands of nano-particles can be placed on a single nano-sensor to rapidly, accurately and efficiently detect the presence of any number of different bacteria or virus and pathogens. A second advantage of nano-sensors is that, given their small size, they can gain access into the tiny crevices where the pathogens often hide.

The bio analytical microsystems that are studied spotlight on the very rapid detection of pathogens in routine drinking water testing, food analysis, environmental water testing and in clinical diagnostics. Nano- scale technology is also being applied to the tagging and monitoring of food items. Nano-technology based anti-counterfeit technologies are in currently in the R&D pipeline of various companies. The tagging of food packages will mean that food can be monitored from farm to fork. The protection of brand authenticity may cause better food safety in certain cases. Another area under development is attempts to characterise biomaterial at the nano-size level. The Atomic Force Microscope in the field of food sciences is currently used to learning the nano-scale structure of foods and other biomaterials. It has the prospective to offer a new way to the tracking and tracing of goods and materials.

Nano- liposomes

Synthetic lipid membranes are considered as promising food-grade nano- vesicles but their function is required to be attributed to empower them as active-nano- vesicle systems. In a study conducted by Niaz et al. (2019), rhamnosomes (RS) were developed by engineering the membrane of nano-liposomes with antimicrobial surfactant viz. rhamnolipids. In the study inherently active rhamnosome nano-vesicles were loaded with bacteriocin to acquire broad-spectrum antimicrobial activity. Addition of rhamnolipids into the lipid bilayer exhibited a boost in the encapsulation efficiency of nisin. Scanning electron microscope and Zeta-sizer disclosed that the surface functionalization with rhamnolipids amplified the vesicles dimension significantly. Whereas, the zeta potential of nisin-loaded rhamnosome nano-vesicles indicated the higher physical stability which not only demonstrated higher antimicrobial activity than liposomes but also augmented the activity of nisin against Gram-positive and Gram-negative resistant food borne pathogens viz. *L. monocytogenes, S. aureus, E. coli* and *P. aeruginosa*. The study also showed 80% reduction in biofilm biomass on treatment with nisin-loaded rhamnosome nano-vesicles which was attributed to improved binding with bacterial surface. The study concluded that surface-active nano-liposomes that are intrinsically active with higher stability can provide an innovative strategy to control the biofilm-forming food associated pathogens.

Silver nano-particles

Recently a novel strain of *Saccaropolyspora hirsuta* was isolated from an insect *Tapinoma simrothi* by Sholkamy et al. (2019) and was subsequently morphologically and physiologically characterized. It was genetically identified using 16S rRNA and sequence similarity percentage in genbank with closely related species as strain ess_amA6 of *Saccaropolyspora hirsuta*. The accession number of strain ess_amA6 was KF996506. Antagonistic activity of strain ess_amA6 against some pathogenic Gram-positive and Gram - negative bacteria, and *Candida albicans,* a unicellular fungus was studied. In addition, star shaped silver nano-particles were biosynthesized using strain ess_amA6. The silver nano- stars were characterized by UV-visible spectrophotometer. The conversion of Ag+ ions to Nano- silver was confirmed by analysis using fourier transform infrared spectroscopy which was attributed to the reduction by capping material of extract. Transmission electron microscopical studies of biosynthesized nano- silver particles showed that they are spherical in shape which ranged from 10 nm to 30 nm in size. Silver atoms were checked in nano- sample by Energy Dispersive X-ray spectroscopy. Bioactivity of biosynthesized nano- silver was observed against some pathogenic microorganisms such as *Staphylococcus aureus, Streptococcus pyogenes, Salmonella typhi, Pseudomonas aeruginosa, Klebsiella pneumonia* and *Candida albicans.* The study showed that these tested microbes were highly sensitive to nano-silver. It also recommended the use of strain ess_amA6 to biosynthesize bio-active nano- silver compounds.

Nano-nutraceuticals and nano-functional food

It is defined that nutraceuticals and functional foods to be food components that might provide demonstrated physiological benefits or reduce the risk of chronic disease and disorders, above and beyond their basic nutritional functions. A functional food is similar to a conventional food, while a nutraceutical is isolated from a food and sold in dosage forms, in powder, capsules or pills. In both cases the active ingredients occur naturally in the food. The bio fortified foods is done with micronutrients, phyto-chemicals, etc. to facilitate health effects.

Bio-engineering and genetics have been thought about as tools to produce more nutritious and functional food. But nano-size technology is moving fast into this area.

Nano- technology into nutrition

One of the oldest and most popular ways to incorporate nano- size technology into nutrition is through micelles, the tiniest of capsules that form naturally when nature requires a fat-soluble substance to be soluble in water. The micelle is fundamentally a cavity that is water-soluble externally, and fat-attracting internally . Micelles support emulsification or dissolution of fat in water solutions example fat globules in milk and absorption in the intestines of fat-soluble micro nutrients such as vitamins D, E and K, Coenzyme Q_{10}, carotenoids and essential fatty acids. Based on this natural nano-technology, many food technologists are developing and incorporating novel micelle nano-technologies in their products. The trends in the development of nano-ceuticals is provided in Fig 3.

Studies are conducted on CoQ_{10} bioavailability via nano-micelle-forming technology. They form a micelle with 25 nm diameter size that has fat-soluble tocopherols as its lipophilic centre. When CoQ_{10} is introduced, it goes straight to this centre, resulting in a clear delivery of CoQ_{10} that is stable at ambient temperature.

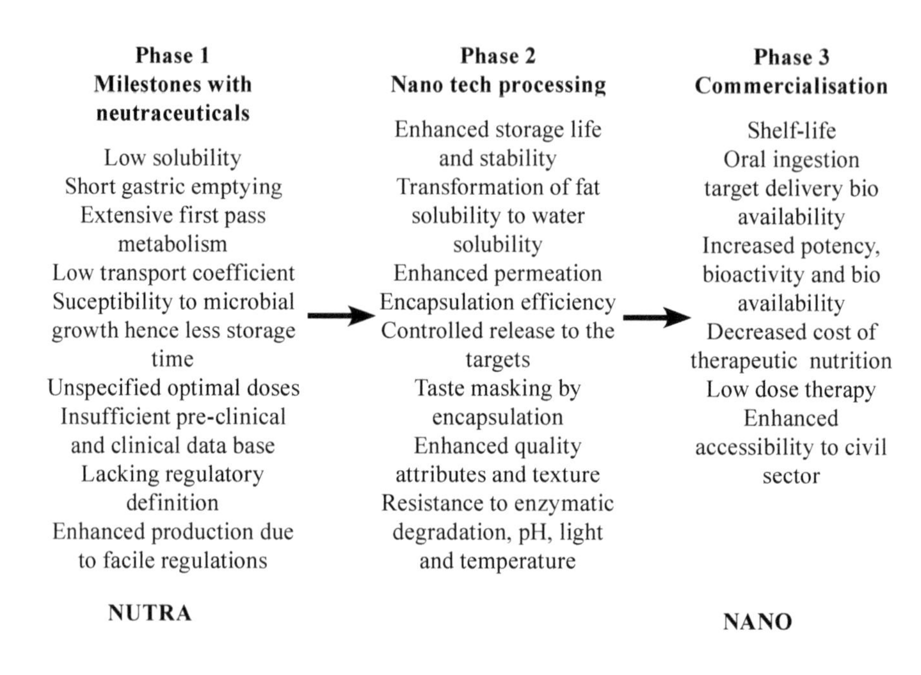

Fig. 3: Recent trends in the development of nano-ceuticals

Research on nano-encapsulation of nutraceutical ingredients, with an emphasis on enhancing body absorption and circulating time of phytochemicals viz. flavonoids, etc. are also on rise. There are abundant dietary supplements available in tablet, powder or pill forms. They have very poor water solubility and oral bioavailability, describing the potential of nano- scopic emulsions and nano- size encapsulation to effectively tackle this issue. The studies are also focussing on safety issues for the formulated nano-emulsions, nano-dispersiosn and polymer micelle-based encapsulated phytochemicals. Undoubtedly, more research is essential thorough pre- clinical/ animal studies on nano-sized phytochemicals. Curcumin, the yellow pigment from turmeric, possess anti-oxidant, anti-inflammatory, and anti-cancer activity, which has yet to be confirmed by human clinical trials. Curcumin being water insoluble very little is found in the bloodstream after digestion (Wang et al., 2008).

It was explored to nano-encapsulate curcumin and measuring its anti-inflammation activity *in vivo* in mice. Many macro and micro nutrients and phytochemicals have poor solubility, which decreases their bioavailability. Accordingly, vitamins, minerals, and phytochemicals are hardly ever used in their pure form, and functional ingredients with demonstrated properties are an element of the delivery system (Tarver, 2006). The delivery system functions to transport the nutrient to its desired site while also influencing the supplement's taste, texture, and shelf life. Theoritically, nano-technology overcomes this puzzle by providing more efficient nano- emulsion forms.

To decline this dilemma, researchers have made oil-in-water nano-emulsions of various sizes to encapsulate curcumin to get better its anti-inflammation activity. The activity was evaluated using a mouse ear inflammation model. The emulsions were prepared using medium-chain triglycerols and polysorbate 20, (a typical molecular-biology-grade detergent) as an emulsifier. On evaluation in rats it showed that 1% curcumin nano-encapsulated in oil-in-water emulsions diminished the induced edema of the ears of mice. Administration of 1% curcumin in polysorbate 20 water solution showed little or no decrease in inflammation (Wang et al., 2008). This suggests that encapsulating curcumin in nano-emulsions increases its bioavailability.

Israel-based Nutralease uses micelles in its nano-sized, self-assembled liquid delivery system for tocopherol, s ω-3 fatty acids, β-carotene, lycopene, isoflavones, CoQ_{10} and lutein (Table 1.) These ingredients are not only used for both dietary supplement to reduce diseases and disorders, but also for cosmeceuticals, in topical applications that are bioavailable, stable and clear. Bio-polymers made using chitosan are put into use for encapsulation and protect antioxidants from the acidic pH in the gut, to make the release in a controlled manner. Usually the biopolymers stick to the walls of the small intestine, where they trickle the antioxidants directly into the cells through which absorption takes place. For improving the absorption rate of catechins, which is only as high as 1.1 percent in regular delivery, the researchers noted this technology could help incorporate ingredients like ω-3 fatty acids in baked foods like bread, biscuits, bun, cakes, etc.

Table 1: Nano- technology based Food Products

Nano-ceuticals slim shake	Assorted Flavor, RBC Life Sciences, Irving, USA
Oat nutritional drink	Assorted Flavor, Toddler Health, Los Angeles, USA
Canola active oil	Shemen, Israel
Nano-tea	Shenzhen Become Industry Trading Co., China
Fortified fruit juice	High Vive.com, USA
Nano-slim beverage	Nano-slim, Canada
Daily vitamin boost	Jamba Juice, Hawaii
Tiptop up bread -Tuna fish oil	Enfield, Australia
Nestle original coffee creamer	Nestle, USA
Trix cereal	General Mills, USA
Mentos fresh mint	Mentos, USA
Nano-technology based supplements	Aquanova (Germany)
Micelle delivery technology	NovaSOL ® line
Micelles for vitamin E, omega-3 fatty acids	Israel based Nutralease
β- carotene, lycopene	Israel based Nutralease
Isoflavone, Co Q_{10}, lutein	Israel based Nutralease

Nano-materials

Naturally occurring and inadvertently produced nano-materials are excluded in this definition. Pharmaceuticals, food and consumer products use nano- materials for a variety of applications in biological filed. Nano-composites, nano-clays, nano-tubes and others are included in nano-materials. Nano- delivery systems will use nano-capsules, nano-cochleates, nano-balls, nano-devices, nano-machines and nano-robots. As on today the nano-technology applications cover all areas of the food chain, from agricultural, bio-medical and enhancing bioavailability of macro and micro nutrients. Nano-technology introduces new avenues for novelty in the food and health sectors at enormous speed, at the same time it faces a long and intensive path towards full realisation and commercialisation.

Bio-active compounds especially active ingredients possess ample health benefits, but they are chemically unstable and susceptible to oxidative damage and degradation. The application of pure bio-active compounds is also very limited in food and drug formulations because of their fast release, low solubility and poor bioavailability. Encapsulation techniques can protect the bio-active materials from environmental stresses, improve physicochemical functionalities, and enhance their health-promoting and antagonistic-disease activities.

Shishir et. al.(2018) have comprehensively discussed the importance and use of biopolymer-based carrier agents and lipid-based transporters with their functional attributes, appropriateness of encapsulation techniques in micro and nano-sized encapsulations, and various forms of better and novel micro and nano-encapsulate systems.

Polysaccharides like maltodextrin in combination with gums or other or proteins have been well studied in the encapsulation formulations. The methods like electro-spinning and electro-spraying are employed while solid lipid nano-particles and nano-structure lipid carriers are revealing themselves as the hopeful and new generation of lipid nano-carriers for bio-actives . Added to this, phytosome, nano-hydrogel, and nano-fibre are also competent and novel nano-vehicles for bio-active compounds. Further studies are warranted for the improvement of existing encapsulate systems and studies on the gastrointestinal systems for industrial application (Shishir et. al., 2018).

Migration of nano-materials from food packaging film was studied. Yang et al (2019) studied the poly lactic acid/nano-TiO_2 composite films with different nano-TiO_2 loading and were prepared and contacted with 50% ethanol solution as food simulant to study the behaviour of nano-TiO_2 migration. The structural changes and intermolecular interactions were overcome by SEM studie, X-ray diffraction, and differential scanning calorimetry. The migration amount increased with the increase of initial nano-TiO_2 content. Scanning electron microscopy images established that the microstructure of poly lactic acid nano-composite films turn into rougher as exposure to ethanol solution in a few days. X-ray diffraction spectra suggested that a decline in the intensity of specific diffraction peak occurred as the decrease in nano-TiO_2content of poly lactic acid nano-composite films during exposure to ethanol simulant. Differential scanning calorimetry analysis ascertained the higher crystallinity % that was obtained during the different degradation times.

Zein-based nano-materials

Zein, a byproduct of corn with renewable resources, unique hydrophobic/hydrophilic character, film/fibre forming and free radical scavenging properties, is a potential biopolymer for diet related functions. The average size of zein nano-particles was reported to be 50–200 nm. The functions of zein nano-materials were several, (a) carrier of food delivery, drinks and micro-nutrient systems (b) a shell or a core of encapsulated systems, (c) or a food ingredient. Zein-based encapsulation in use of food and nutrient components viz. lipids; essential oils, vitamins A,D,E and K, food colorants, food flavours and naturally occurring anti-oxidants. The improvement in the bioavailability of food and nutrient components viz. folic acid, vitamin D_3, curcumin, β-carotene, and resveratrol was observed by employing the zein-nano-particles in comparison with the bulk counterparts. Kasaai (2018) reported that the bio-active substances with potential applications intended for food and nutrition industries were stabilized by zein/zein-based nano-materials.

Composite films, a likely potential alternate to packaging materials, were prepared by extrusion technology using zein, poly propylene carbonate (PPC) and nano-TiO_2 (Li, et al., 2018). The samples were characterized by microscopic and calorimetric techniques and mechanical properties. The scanning electron microscopy study revealed that zein and PPC could form a homogeneous network structure and uniform distribution of nano-TiO_2 in films. Further, the composite films demostrated a good thermal stability. The work on antibacterial effect indicated that zein/PPC/nano-TiO_2 film had superior activity against the ***Escherichia coli***, *Staphylococcus aureus* and *Salmonella*. This composite film may be a potential alternative for food packaging (Li, et al., 2018).

Nano- selenium and garlic oil

Selenium is an important trace mineral that supports with cognitive function and fertility. Selenium may support to prevent cardiovascular disease, thyroid problems, cognitive decline (disorders related to thinking), cancer, and others. Brazil nuts are one of the finest sources of selenium (approximately 68 to 91 μg per nut). Among fish yellow fin tuna contains about 92 μg of selenium per 3 ounces, making it an excellent source of selenium. The other sources are ham, pork, beef, chicken and turkey. Garlic (*Allium sativum* L.) is a widely consumed spice in the world. Garlic contains varied bio-active compounds, such as allicin, alliin, diallyl sulfide, diallyl disulfide, diallyl trisulfide, ajoene, and S-allyl-cysteine. There are substantial studies showing that garlic and its bio-active constituents exhibit free radical scavenging (Khanum et, al., 1998), anti-inflammatory, anti-bacterial, anti-fungal, immunomodulatory, cardio vascular protective, anti-cancer(Khanum et, al., 2004 and Nagaraj et al., 2010), hepato-protective, digestive system protective, anti-diabetic, anti-obesity, neuro-protective, and renal protective effects. There are at least 30 selenoproteins that are invented which shows the tremendous interest in researching garlic. Moreover in the form of selenoproteins, Selenium can subsist in many different chemical forms in biological materials either as organic Se compounds, such as selenomethionine and dimethylselenide, and inorganic selenites and selenates. In foods, Se is predominantly present as selenomethionine, which is an important source of dietary Se in humans, and also as a chemical shape that is commonly used for Selenium supplements in clinical trials (Tinggi, 2008).

Abdel-Wareth et. al. (2019) worked on the result of nano-selenium (nano-Se), garlic oil, and their combination on nutrients digestibility, semen quality, serum testosterone and metabolites of male Californian rabbits. The results suggested that improvements were observed in body weight gain and feed conversion ratio in nano-Se treated group than non-supplemented group, and an interaction effect was observed on final body weight of rabbits. The digestibility of all measured nutrients were significantly higher for animals fed nano-Se, garlic oil, and the combination group compared to their non-supplemented counterparts. The study also showed that semen characteristics were significantly improved in nano-Se and garlic oil supplemented groups compared to control group of rabbits. Likewise, supplementation of nano-Se, garlic oil, and their combination significantly improved liver and kidney functions as indicated from serum metabolites, and increased serum testosterone hormone levels. This pre-clinial study showed that nano-Se or oil of garlic and its combination improves the nutrients digestibility, semen quality, kidney and liver functions, and testosterone levels in male rabbits.

Hebat-Allah et al., (2019) worked on the effect of nano-selenium that were applied to three different Egyptian groundnut (*Arachis hypogaea* l.) cultivars; under sandy soil conditions at vegetative growth stage to study their effects on yield components, signatures of protein and fatty acids, antioxidant levels and cytotoxicity of yielded seeds. The work showed that the tested nano-selenium concentrations improved yield components and seeds oil. However, nano-selenium altered protein marks and fatty acids composition by increasing unsaturated fatty acids and/or decreasing saturated fatty acids as compared with control. The cytotoxicity assessments proved safety of the yield for human health (Hebat-Allah et al., 2019).

Ginkgo biloba seeds

Ginkgo biloba has many wellbeing benefits. It's often used to treat mental health conditions, Alzheimer's disease and fatigue. It came on the Western culture scene a few centuries ago, but has enjoyed a gush of reputation over the last few decades. It is also used in conditions like: anxiety and depression, schizophrenia, inadequate flow of blood to brain, hypertensive problems, altitude sickness, erectile dysfunction, asthma, neuropathy, cancer, premenstrual syndrome, eye lens degeneration, etc. Like many natural remedies, ginkgo isn't well-studied for many of the conditions it's used for. The preservation of *Ginkgo biloba* (common names are Eun-haeng, fossil tree, Japanese silver apricot, kewtree, maidenhair tree, salisburia, silver apricot, etc.) seeds coating with these films were investigated during storage. In a study conducted by Tian et al. (2019), the chitosan, chitosan/nano-TiO_2 and chitosan/nano-SiO_2 coating films were prepared, their physico-chemical properties were studied. The amalgamation of nano-TiO_2 and nano-SiO_2 particles of the most exceptional formula enhanced the mechanical properties of the composite films, which enhanced the water-vapour and gas permeability. Chitosan coating incorporation of nano-TiO_2 and nano-SiO_2 particles could significantly lessen the rates of decay, shrinkage, respiration, ethylene production, electrolyte seepage, and superoxide anion ($O_2\cdot-$) production and the accumulation of malondialdehyde content. After storage, the firmness of *G. Biloba* seeds coated with chitosan alone and chitosan/nano-composite were markedly higher than those of control. Higher levels of peroxidase, superoxide dismutase, and catalase were enhanced in the chitosan/nano-TiO_2 and chitosan/nano-SiO_2 coating treatments during storage time. Either the combined treatment of chitosan with nano-TiO_2 or chitosan/nano-SiO_2 demostrated desirable presentation in inhibiting mildew occurrence, shrinkage and maintaining the firmness of *G. biloba* seeds, positively have an effect on the antioxidant activity in *G. biloba* seeds, thereby leading to the enhancement of seed quality. Coating *G. biloba* seeds with chitosan/nano-TiO_2 or chitosan/nano-SiO_2 is a potential method for commercial preservation.

Catechin nano-particles

Catechins are natural antioxidants that help prevent cell damage and provide other benefits. Tea is rich in polyphenols that have effects like reducing inflammation and helping to fight cancer. Green tea is about 30% polyphenols by weight, including large amounts of a catechin called epigallocatechin gallate. Novel starch-based nano-particles from three sources: horse chestnut, water chestnut and lotus stem were prepared for nano-encapsulation of catechin (Ahmad et al., 2019). Average particle size of horse chestnut, water chestnut and lotus stem based nano-particles were found out with encapsulation efficiency and negative zeta potential. Structural and physical properties including thermal were differentiated by fourier transform infra-red spectroscopy, scanning electron microscopy, X-ray diffraction and differential scanning calorimetry. The studies of scanning electron microscopy revealed capsule formation with entrapped catechin, depicting encapsulation of catechin in starch nano-particles without any evident interaction. X-ray diffraction work showed loss of crystallinity after encapsulation. Higher content of catechin in intestinal juice ensured controlled release in intestine. Bio-active properties were retained at higher level in encapsulated catechin compared to free catechin upon *in-vitro* digestion.

Mentha longifolia L. essential oil

In the present-day period, various pharmacological activities have been confirmed for *M. longifolia*, (Horse Mint French) such as anti-parasitic, anti-microbial, anti-insect, anti-mutagenic, anti-nociceptive, anti-inflammatory, anti-oxidant, keratoprotective, hepatoprotective, anti-diarrhea, and spasmolytic effects. Restrictions in the conventional medical management of diseases suggest a clear-cut requirement for safe and efficacious treatments. Rezaeinia et al. (2019) optimized encapsulation of *Mentha longifolia* L. essential oil into Balangu (*Lallemantia royleana*) seed gum nano-capsules, to increase their usefulness as flavoring and bio-active agents in foods and functional beverages. Essential oil emulsions with Balangu seed gum and various polyvinyl alcohol concentrations combined with Tween-20 were electro-sprayed. Increasing the concentration of polyvinyl alcohol enhanced the emulsion viscosity and enhanced loading capacity. Field emission scanning electron microscopy suggested that on increasing the gum amount and poly vinyl alcohol the process could be made to produce nano-fibers. The *Mentha longifolia* L. essential oil was entrapped in nano-structures without any chemical interaction with encapsulated material. This was demonstrated by fourier transform i-r spectroscopy and differential scanning calorimetry. The release mechanisms and kinetics of loaded *Mentha longifolia* L. essential oil were evaluated in different simulated food models viz. aqueous, acidic, alcoholic or alkaline and oily food models. The release profiles data were fitted to first order, Kopcha, Korsmeyer-Peppas, and Peppas-Sahlin models. The essential oil release profiles fitted well to the Peppas-Sahlin model for a range of simulated foods. The release mechanism of the essential oil from the nano-structure of the Balangu seed gum was reported to be mainly controlled by the Fickian diffusion phenomenon.

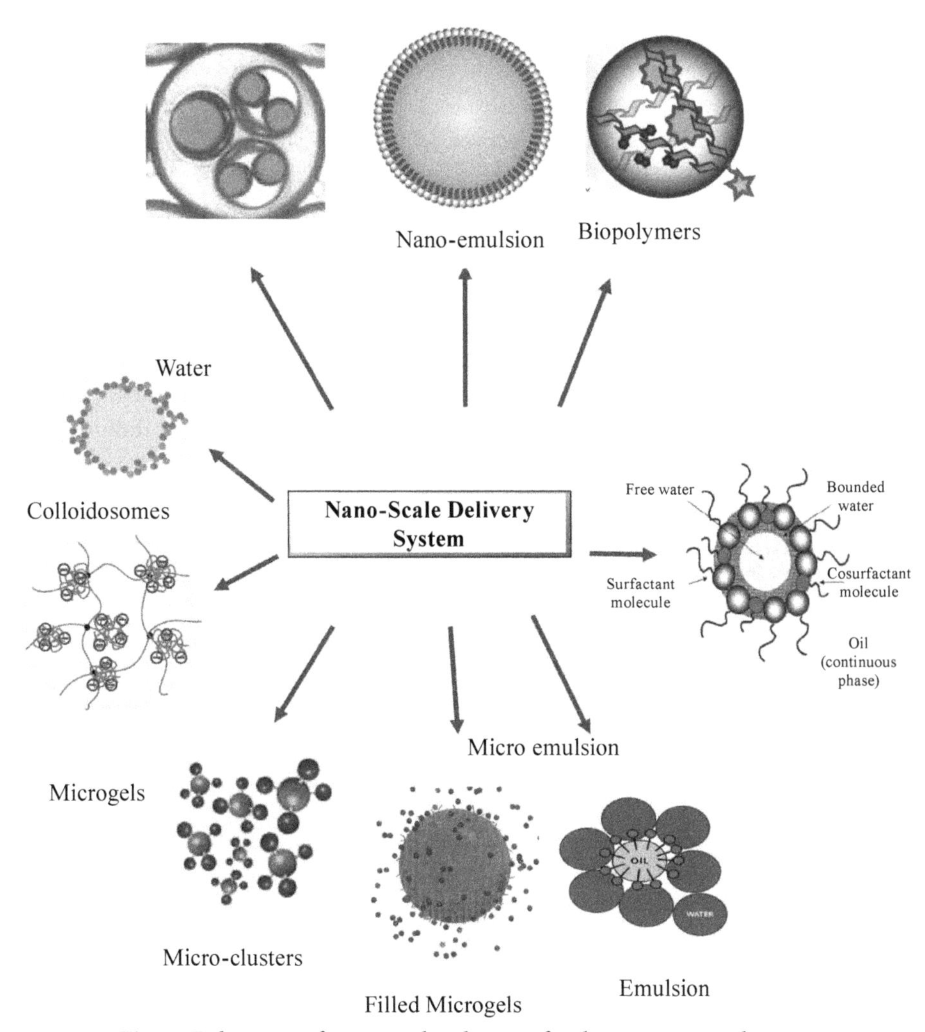

Fig. 4: Relevance of nano-technology in food processing industries

Nano-delivery systems

The encapsulation system has numerous benefits, viz. ease of handling, enhanced stability, protection against oxidation, taste masking, retention of volatile ingredients. moisture-triggered controlled release, pH-triggered controlled release, heat-triggered release, consecutive delivery of multiple active ingredients, long-lasting organoleptic perception and enhanced bioavailability and efficacy. The system can be utilised to transform volatile liquids such as flavours into a powder, which are in many cases easier to handle. The various delivery systems in nano- size are summarized in Fig 4.

Nano-ceutical delivery system, can enable taking vitamin B_{12}, children will love the taste, like candy, benefitting elder people who can be more alert and sleep better. Vitamins can be sprayed it into the mouth, not only does it taste good but it is completely nutritional. The pleasant, convenient way of delivery, is easily absorbed up to 90%, by tissues beneath the tongue and lining of the cheeks.

Simões et. al. (2017) provided a summary of the bio-based substances that are being used for encapsulating bio-active compounds. Assessments between various *in vitro* digestion systems the merits and de-merits of currently used systems, and correlations between the behaviour of micro- and nano-systems studied via *in vitro* and *in vivo* systems were also summarised. Finally, examples of micro- and nano-systems that are bio-active for authentication of food matrices are provided, together with a reconsideration of the main challenges for their safe commercialization, the regulatory aspects involved and the main legislation features.

Chitosan

Functional food components have attracted extensive attention for their nutritional and physiological properties. Chitosan polymer has been used extensively in the medical ground. It is either partially or fully deacetylated chitin. As chitin occurs naturally in fungal cell walls and crustacean shells, chitosan is a fully biodegradable and bio-attuned and can be used as an anti-bacterial and anti-fungal agent. However, the instability and low bioavailability seriously restricted their uses in the food and pharmaceutical industries. Sun et al (2019) successfully manufactured a multifunctional three-phase nano-delivery system of polysaccharide-polyphenol-protein, viz. chitosan-tannin-lysozyme nano-particles (CS-TA-L NPs). In this study, employing chitosan as the wall material, tannin and lysozyme were incorporated into the chitosan matrix by ion cross-linking and ultrasonic assisted methods, forming the nano-particles with core-shell structures. As a result of this, this uncomplicated, reasonable and green nano-delivery system protected the biological activities of tannin and lysozyme, improved the stability of chitosan-tannin-lysozyme nano-particles. This was endowed with chitosan nano-particles with anti-oxidant and anti-bacterial effects resulting in providing a theoretical basis for constructing the multifunctional nano-delivery systems.

Oligo-hyalurosomes nano-delivery system

Unlike water soluble antioxidants, the lipid soluble version do not dynamically go out seeking to wipe out rogue cells in the human body. These are the antioxidants which really have a much more passive role in keeping the human body healthy. Basically, these antioxidants work by adhering to damaged cells, and injecting valuable micro-nutrients which hold up the replenishment and strength of that particular individual cell. In this way, antioxidants are able to uphold the health of cells on a celllular level, by working in synchronisation with the cells themselves. Examples are: carotenes and ubiquinol, the reduced coenzyme Q_{10}. A deficiency of antioxidants could cause, or support in causing

Alzheimer's disease, cancer, cardiovascular disease, cataract, diabetes, hypertension, infertility, eye lens degeneration , psychosis and respiratory tract infection. By adding adequate antioxidants to the diet, there is less oxidative stress, and aging is also slowed down. In a work by Guo et al, (2018), in order to enhance the stability, bioavailability and antioxidant activity of such insoluble antioxidants used into juice, yoghourt and nutritional supplements, the oligo-hyalurosomes nano-delivery system based on oligo-hyalurosomes nano-delivery system - curcumin polymer loaded curcumin and resveratrol was fabricated with nano-scopic technolgy. The rosy biodegradable amphiphilic oligo-hyalurosomes nano-delivery system polymer was successfully synthesized and used to fabricate the hyalurosomes containing both curcumin and resveratrol, called oligo-hyalurosomes. The hyalurosomes containing both curcumin and resveratrol, can spontaneously self-assemble into nano-sized spherical shape of average particle size. *In vitro* gastrointestinal release analysis showed stability and outstanding sustained release character. Oligo-hyalurosomes nano-delivery system showed a dose-dependent behaviour with a higher free radical scavenging ability in comparison with single formulations and liposomes. This novel nano-food delivery system manifested the hopeful properties for the new effectual gastrointestinal formulation and promising new nano-food delivery system in the use of juice, yoghurt and nutritional supplements.

Pectin polymers as carrier

Pectin is a naturally occurring polysaccharide frequently found in berries, apples and other fruits. When heated together with sugar, it causes a solidifying effect that is distinctive of jams and jellies. It has been put into use as a carrier for the protection and targeted delivery of bio-active compounds and for enhancing their shelf life and stability. The process of nano-encapsulation is one of the procedures that have been used for the efficient protection of bio-active compounds. The various sources and depiction of pectin along with various encapsulation processes of different bio-active compounds was described by Rehman et al. (2019). Pectin is applied for nano-encapsulation, where nano-capsules can be formed through diverse methods viz. spray drying, emulsion and through the creation of hydrogel, liposomes, etc. Additionally, the use of pectin in combination with other compounds such as proteins and lipids were established to be the most potential wall material for the bio-actives .

Ferritin based nano-carrier

Ferritin protein has received more and more attention in encapsulation and delivery of the bio-active compounds due to its nano-sized shell-like structure and its reversible self-assembly character. Ferritin is a blood cell protein that contains iron. A ferritin test helps us to understand how much iron is storing in our body. If a ferritin test exposes that your blood ferritin level is lower than normal, it indicates that our body's iron stores are low and we have iron deficiency. The low solubility, instability and low bioavailability of bio-active compounds from food such as polyphenols and flavonoids, restrict their use in the fields of life science including nutrition. After encapsulation, bio-actives can be functionalized using the ferritin vehicle to accomplish stabilization, solubilization and targeted release. The outer interfaces and the porous structure of ferritin are also artfully connected for encapsulation. Liu et al (2019) focused on the latest advances in the manufacture, characterization, and function of ferritin-based nano-carriers for bio-active compounds by the methods like reversible self-assembly, outer-interface decoration and the channel-directed strategy. The functional advances of food bio-active compounds, viz. their solubility, stability and cellular uptake, are highlighted. The limitations that affect ferritin encapsulation are also examined.

Nano-emulsions

Nano-emulsions are ever more used in a diversity of products so as to incorporate easily degradable bio-active compounds, defend them against oxidation and augment their bio-availability. However, accomplishing a stable emulsion formulation with desirable properties, causes a challenging task. In a work conducted by Katsouli et al. (2018) the composition of water in extra virgin olive oil nano-emulsions incorporating two bio-active compounds, ascorbic/gallic acid, was optimized using Response Surface Methodology (RSM). The selected composition variables were the emulsifier and the ratio of bio-active compounds and the scrutinized properties were the emulsion stability, mean droplet diameter, poly-dispersity and turbidity. The features in the water-in-oil nano-emulsions were considerably influenced by both composition variables. The best possible formulation for producing water-in-oil nano-emulsion with desirable characteristics incorporating ascorbic and gallic acid was provided by the same combined levels of emulsifier and bio-active compound ratio i.e. 1% w/w bio-active compound and 4% w/w Tween 20. Experimental values were in close agreement with predicted values.

Carotenoids are added to lemonades, fruit juices, squashes, crushes and margarines. The capability to fabricate materials at the atomic or molecular level gives an impact on the food segment if it is possible to develop a sort of coatings, barriers, release materials, devices and new packaging substances. The branch of science called 'Food colloid science' include fabrication and characterisation of nano-particles in foods and it has remarkable scope in this field upon which nano-technology may draw upon. The highly specialised ultrasound spectroscopy offers the best outlook for the characterisation of concentrated systems of nano-particles and nano-emulsions for the beverage sector. Nano-emulsions have recently received lots of attention from the food segment because of their high clearness and adherence. The addition of nano-emulsified bio-actives and flavors to a cocktail/drink makes possible product appearance without any transformation. The type of surfactants is highly challenging in the emulsification process and the nano-emulsion supports to overcome this limitation. Among various advances that have been utilized include micro-emulsion and nano-emulsion. Nano- and micro-engineering techniques are employed for microfiltration and nano-sensing applications. With micro-engineering techniques it is likely to assemble very specific microsieves. The pores, which are well defined by employing photolithographic methods permit accurate separation of particles by size. In this case, the membrane thickness is typically less than the pore size which results in operational progression fluxes. This supersedes the one to two decades old conventional filtration methods. Various advantages of nano- emulsions are provided in Fig 5.

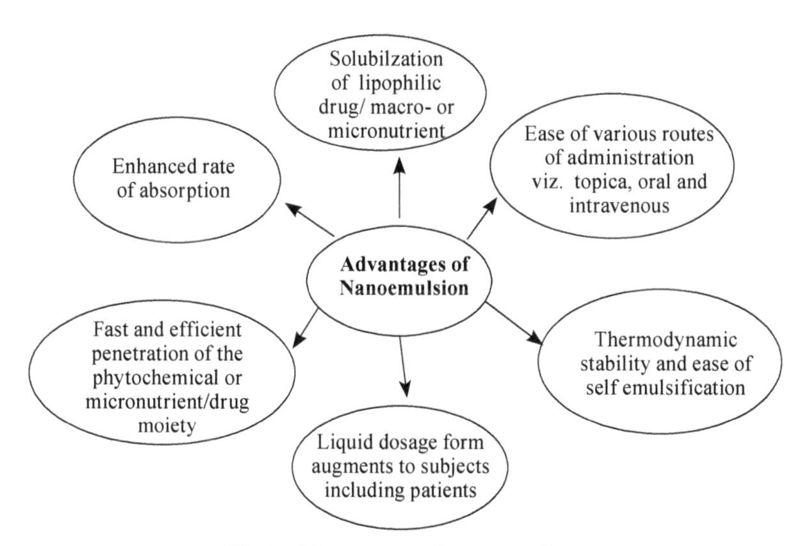

Fig. 5: Advantanges of nano-emulsions

Capsaicin

Capsaicin, found in hot chilli peppers is reported to have variety of pharmacological effects such as effective role in pain control and may be used as an effectual analgesic agent. Nigam e al. (2019) prepared capsaicin extracts by means of techniques viz. liquid-liquid extraction and solid-liquid extraction. The extracts were then characterized via qualitative and quantitative chromatographic experiments, TLC and HPLC. Plant extracts from dried hot chilli peppers showed better yields and higher alkaloid and capsaicinoids contents as compared to fresh capsaicin extracts. Clear, uniform and stable oil-in-water nano-emulsions of capsaicin were made using oleic acid/labrasol, Tween-20, glycerol and water by means of high-speed homogenizing and ultra-sonication. The developed capsaicin nano-emulsion was characterized for particle size, poly dispersity index and zeta potential. Percentage transmittance suggested clear nano-emulsion. Further cytotoxicity was evaluated on Neuro-2a cell using MTT assay. The results of cell viability test indicated viablility with nano-emulsion prepared from capsaicin.

New trends in nano- food processing

Nano- scale powder processing system

One of the most notable qualities of nano-powders is the specific surface area. Nano- powders may be produced by diverse kinds of processes: physical processes that include laser ablation, evaporation/condensation, etc., mechanical processes that include mechano-synthesis, high deformation at low temperature, etc. and sol-gel, hydrothermal, plasma synthesis (chemical processes). Nano-powders are composed of nano-particles with an average diameter below 50 nm. Most of the manufacturers of nano-powder produce micro powders that are accumulated of nano-particles, but the powder itself is atypically a nano- powder. Such compounds have two or three different cations in their chemical formula. The best example of chemical compound in this respect is calcium titanate ($CaTiO_3$). As the size reduces to nano-scale, the properties of atoms present on the outer boundary of particles become dominant (Wang et al., 2019).

To enhance solubility and dispersibility of nutritional elements of barley grass for rehabilitation therapy, high-energy ball-milling was conducted to obtain nano-scale powder of barley grass (Cao et al., 2018). Nano-scale barley grass powder manifests a fascinating performance of quality, hydration and nutritional properties. Ball-milling is a processing to make barley grass easy digestion for human body.

Nano- phytosome technology

Flavonoids have many nutritional-therapeutic and preservative features. Hence, using them for fortifying food products or developing supplements could provide new horizon in developing new functional foods. However, there are several concerns connected with the use of these phyto-actives in foods and drinks. This particular nano-some is one of the latest lipid based nano-carrier and rapidly growing attractive means of delivering plant based nutraceuticals. Although it has been developed for pharmaceutical applications, its application in food products for inventing novel functional foods and beverages could be explored. Ghanbarzadeh (2016) reviewed physicochemical properties of nano-phytosome to be used as carrier of food bio-actives in development of nano-ceuticals.

Nano-pesticides

In order to progress the yield and dominance of food and lighten the pressure of traditional pesticides on the environment, pesticides in nano encapsulated form are on developmental stages and have been used. This would add to the yield of agricultural products due to their immense role in saving farming costs, the nutritional characteristics and keeping quality of foods, and achieving precision agriculture. Still, the impact of nano-pesticides on biodiversity is still lacking a thorough investigation. Sun et al. (2019) examined whether the application of nano-pesticides is a great challenge for biodiversity.

A key dispute in the agricultural sector is the need to address problems associated with pesticide use that leads to environmental contamination, bio-accumulation and increasing pest resistance. This demands a declining effect in the measure of pesticide applied for protection of the stored products. Nano-scopic technology is promising as a highly attractive means to achieve this goal. This often offers new means for the formulation and delivery of pesticides and novel dynamic ingredients (Hayles et al., 2017).

The development of rapid, sensitive, specific, and economically viable methods for pesticide detection to ensure food safety and security has been highlighted in the recent past. Nano-technology may be applied to develop inexpensive, portable nano-sensors combining a biological recognition mechanism with a physical transduction technique for rapid and real-time detection of pesticides at dispersed locations. Saini et al. (2017) provided a critical evaluation of newer trends in biosensors, such as tailor-made biomolecules that include aptamers and molecularly imprinted polymers, optical, electrochemical, and piezoelectric materials, and so forth. Acetamiprid is an insecticide that can generate potential health risk of humans. Aptamers have attracted as a novel sensing elements. The combined nano-material and aptamer technology has opened a new opening in the monitoring of pesticide residues (Verdian, 2018).

Controversies in the application of nano- techniques for foods

The rapid propagation in the area of nano-technology has led to a growing function of nano-scale materials in a vast group of industrial and consumer products. This includes beverages and food products, dietary supplements, and food contact with substances. On the other hand, the widespread use of nano-materials, and research in this field that are basically not subjected to tests for its toxicity has also led to a number of uncertainties and worries. The regulatory aspects related to nano-technology based food products is absolutely necessary along with public attentiveness and understanding of the role of nano- size technology.

A revolution with jeopardy

Yesterdays “micro” is replaced by Today’s “nano”. It may be note that a micrometer is one millionth of a meter; whereas a nano-meter is equal to one billionth of a meter. It is roughly equal to 10 hydrogen atoms. Nano- products are rapidly being discharged into the marketplace without being properly tested or experimented. This is predominantly in relation to skin products, physiologically active foods and beverages coupled with food packaging and dietary supplements. Many consumers have modest or no knowledge about nano-technology and are probably unaware that some products they use are made with nano- sized particles. It is hardly anything is known about the use of engineered nano- particles in the dietary supplement market (Schultz and Barclay , 2009). Although the nano- sized materials brings many potential advantages to food production, such as enhanced shelf life and resistance to pathogens, its progress must be monitored by appropriate safety assessments and regulation to minimise the risk associated with it. Additionally, it is important to validate, quick tests in place to evaluate the potential toxicity of new nano-materials. Two new emerging technologies highlighted for the toxicological assessment of nano-materials in food applications are high content screening and what is called the "zebrafish model". The high content screening technique appraises the biological effects of chemical substances in *in vitro* cell based assays. The Zebra fish (*Danio rerio*) has been an important model vertebrate in many of the biological disciplines. It can endow with valuable experimental toxicity information.

Nano- food market

Just as in the case of safety data which is scanty and inconclusive, solid market data on nano-technology based foods or dietary supplements may be intangible. It is seen that many companies simply do not report the presence or use of nano-materials in their products. The subjective application of a 100 nm limit in size included in most definitions of nano-technology exclude products that may still act nano-like.

Segmenting the nano-sized materials incorporated into functional foods and beverages, dietary supplements, personal care and cosmetics, providing market numbers make matters worsened. Nano-based foodstuffs and the companies are provided in Tab 1.

Several kinds of food-related applications are in different developmental stages, with a few such applications that are already on the market. More than 200 companies around the world are on the active mode in R&D. Oxygen and free radical scavengers (for sliced processed meat, ready-to-eat meals and beer), moisture absorbers (for fresh meat, poultry and fresh fish) and barrier packaging (for packaging of fruit and vegetables) and ethylene-scavenging bags are among the tops among active technologies, with more than 80 % of the current market. Among these 60 % (antioxidant) and 40% (antimicrobials) are on dominate controlled release.

The application of active packaging is already successful in the United States. Japan and Australia relate nano-technology to make longer shelf-life ensuring maintenance of nutritional quality and microbiological safety.

The future perspectives

Nano-encapsulation is the most noteworthy encouraging methods with the possibility to entangle biologically active chemicals (Thiruvengadam et al., 2018). As an active area of research, nano-technology is witnessing fast commercialisation. Food wrapping has also been considered as a potential beneficiary of nano-technology.

Research is being done to develop nano-encapsulated materials containing macro and micro-nutrients that would be released when nano-sensors detect a deficiency in our body. Materials in nano size are being developed to get better the colour, texture and taste, of foods. Now a day "interactive" foods are being experimented that would allow the consumers to prefer which flavour and colour a foodstuff has. Modern biotechnologies involve making useful products from whole organisms or parts of organisms, such as molecules, cells, tissues and organs. Recent developments in biotechnology include genetically modified plants and animals, cell therapies and nano-technology. These products are not in everyday use but may be of benefit to the purchaser in the future.

Nano-technology is intangible, defined and quantified in a market intellect. Nano-technogy materials such as micelles and liposomes may seem harmless for use in making certain micronutrients, ω-3 fatty acids and other antioxidants more bioavailable and stable. All the more, when the nano-sized materials are applied to more potent dietary supplement ingredients, where these nano-enabled substances will go in the body and how they will have an consequence on health is still largely unknown. This lends to a fear by regulators and consumers, both which have been slow to earn about nano-technolgy. It is in fact an infant frontier that warrants ample scope for future research.

References

Abdel-Wareth A. A. A., Ahmed, Hassan H. A, Abd El-Sadek M. S.and Lohakare J.(2019). Nano-selenium, garlic oil, and their A. E combination on growth and reproductive performance of male Californian rabbits. *Animal Feed Science and Technology*, 249: 37-45.

Ahmad M., Mudgil P., Gani A., Hamed F. and Maqsood S. (2019). Nano-encapsulation of catechin in starch nanoparticles: Characterization, release behavior and bioactivity retention during simulated in-vitro digestion. *Food Chemistry*, 270 (1): 95-104.

Cao X., Zhang M., Mujumdar A. S., Zhong Q, and Wang Z. (2018). Effect of nano-scale powder processing on physicochemical and nutritional properties of barley grass. *Powder Technology*, 336, , 161-167.

Ghanbarzadeh B., Babazadeh A. and Hamishehkar, H. (1 September 2016). Nano-phytosome as a potential food-grade delivery system. *Food Bioscience*, 15:126-135.

Guo C., Yin J. and Chen D. (2018). Co-encapsulation of curcumin and resveratrol into novel nutraceutical hyalurosomes nano-food delivery system based on oligo-hyaluronic acid-curcumin polymer. *Carbohydrate Polymers*, 181(1):1033-1037.

Hayles J., Johnson L., Worthley C.and Losic D. (2017).5 - Nanopesticides: a review of current research and perspectives. *New Pesticides and Soil Sensors.*, 193-225.

Hebat-Allah A., Osama M.H., Bahaa B.D., Salwa M.M. and and El-Hallouty (2019). Evaluation of cytotoxicity, biochemical profile and yield components of groundnut plants treated with nano-selenium. *Biotechnology Reports*, 24

Kasaai M.R. (2018). Zein and zein -based nano-materials for food and nutrition applications: A review. *Trends in Food Science & Technology*. 79:184-197.

Katsouli M., Polychniatou V. and Tzia C. (2018). Optimization of water in olive oil nano-emulsions composition with bio-active compounds by response surface methodology, *LWT*, 89:740-748.

Khanum F., Anilakumar K.R., Viswanathan K.. R. and Sudarshanakrishna. K. R. (1998). Effect of feeding fresh garlic and garlic oil on detoxifying enzymes and micronuclei formation in rats treated with azoxymethane. *Int. J. Vitamin Nutr Res.*, 68:208-213.

Khanum F., Anilakumar K R, and Vishwanathan K R. (2004). Anticarcinogenic properties of garlic-a review. *Crit Rev Food Sci Nutr* .44:479-488.

Li S, Zhao S, Qiang S, Chen G and Chen Y. (2018).A novel zein/poly (propylene carbonate)/nano-TiO_2 composite films with enhanced photocatalytic and antibacterial activity. *Process Biochemistry*, 70:198-205.

Liu Y., Yang R., Liu J., Meng D. and Blanchard C. (2019). Fabrication, structure, and function evaluation of the ferritin based nano-carrier for food bio-active compounds. *Food Chemistry,* 299, 30,

Nagaraj N. S., Anilakumar K. R., and Singh O.V. (2010). Dially disulfide causes caspase-dependent apoptosis in human cancer cells through a Bax-triggered mitochondrial pathway. *J Nitr. Biochem.* 21(5):405-412,

Niaz T. , Shabbir S., Noor T. and Imran M. (2019). Antimicrobial and antibiofilm potential of bacteriocin loaded nano-vesicles functionalized with rhamnolipids against foodborne pathogens. *LWT,* 116, .

Nigam K., Gabrani R., and Dang S. (2019). Nano-emulsion from Capsaicin: Formulation and Characterization. *Materials Today: Proceedings,* 18:869-878.

Rehman A., Ahmad T., Aadil R. M, Spotti M. J., and Quny. (2019). Pectin polymers as wall materials for the nano-encapsulation of bio-active compounds. *Trends in Food Science & Technology,* 90:35-46.

Rezaeinia H., Ghorani B., Emadzadeh B. and Tucker N. (2019). Electrohydrodynamic atomization of Balangu (*Lallemantia royleana*) seed gum for the fast-release of Mentha longifolia L. essential oil: Characterization of nano-capsules and modeling the kinetics of release. *Food Hydrocolloids,* 93:374-385.

Saini R. K., .Bagri L. P. and Bajpai A. K. (2017) 14 - Smart nanosensors for pesticide detection. *New Pesticides and Soil Sensors.,* 519-559.

Schultz W.B. and Barclay L. (January 2009). A Hard Pill to Swallow: Barriers to Effective FDA Regulation of Nanotechnology-Based Dietary Supplements. *Available at: http://www.nanotechproject.org/process/assets/files/7056/pen17_final.pdf*

Shishir M.R.I., Xie L, Sun C., Zheng X. and Chen W. (2018).Advances in micro and nano-encapsulation of bio-active compounds using biopolymer and lipid-based transporters. *Trends in Food Science & Technology,* 78:34-60.

Sholkamy E. N., Ahamd M. S., Yasser M.M. and Eslam N. (2019). Anti-microbiological activities of bio-synthesized silver Nano-stars by *Saccharopolyspora hirsuta. Saudi Journal of Biological Sciences,* 26(1):195-200.

Simões L. D., a Madalenaa D A., Pinheiro A., C. Teixeira J.A., Vicente A., A. and Ramos O. L. (2017). Micro- and nano bio-based delivery systems for food applications: *In vitro* behavior. *Advances in Colloid and Interface Science,* 243:23-45.

Sun X., Jia P., Zhe T., Bu T., Wang L. (2019). Construction and multifunctionalization of chitosan-based three-phase nano-delivery system. *Food Hydrocolloids,* 96:402-411.

Sun Y., Liang J., Tang L., Li H. and Zeng G. (24 July 2019) .Nano-pesticides: A great challenge for biodiversity? *Nano Today,* In press, corrected proof, Available online.

Tarver T. (2006). Food nanotechnology. *Food Technology.* 60(11):22-26.

Thiruvengadam, M., Rajakumar, G. and Chung, I.M.(2018). Nanotechnology: current uses and future applications in the food industry. 3 *Biotech* 8: 74. *https://doi.org/10.1007/s13205-018-1104-7*

Tian F , Chen W., Wu C., Kou X. and Wu Z. (2019). Preservation of Ginkgo biloba seeds by coating with chitosan/nano-TiO2 and chitosan/nano-SiO2 films. *International Journal of Biological Macromolecules,* 126, (1):917-925.

Tinggi U. (2008 Mar). Selenium: its role as antioxidant in human health. *Environ Health Prev Med;* 13(2):102–108.

Verdian A. (2018). Apta-nanosensors for detection and quantitative determination of acetamiprid – A pesticide residue in food and environment. *Talanta,* 176(1):456-464.

Wang J., Wu S. and Liao H.(2019). Advanced Nanomaterials and Coatings by Thermal Spray Micro and Nano Technologies. *Chapter 2 - The Processes for Fabricating Nanopowders.* 13-25.

Wang X., Jiang Y, Wang Y., Huang M.T., Ho C.T.and Huang Q. (2008). Enhancing anti-inflammation activity of curcumin through O/W nanoemulsions. *J. Food Chem.* 108(2): 419-424.

Yang C., Zhu B., Wang J. and Qin Y.(2019). Structural changes and nano-TiO_2 migration of poly (lactic acid)-based food packaging film contacting with ethanol as food simulant. *International Journal of Biological Macromolecules*, 139(15): 85-93.

5

Recent Trends in Functional and Nutraceutical Foods

Rashim Kumari[1], ***Vikas Dadwal***[1,2], ***Himani Agrawal***[1,2] **and** ***Mahesh Gupta***[1]

[1]*Food and Nutraceutical Division, CSIR-Institute of Himalayan Bioresource Technology, Palampur, Himachal Pradesh, India*
[2]*Academy of Scientific and Innovative Research (AcSIR), India*

Introduction

For human beings, the three most important necessities for their survival are food, clothing and shelter. among all of these food is the top most priority. Food safety and Standards defines food into 8 categories (Nutraceutical, Heath Supplements/Dietary Supplements, Food for special medical Purpose, food for Special Dietary Use, Novel Food, and Functional food) in regulation, 2016. A growing trend of health products, concept of food as medicine, associated public interest and consumers demand leads to a progressive research interest in functional foods and nutraceuticals development. Abnormal lifestyle, food habits (consumption of junk food) and low physical activity creates an exponential growth in a number of chronic diseases; obesity, cardio-vascular syndromes, osteoporosis, arthritis and cancer (Das et al., 2012) among the developing and developed countries. While nutraceuticals and functional foods are becoming an alternative affordable system to reduce the risk of disease through prevention in spite of high tech therapies and costly medicines.

The current survey showed that worldwide revenue of functional foods is projected to increase from about 300 billion U.S. dollars in 2017 to over 440 billion dollars in 2022. Also according to the International Food Information Council (IFIC), health benefits achieved from functional food components may provide benefits far beyond than basic nutrition. Products like dietary supplements containing vitamins, functional foods such as sports drinks or products fortified with probiotic play a major role in the market rise (www.statista.com).Indian and Chinese traditional foods, recipes, spices and diversified ingredient market is well known around the world from centuries. Hence, a double-digit growth rate is also experienced in Asian nutraceutical and functional food market. Another reason for the growing trend in consuming functional foods is education and proper health awareness. Furthermore, people are more concern about the optimization of health-promoting capabilities in their diet using dietary supplement and fortified foods. While, food product based industries are marketing health supplements using scientific facts and the relationship between diet and health. Government, non-government organizations (NGOs) and web-based online information also play a key role in promoting and trending food based health products.

Current scientific knowledge revealed that prime bioactive ingredient used in fortified or functional foods were isolated from different plant sources including fruits, vegetables, whole grains, cereals, nuts and marine foods (a rich source of omega 3 fatty acids) (Shahidi et al., 2009). In addition, residual waste (pulp, peel, seeds) of fruits and vegetable were reprocessed to recover bioactive compounds for possible utilization in other food products (Sagar et al., 2018). Studies also showed that micronutrients including vitamins and minerals (Hoeft, B et al., 2012), polyphenolic compounds (Khurana et al., 2013; Zhou et al., 2016), unsaturated fatty acids (Ruxton et al., 2004), plant sterols and probiotics (Nagpal et al., 2012) are the major bioactive components used in nutraceuticals and functionalized foods. While, the present trend of utilizing milk-based beverages becomes an ideal delivery system to transport these bioactive compounds. Recent years, rather than some single bioactive molecules, a physical mixture of different bioactive compounds in a single functional food product is also in trend because of its associated health promoting synergistic effects.

The following sections are demonstrating the latest trends in different categories of nutraceutical and functional foods, its associated health benefits, market and future perspective.

Nutraceutical, dietary supplements and functional foods

Nutraceuticals

Nutraceutical, Dietary Supplements and Functional foods are seems to be a thin line difference terms which are often used interchangeably. Nutraceutical is a hybrid term of 'nutrition' and 'pharmaceutical'. The term nutraceutical was coined in 1989 by Stephen DeFelice MD, founder and chairman of Foundation for Innovation in Medicine (Aronson, 2017). Any food or part of food providing medical or health benefit including prevention and treatment of diseases or a part of food that can be modified into capsules, pills or liquid to deliver and promotes human health refers to as nutraceuticals. Nutraceuticals not only supplement the food but also prevent and/or treat disease. They are also used as a conventional food or single item of meal. While previous reports showed that traditional cuisines or food products were also exerts a wide range of physiological effects including gastrointestinal health, metabolic homeostatic, signaling pathways, antioxidative defenses and other detoxifying processes (Bhaskarachary, K., 2016). While, a variety of food products, herbal products, processed foods such as soups, cereals and beverages are referred to as functional foods as they are also responsible for providing health related benefits. These food products also show resemblance with traditionally consume foods but enhanced by more nutrients to establish physiological benefits to maintain human health. Whereas traditional foods itself described as the foods eaten from last many decades assumed that containing biological active molecules exhibiting beneficial effects also included in functional foods (Bhaskarachary, K., 2016). Despite from traditional knowledge present scientific expertise also leads to the formulation of a wide range of edible products containing extracted molecules from different food sources.

Still, at present there is no clear definition of nutraceutical and functional foods and most of them are provided by professional or regulatory organizations. According to American Dietetic association, whole or fortified, enriched foods having potentially beneficial effects on human health when consumed as part of a varied diet or in regular basis at an effective level refer to as functional foods (Hasler et al., 2004). While, Nutrition Business Journal states that any food that can be consumed particularly for health reasons refer to as nutraceuticals (www.nutritionbusinessjournal.com) At the same time, Health Canada states that a nutraceutical is a product "prepared as food, but sold and supplemented in the form of pills or powders or in the other medicinal forms and provides protection against chronic diseases" (www.canada.ca), according to Functional food science (FFC) a "functional food is an inexpensive and convenient solution for chronic health problems". (Martirosyan, D.M. and Singh, J., 2015). Hence, as per above, functional food can be included into nutraceuticals.

Dietary supplements

Worldwide, there are various definitions for "dietary supplements". As per Dietary Supplement Health and education Act of 1994 (DSHEA), dietary supplements are defined as "a product (other than tobacco) intended to supplement the diet that bears or contains one or more of the following dietary ingredients: a vitamin, a mineral, an herb or other botanical, an amino acid, a dietary substance for use by man to supplement the diet by increasing the total dietary intake, or a concentrate, metabolite, constituent, extract, or combination of these ingredients". These dietary supplements are in the form of drug, liquid and not used as conventional food or single item of a meal like nutraceuticals.

It is an innovative idea which is comparatively different from nutritional supplements, which correct specific deficiency diseases, such as vitamins and minerals but sometimes, there is not enough or scientific data proof to support these claims. Although, the dietary supplement industries are vastly growing due to the huge demand but simultaneously, people are also in dilemma because these supplements do not have any scientific data on some health claims, plus there are data on the adverse effects of these health supplements. Dietary supplements are marketed as semi "drugs," but they fully evade the regulations and official process. Therefore, it is consumer responsibility to check and analyze the status and benefits of dietary supplements, and also to check at the possibility of solution to the problems through traditional knowledge process, like Ayurveda.

Functional food

Almost every food has some effect which makes the debate on functional food more complex. Japan in 1980 was the first country to use the term "functional food". Functional food contains fortified and enriched foods in their natural form which prevent and control diseases. The Academy of Nutrition and Dietetics explains functional foods as: "whole foods along with fortified, enhanced or enriched foods that have a potential health benefits when taken as part of a varied diet on a regular basis at effective levels based on significant standards of evidence (Table 1).

Table 1: Classification of Functional foods and its health benefits

Food	Key Ingredients in functional food	Health benefit	Claim on Label
Whole Food	Oats (Total dietary fiber and beta-glucan)	Reduces the cholesterol and constipation. Reduces the risk of heart disease	May reduce the risk of heart disease.
	Soy (Isoflavones)	Reduces cholesterol. Reduces the risk of osteoporosis and certain cancers.	May reduce the risk of some cancers.
	Fruits and vegetables (Dietary Fiber, Lycopene, Lutein, Catechins, Plant Sterols, Flavanones, Polyphenols)	Reduces the heart diseases and the risk of certain cancers. Reduces hypertension.	May reduce the risk of heart disease
	Garlic (Allicin)	Reduces the risk of heart disease and certain cancers. Reduces cholesterol level	None
	Flaxseed(γ-tocopherol, PUFA)	Reduces the risk of heart disease and certain cancers. Reduces triglycerides.	None
	Nuts (MUFA, Fiber, magnesium)	Reduces the risk of heart disease	None
Fortified foods	Calcium fortified juices	Reduces the risk of osteoporosis. Reduces hypertension.	Helps in maintaining healthy bones and may reduce the risk of osteoporosis
	Fortified fibrous grains	Reduces the risk of certain cancers and heart disease. Reduces cholesterol level and constipation. Increases blood-glucose control	May reduce the risk of some cancers; May reduce the risk of heart disease
	Iron rich weaning food	Reduces risk of iron deficiency in children.	None
	Fortiofied fiber juices	Reduces the risk of certain cancers and heart disease. Reduces the level cholesterol, hypertension, and constipation.	May reduce risk of some cancers
	Vitamin D fortified milk	Reduces risk of osteomalacia and osteoporosis	Helps maintain healthy bones and may reduce risk of osteoporosis
	Grains fortified with folic acid	Reduces the risk of heart disease and neural tube birth defects.	May reduce risk of brain and spinal cord birth defects

Enhanced foods	Probiotic dairy products	Reduces the risk of colon cancer and candidal vaginitis. Controls inflammation. Treats respiratory allergies, diarrheal disorders, and eczema.	Structure/function claim
	Sports bars	Depends on ingredients	Structure/function claim
	Herabl foods and beverages	Depends on ingredients	Structure/function claim
	sugar alcohols foods instead of sugar	Reduces the risk of tooth decay	May reduce risk of tooth decay
	Stanol esters spreads	Reduces cholesterol level.	Structure/function claim
	Salad dressing beverages with antioxidants	overall health	Structure/function claim
	omega-3 fatty acid egg	Reduces the risk of heart disease	Structure/function claim
Enriched foods	Enriched grains (phenolic acids, flavones, flavanoids, coumarins, terpenes)	Reduces the risk of certain cancers, heart disease, and nutritional deficiency	May reduce the risk of some cancers; May reduce the risk of heart disease

Functional foods as disease preventive

Functional foods have health benefits along with basic nutrition because of certain active components, which later may or may not have been altered to enhance their bioactivity. These foods either may help to prevent disease or reduce the risk of developing disease. During the late twentieth century consumer awareness and interest in functional foods has been increased due to its health benefits. They have become more aware of the health benefits associated with specific foods and are incorporating elements such as fiber, calcium, and soy into their diets. Clearly most foods have functional property in some way. Functional foods varied from cereals, fruits, vegetables etc, which are naturally rich in phytochemicals. They affect biological responses in the body and promote health benefits (Fig.1):

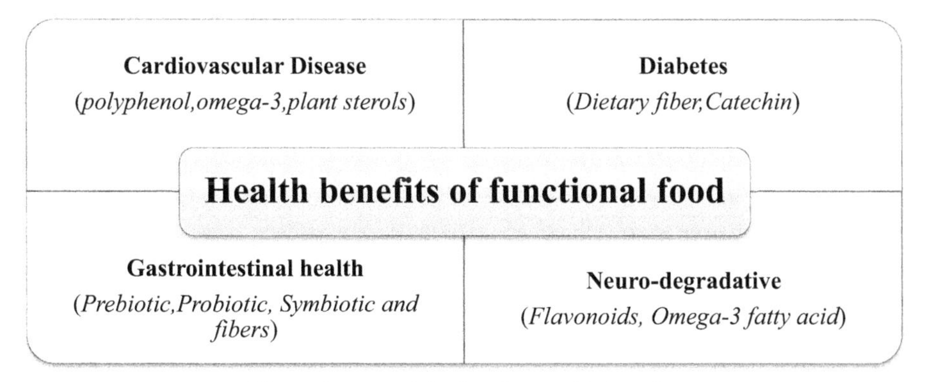

Fig. 1: Functional food and its health benefits

Diabetes

Diabetes is one of the world's oldest diseases and current metabolic syndrome which is speedily increasing (Lakhtakia, 2010). The most common symptom of diabetes is hyperglycemia and if it is increased can cause major complications (Forbes and Cooper, 2013). Generally, foods that help to control diabetes are divided into three food groups. Dietary fiber is one of them which lower the risk of diabetes. As fiber affects the absorption of nutrient in the gastrointestinal tract and decrease the postprandial blood sugar level (American Diabetes Association, 2008). In recent years the anti-diabetic activity of pumpkin came into lime light. From studies it is revealed that its pulp, seeds, and oil may contain hypoglycemic property, which means it is a good food for diabetic patients (Williams et al., 2013). Cinnamon is also related with diabetes control. In a study it is observed that people suffer from metabolic syndrome added cinnamon in their diet due to which a drop in their fasting glycemia and an improvement in body composition recorded (Ziegenfuss et al., 2006). Similarly in-vivo studies showed lower postprandial blood glucose in rats after giving cinnamon extract as it suppressed maltose and sucrose (Shihabudeen et al., 2011). Reports showed that millets grain plays a vital role for the management of diabetes specially finger millet and foxtail millet help to control blood sugar level (Kumari and Sumathi, 2002). Moreover, it has been recommended that cinnamon has a key role in the prevention of insulin resistance (Qin et al., 2010). Aditionally, catechin present in green tea and black tea have major contribution in maintaining blood sugar (Beidokhti and Jäger, 2017).

CVD

With an increasing prevalence, cardiovascular disease (CVD) is still the major reason for mortality and morbidity worldwide. CVD is a group of diseases that includes myocardial infarction, atherosclerosis and heart strokes (Martínez-Augustin et al., 2012). For the treatment and prevention of this disease a different approach in lifestyle such as normal body weight, regular exercise, smoking cessation, healthy diet etc along with medication is required (Mente et al., 2009). Additionally, functional food should be taken in daily diet because they act as supplementary treatment to CVD patients (Tomé-Carneiro and Visioli, 2016). Food rich in antioxidant, mainly plant flavonoids plays potential role in the prevention of cardiovascular disease. Medical studies showed that the polyphenols present in fruits and tea helps in the prevention of CVD (Sarriá et al., 2015). Occurance of reactive oxygen species (ROS) and oxidative stress lead to damage of endothelial cell, myocardial infarction, development of atherosclerosis, ischemia and inflammation (Dhalla et al. 2000; Raedschelders et al., 2012). Defenses of antioxidants enhances the property of flavonoids which helps in anti-inflammation, lowering blood pressure, oxidation of low-density lipoproteins (LDL) and improving the functionality of endothelial cell (Desch et al., 2010). Results from latest meta-analyses and cohort studies have shown that intake of total flavonoid lower the risk of CVD, heart problems, and mortality rate (Ivey et al., 2015). Studies revealed a link among level of plasma triglycerides and cardiovascular disease (Hokanson and Austin, 1996). The triglyceride-lowering effects of long-chain omega-3 fatty acids (O3FAs) include docosahexaenoic acid (DHA) and eicosapentaenoic acid (EPA), and are well documented. Fatty fish is the major source of EPA and DHA, such as mackerel, tuna, salmon etc (Tur et al., 2012). In addition, O3FAs reduces oxidative stress, eicosanoids, cytokines, leukotrienes and altering endothelial and immune cell function as it has antiinflammatory and immunomodulatory property (Calder, 2013). Plant sterols (PS) are natural group substances of plant cell membranes which contains phytosterols and phytostanols. It is well documented that phytosterol not only helps in reducing the level of LDL and absorption of cholesterol but also reduces the occurance of CVD (Alemany et al., 2014).

Neuro-degenerative disorders

Neurodegenerative diseases (NDs) are disorder of nervous system which caused by factors like aging, hereditary, environmental etc. ND is a progressive loss of neurons which leads to memory loss, Bone dysfunctioning, defect in cognitive skills, emotional and behavioral disturbance. Diseases like Alzheimer's disease (AD), Huntington's disease, Parkinson's disease, multiple sclerosis are result of neuronal degeneration (Amor et al., 2010). In human body, energy is produced by nervous system as it contains the highest amount of oxygen, so it is likely very prone to the effects of ROS and reactive nitrogen species. Clinical evidences showed that phenolic rich diet is very effective in evading the neurodegenerative diseases especially AD (Sofi et al., 2010). In-vivo studies of flavonoid rich have shown good results as it stimulates neuron regeneration and protects (Casadesus et al., 2004; Galli et al., 2002). Through different mechanism flavonoids can produce neuroprotective which have antioxidant property to protect neurons by creating a barrier in the reaction of nicotinamide adenine dinucleotide phosphate (NADPH) oxidase (Solanki et al., 2016; Vauzour et al., 2008). O3FAs, such as DHA, are allied with superior act and helps in avoiding age-related problems. DHA with low plasma concentration turn down the coginitve health in AD patients (Beydoun et al., 2007; Heude et al., 2003). Studies have shown intake of O3FA lowers the risk of decline in memory ability (Morris, 2016), enhance the cognitive ability (van Duijn et al., 2016), and lowers neuroinflammation. The essential mineral selenium (Se) has also been recognized as playing a vital role in NDs. Availability of selenoproteins, such as amino acid residue and selenocysteine mainly depend on the amount of Se in the body. Though the studies of Se intake in ND patients are unsure because the data available on the Se intake and its potential benefit for prevention and/or treatment are uncertain and insufficient (Cardoso et al., 2010; Loef et al., 2011). Future studies on Se is needed to know the benefits of supplementing Se for the prevention and delay the occurrence of NDs. Food and food habits play a key role and one of the most promising variable risk factor for NDs. The literature shows that for the prevention of ND antioxidants, O3FAs, Se and polyphenols rich diet should be consumed to get potential health benefits.

Gastrointestinal health

The largest internal organ of the body is gut which acts as a barrier against pathogens and antigens of intestinal lumen (Gatt et al., 2007). Nowadays, research interest in the area of human microbiota has increased, because it produces metabolites which with series of chemical interactions and signaling pathways strengthen the immune system (Sawicki et al., 2017). The microbiota of intestine was developed after birth and gets affected by various factors, like the method of birth, use of antibiotic, infant nutrition, and age (Montalto et al., 2009). In the gut the partition of various strains of bacterial species will determine the metabolic profile of the microbiota, which could have possible health benefits (Flint et al., 2015).Eating habits, diet and lifestyle of every individual have health impacts. So, many gut diseases occur because of improper diet which creates an imbalance of intestinal microbiota, (Cencic and Chingwaru, 2010). Functional foods having prebiotics, probiotics, symbiotic, and fibers helps to get better microbiota that enhance gut function (Tur and Bibiloni, 2016).Probiotics are live microorganisms which have certain health benefits. It is the bacteria of natural biota having less or no pathogenicity. The most commonly used bacteria as a probiotic in different products are lactobacilli and bifidobacteria. These are mostly used in dairy products, such as cheeses, traditional buttermilk, probiotic milk drying, yogurts etc. Due to the health benefits of these microorganisms, the food industries are interested to use/add probiotics in fermented food products also, rather than only in milk and milk products, such as fermented fruits and vegetables, drinks, and cereals (Hittinger et al., 2018). By using probiotics in our daily diet, can have positive effect on human health because it modifies the gut microbiota by correcting the imbalance of bacteria. Some probiotics have the capability to lower the pH of the lumen of intestine and boost the formation of lactate and short-chain fatty acids (SCFAs), which decreases the intestinal transit time and increases peristaltic movements (Dimidi et al., 2014).

Clinical studies carried out on the effect of probiotic showed that probiotic are dependent on different factors such as starin of probiotic, quantity of dose used, infection type and time taken for treatment but, normally the effective dosage is 10^7–10^9 colony-forming units/mg per day in humans (Minelli and Benini, 2008). Prebiotics are undigestible fermented substances that result in the formation or activity of gut microbiota by creating a particular change in the intestine. are inulin-type fructans are used as most common prebiotics. It includes of enzymatically hydrolyzed inulin, native inulin, oligofructose and synthetic fructooligosaccharides (Roberfroid and Delzenne, 1998) and wheat, onion, banana, garlic, and leek are the most common sources (Van Loo et al., 1995). Probiotics and prebiotics have exclusive role to play in human health, mainly in the formation and balancing themicrobiota that colonize the human gastrointestinal tract (Douglas and Sanders, 2008). Intake of probiotics or prebiotics in our daily life results in health benefits which include better function of immune system, improvement in colonic integrity, low regulated allergic response, better digestion and elimination and body less prone to intestinal infections (Cencic and Chingwaru, 2010).

Furthermore, dietary fiber, which includes non-starch polysaccharides (cellulose, pectins, dextrins, beta-glucans, lignin, mucilages) not only modify the transit time from the gut but also provide positive effects to inulin-type fructans. Its rich sources are cereals, oats, chia, nuts etc. They also lead to fermentation of bacteria that alter the bacterial growth, mainly in the population of lactobacilli and bifidobacteria. In addition, these dietary soluble fibers also show other health benefits, which could be partially a result of their effect on the composition of microbiota (Laparra and Sanz, 2010).After fermentation of dietary fiber butyrate, acetate, and propionate formed which may play a vital role in the homeostatic control of energy, preventing diseases, improving immune function, and microbe signaling (Sawicki et al., 2017). Therefore, interest has gained in the fiber-induced modulation of the gut microbiota for its significant effect on the health of host (Flint et al., 2012).

Permissible limits for nutraceuticals

According to FSSAI (HSN) 2016, the amount adding nutrients should not exceed the recommended daily allowance (RDA) provided by the Indian Council of Medical Research (ICMR) in nutraceuticals and health supplements. The limit of nutrient used in the composition of nutraceuticals are given in the regulatory clause FSSA (HSN) [6(1)(i-iii) and [7(1)(i-iii)]: Schedule I for vitamins, minerals and their forms, II for amino acids, IV for Plant or botanical ingredients, VI for Nutraceuticals as ingredients VII for Probiotic and VIII for Prebiotics.

Bioaccessibility of Nutraceutical and Functional Food

Over recent years, chronic diseases including diabetes, cancer, cardiovascular disease (CVD) and obesity growing in the society with an exponential rate. A WHO report states that, 60% of related aspects are correlated between an individual's health and their lifestyle. Hence, major aspect includes the unhealthy diets that lead to the metabolic syndromes, disabilities, and illness among the millions. For the delivery of affordable health benefits, scientific communities are consistently working on the identification of major nutrients and bioactive molecules from various sources. While, these bioactives were finally used to develop health promoting functional foods and nutraceuticals. Fruits, vegetables, cereals, crops and medicinal herbs mention as the good source of minerals, vitamins, fibers and potential bioactive compounds refer to as phytochemicals or phyto-constituents. These phytochemicals help to fight against the metabolic stresses or the components that involves in the development of chronic diseases. Bioaccessibility is becoming the major concern, as these phytochemicals are highly unstable and less bioactive at extreme environments which finally promotes higher dose at target tissues (Manach et al., 2005). Several factors inhibit the bioaccessibility of the phytochemicals including food source, gastrointestinal conditions and chemical modifications due to the interaction of other biomolecules in the food matrix (Palafox-Carlos et al., 2018).

Parameters affecting bioaccessibility of nutraceuticals

Previous studies conducted by Aggett, P.J., 2010 and Hurrell, R. and Egli, I., 2010 demonstrated the different components in the metabolic pathway where changes in nutrient bioavailability can be studied including: nutrient release from the orally administered matrix, effects of digestive enzymes, intestinal response including binding and uptake, transfer from the gut wall to the lymphatic circulation, systemic distribution, deposition (stores), functional and metabolic use.

Gastrointestinal release of nutraceuticals

Nutraceuticals refer to as the bioactive constituents that is held to provide the medical and health benefits including prevention and treatment of disease other than basic nutritive value. Nutraceuticals consisting a wide range of biomolecules including bioactive peptides, vitamins, phenolic acids, carotenoids, lipids. While, incorporation of nutraceuticals with food stuff can help to develop variety of functional food products (Gonçalves et al., 2018). From last many decades, the most favored route for the delivery of bioactives for the prevention of chronic disease is the oral administration. The major factor influencing the nutraceutical delivery is the physiochemical environment of gastro-intestine, where the delivery is highly dependent on efficacy, solubility and stability of the delivery vehicles. These vehicles or delivery systems were designed and capable to entrap and controlled release of bioactive molecules for the of bioactives (Ting, Jiang, Ho & Huang, 2014). Currently, scientists are mainly focused on physiochemical behavior of the delivery systems for the enhancement of bioactivity, nutraceutical concentration and shelf life.

Development of new functional foods and nutraceutical depends on the type of delivering nutrient molecule and material used in matrix designing for the systematic or controlled release the nutrient component unaffected by external environment. There are various definitions for nutrient bioavailability, but mainly it describes as a proportion of a nutrient in the diet which is absorbed and used for normal body functions. Change in nutrient bioavailability is described by the different steps of the metabolic pathway (Fig. 2):

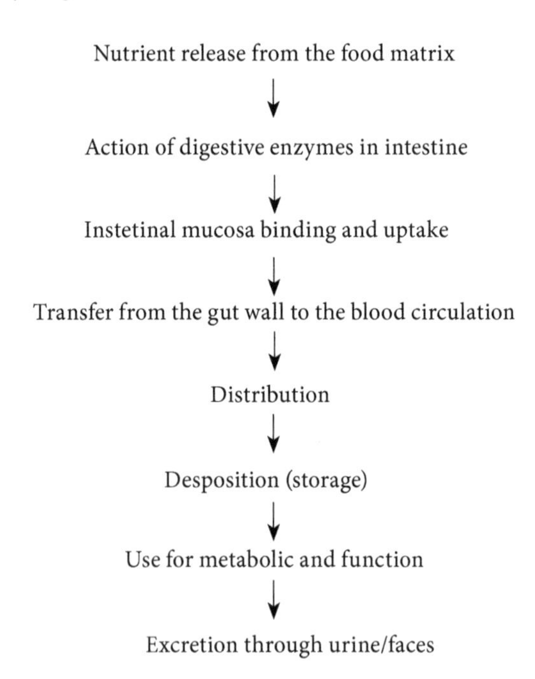

Fig. 2: Metabolic pathway of nutrient bioavailability

Market potential of nutraceuticals and functional foods

In the late 1980s, functional foods marketing got initialized to promote and develop the fortified foods with special ingredients capable of generating health benefits.

Such modification in the consumption of food considered as functional if it targets functions in the body by the reducing the risk of diseases (Figueroa-González et al., 2011). While such an awareness not only creates and linkage between the diet, nutrition and health in current society, a rapid growth is also observed in the coming years in functional food sector exploring an exponential growth of 67% in between the years of 2017-2022 globally (*https://www.statista.com/statistics/252803/global-functional-food-sales/*). Nutraceutical market was noted to be $117 billion in 2008 world widely, in which contribution of India's was only 0.9% (*http://www.nutraceuticals world.com/issues/2012-11/view_features/tapping-indias-potential/*). Quickly moving pharmaceutical and consumer goods companies are the main leads in the Indian nutraceutical market because of which in the next 10 years' nutraceutical market will experience remarkable expansion. Countries are now become very health conscious and consumers nowadays are ever more aware of the benefits of these products. However, in many developing countries, including India, general population's disposable income has also grown exponentially.

Demand scenario: With CAGR of 8.8% the nutraceuticals market is expected to reach $578.23 billion by 2025, globally (Fig. 3). Increasing health problem may raise the demand of these products in next upcoming years.Factors such as easy availability and low price as compared to prescribed drug expected to increase over the next few years. One can get them in different forms, including tablet, capsule, syrups etc. Global participants rely on an extensive distribution network to ensure sales across regions. Manufacturers shows interest in scientific research and development so that they can develop new products and can compete in the market.

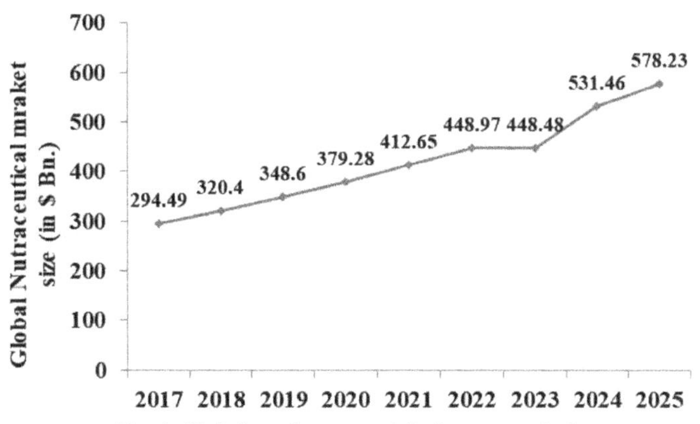

Fig. 3: Global market potential of nutraceuticals

Indian Market Size & Growth

Each year with an increase of 21%, the Indian market of nutraceuticals is predicted to rise from $4 billion to $18 billion in the year 2017 to 2025 (Fig. 4). Due to low physical activity and change in the food habits made Indian population more susceptible to health ailments but simultaneously about half Indian population is becoming alert about health & fitness. This consciousness among people gives a lift to growth opportunity for Nutraceuticals.

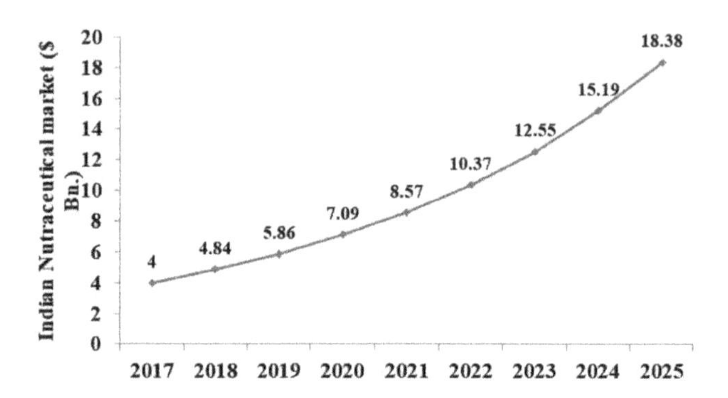

Fig. 4: Indian market potential of nutraceuticals

Key Players in Indian nutraceutical market

In Indian nutraceutical market, Pharmaceutical and FMCG companies are more in number than companies of nutraceutical product. Nowadays in India people are more conscious and aware about their health which increases the market of these nutraceutical companies, making it one of the fastest rising supplements. The main players ruling the Indian nutraceuticals markets are shown below in fig. 5:

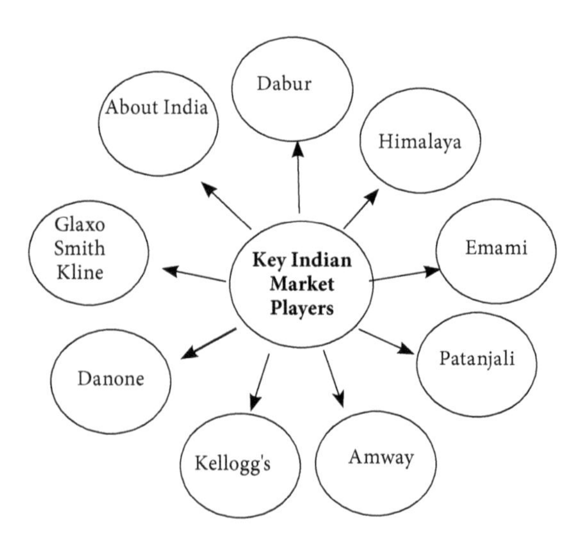

Fig. 5: Key Players of Indian market

Safety regulations

In every country food laws are the basis of regulations of all types of food including health food, dietary supplements, functional food and nutraceuticals with specific guidelines and regulations that framed to regulate health food and to prevent people from exploitation, who might rely on the vague claims of these product which are made by the producer to increase its marketing. United States Food and Drug Administration (FDA) regulate the functional foods under the authority of two laws. The Federal Food, Drug, and Cosmetic Act (FD&C) 1938 provides for the regulation of all foods and food additives. For dietary supplements and ingredients of dietary supplements, The Dietary Supplement Health and Education Act (DSHEA) 1994 were amended. Functional foods are categorized in different foods as describes above such as whole foods, fortified foods, enhanced foods and enriched foods. Claims on the label of functional foods are of two types:

(1) Functional claims, do not claims that the food can prevent, diagnose, treat or cure a particular disease but explains the effects on the normal body functioning; and
(2) Disease risk reduction claims, which explains the link between nutritional components and a health state.

Functional claims do not require much more scientific data as in disease risk reduction claims and they also do not need any preapproval from FDA. Under the FD&C Act, functional claims cannot be wrong or confusing. Under this act there is no level of confirmation to support these claims is required. To create confusion, the data available to support functional claims varies widely.

Before claiming, disease-risk reduction claims needs an approval from FDA and must show scientific facts. For example, the health claim in relation to any food product and CVD reads: "Low in saturated fat and cholesterol that include 25 grams of soy protein a day may reduce the risk of heart disease. One serving of (name of food) provides grams of protein." This claim is only on those food products who give minimum 6.25gram protein in one serving. Other FDA-approved health claims are of fruits and vegetable that claims to lower the risk of certain cancer; high sodium and increased risk for hypertension; intake of saturated fat and an increased risk of heart disease; and food fortified with folic acid and lower risk of birth defects. Many functional foods may appear to have certain benefits to host health. For example, orange juice fortified with calcium gives almost the same quantity of calcium present in milk. In case of calcium requirement around 85 percent adolescent women and almost half of children under the age of 5, do not get the required Dietary Reference Intake (DRI), therefore orange juice fortified with calcium may help appreciably to meet the daily calcium requirement. Secondly, it is very difficult to create a good impact on health by other developed functional foods. These consist of food comprise with listed herbal ingredients on the label, or inadequate amount of these ingredients to create the claimed effect. Moreover, few herbal sources can be harmful, such as kava, which has been linked with liver damage, and belladonna, which is toxic.

Conclusion

At the present time, there is a various number of scientific evidence signifying the health benefits of consumption of functional foods and nutraceuticals, particularly in the reduction of risk of a specific disease. Frequently to evaluate the positive health effects of these compounds on certain biomarkers of the host health, studies have been performed on in-vitro with model systems. Nevertheless, the process of the some of these compounds is still not well understood. In addition, they have other bioactivities that could be exploited other than their main functionality. For this reason, further scientific research after the ingestion of these compounds needed to evaluate their whole physiological activity. On the other hand, the functional effects are highly depending on food composition and vary in every individual customer and, consequently, results obtained after in-vivo studies were also not consistent. Thus, to encourage the development of new functional foods and nutraceutical and to raise their market it will be necessary to understand the link among functional food and their ingredient during food processing, as well as performing more certain clinical studies to assess the claims correctly.

References

Aggett, P. J. (2010). Population reference intakes and micronutrient bioavailability: a European perspective. *The American Journal of Clinical Nutrition*, 91(5), 1433S-1437S.

Alemany, L., Barbera, R., Alegría, A., & Laparra, J. M. (2014). Plant sterols from foods in inflammation and risk of cardiovascular disease: a real threat?. *Food and Chemical Toxicology*, 69, 140-149.

American Diabetes Association. (2008). Nutrition recommendations and interventions for diabetes: a position statement of the American Diabetes Association. *Diabetes care*, *31*(Supplement 1), S61-S78.

Amor, S., Puentes, F., Baker, D., & Van Der Valk, P. (2010). Inflammation in neurodegenerative diseases. *Immunology*, 129(2), 154-169.

Aronson, J. K. (2017). Defining 'nutraceuticals': neither nutritious nor pharmaceutical. *British Journal of Clinical Pharmacology*, 83(1), 8-19.

Beidokhti, M. N., & Jäger, A. K. (2017). Review of antidiabetic fruits, vegetables, beverages, oils and spices commonly consumed in the diet. *Journal of Ethnopharmacology*, 201, 26-41.

Beydoun, M. A., Kaufman, J. S., Satia, J. A., Rosamond, W., & Folsom, A. R. (2007). Plasma n–3 fatty acids and the risk of cognitive decline in older adults: the Atherosclerosis Risk in Communities Study. *The American Journal of Clinical Nutrition*, 85(4), 1103-1111.

Bhaskarachary, K. (2016). Traditional foods, functional foods and nutraceuticals. *Proceedings of the Indian National Science Academy*, 82(5), 1565-1577.

Calder, P. C. (2013). Omega-3 polyunsaturated fatty acids and inflammatory processes: nutrition or pharmacology?. *British Journal of Clinical Pharmacology*, 75(3), 645-662.

Cardoso, B. R., Ong, T. P., Jacob-Filho, W., Jaluul, O., Freitas, M. I. D. Á., & Cozzolino, S. M. F. (2010). Nutritional status of selenium in Alzheimer's disease patients. *British Journal of Nutrition*, *103*(6), 803-806.

Casadesus, G., Shukitt-Hale, B., Stellwagen, H. M., Zhu, X., Lee, H. G., Smith, M. A., & Joseph, J. A. (2004). Modulation of hippocampal plasticity and cognitive behavior by short-term blueberry supplementation in aged rats. *Nutritional Neuroscience*, 7(5-6), 309-316.

Cencic, A., & Chingwaru, W. (2010). The role of functional foods, nutraceuticals, and food supplements in intestinal health. *Nutrients*, 2(6), 611-625.

Das, L., Bhaumik, E., Raychaudhuri, U., & Chakraborty, R. (2012). Role of nutraceuticals in human health. *Journal of Food Science and Technology*, 49(2), 173-183.

Desch, S., Schmidt, J., Kobler, D., Sonnabend, M., Eitel, I., Sareban, M., & Thiele, H. (2010). Effect of cocoa products on blood pressure: systematic review and meta-analysis. *American Journal of Hypertension*, 23(1), 97-103.

Dhalla, N. S., Temsah, R. M., & Netticadan, T. (2000). Role of oxidative stress in cardiovascular diseases. *Journal of Hypertension*, 18(6), 655-673.

Dimidi, E., Christodoulides, S., Fragkos, K. C., Scott, S. M., & Whelan, K. (2014). The effect of probiotics on functional constipation in adults: a systematic review and meta-analysis of randomized controlled trials. *The American Journal of Clinical Nutrition, 100*(4), 1075-1084.

Douglas, L. C., & Sanders, M. E. (2008). Probiotics and prebiotics in dietetics practice. *Journal of the American Dietetic Association*, 108(3), 510-521.

Figueroa-González, I., Quijano, G., Ramírez, G., & Cruz-Guerrero, A. (2011). Probiotics and prebiotics—perspectives and challenges. *Journal of the Science of Food and Agriculture*, 91(8), 1341-1348.

Flint, H. J., Duncan, S. H., Scott, K. P., & Louis, P. (2015). Links between diet, gut microbiota composition and gut metabolism. *Proceedings of the Nutrition Society*, 74(1), 13-22.

Flint, H. J., Scott, K. P., Louis, P., & Duncan, S. H. (2012). The role of the gut microbiota in nutrition and health. *Nature Reviews Gastroenterology & Hepatology*, *9*(10), 577.

Forbes, J. M., & Cooper, M. E. (2013). Mechanisms of diabetic complications. *Physiological Reviews*, 93(1), 137-188.

Galli, R. L., Shukitt-Hale, B.A.R.B.A.R.A., Youdim, K.A., & Joseph, J.A. (2002). Fruit polyphenolics and brain aging: nutritional interventions targeting age-related neuronal and behavioral deficits. *Annals of the New York Academy of Sciences*, 959(1), 128-132.

Gatt, M., Reddy, B. S., & MacFie, J. (2007). bacterial translocation in the critically ill–evidence and methods of prevention. *Alimentary Pharmacology & Therapeutics*, 25(7), 741-757.

Hasler, C. M., Bloch, A. S., Thomson, C. A., Enrione, E., & Manning, C. (2004). Position of the American Dietetic Association: functional foods. *Journal of the American Dietetic Association*, 104(5), 814-826.

Heude, B., Ducimetière, P., & Berr, C. (2003). Cognitive decline and fatty acid composition of erythrocyte membranes—The EVA Study. *The American Journal of Clinical Nutrition*, 77(4), 803-808.

Hittinger, C. T., Steele, J. L., & Ryder, D. S. (2018). Diverse yeasts for diverse fermented beverages and foods. *Current Opinion in Biotechnology*, 49, 199-206.

Hoeft, B., Weber, P., & Eggersdorfer, M. (2012). Micronutrients—A global perspective on intake, health benefits and economics. *Int. J. Vitam. Nutr. Res*, 82(5), 316-320.

Hokanson, J. E., & Austin, M. A. (1996). Plasma triglyceride level is a risk factor for cardiovascular disease independent of high-density lipoprotein cholesterol level: a metaanalysis of population-based prospective studies. *Journal of Cardiovascular Risk*, 3(2), 213-219.

Hurrell, R., & Egli, I. (2010). Iron bioavailability and dietary reference values. *The American Journal of Clinical Nutrition*, 91(5), 1461S-1467S.

Ivey, K. L., Hodgson, J. M., Croft, K. D., Lewis, J. R., & Prince, R. L. (2015). Flavonoid intake and all-cause mortality. *The American Journal of Clinical Nutrition*, 101(5), 1012-1020.

Khurana, S., Venkataraman, K., Hollingsworth, A., Piche, M., & Tai, T. (2013). Polyphenols: benefits to the cardiovascular system in health and in aging. *Nutrients*, 5(10), 3779-3827.

Kumari, P. L., & Sumathi, S. (2002). Effect of consumption of finger millet on hyperglycemia in non-insulin dependent diabetes mellitus (NIDDM) subjects. *Plant Foods for Human Nutrition*, 57(3-4), 205-213.

Lakhtakia, R. (2013). The history of diabetes mellitus. *Sultan Qaboos University Medical Journal*, 13(3), 368.

Laparra, J. M., & Sanz, Y. (2010). Interactions of gut microbiota with functional food components and nutraceuticals. *Pharmacological Research*, 61(3), 219-225.

Loef, M., Schrauzer, G. N., & Walach, H. (2011). Selenium and Alzheimer's disease: a systematic review. *Journal of Alzheimer's Disease*, 26(1), 81-104.

Manach, C., Williamson, G., Morand, C., Scalbert, A., & Rémésy, C. (2005). Bioavailability and bioefficacy of polyphenols in humans. I. Review of 97 bioavailability studies. *The American Journal of Clinical Nutrition*, 81(1), 230S-242S.

Martínez-Augustin, O., Aguilera, C. M., Gil-Campos, M., Sánchez de Medina, F., & Gil, A. (2012). Bioactive anti-obesity food components. *International Journal for Vitamin and Nutrition Research*, *82*(3), 148.

Martirosyan, D. M., & Singh, J. (2015). A new definition of functional food by FFC: what makes a new definition unique?. *Functional Foods in Health and Disease*, 5(6), 209-223.

Mente, A., de Koning, L., Shannon, H. S., & Anand, S. S. (2009). A systematic review of the evidence supporting a causal link between dietary factors and coronary heart disease. *Archives of Internal Medicine*, 169(7), 659-669.

Minelli, E. B., & Benini, A. (2008). Relationship between number of bacteria and their probiotic effects. *Microbial Ecology in Health and Disease*, 20(4), 180-183.

Montalto, M., D'onofrio, F., Gallo, A., Cazzato, A., & Gasbarrini, G. (2009). Intestinal microbiota and its functions. *Digestive and Liver Disease Supplements*, 3(2), 30-34.

Morris, M. C. (2016). Nutrition and risk of dementia: overview and methodological issues. *Annals of the New York Academy of Sciences*, 1367(1), 31-37.

Nagpal, R., Kumar, A., Kumar, M., Behare, P. V., Jain, S., & Yadav, H. (2012). Probiotics, their health benefits and applications for developing healthier foods: a review. *FEMS Microbiology Letters*, 334(1), 1-15.

Palafox-Carlos, H., Ayala-Zavala, J. F., & González-Aguilar, G. A. (2011). The role of dietary fiber in the bioaccessibility and bioavailability of fruit and vegetable antioxidants. *Journal of food science*, *76*(1), R6-R15.

Qin, B., Panickar, K. S., & Anderson, R. A. (2010). Cinnamon: potential role in the prevention of insulin resistance, metabolic syndrome, and type 2 diabetes. *Journal of Diabetes Science and Technology*, 4(3), 685-693.

Roberfroid, M. B., & Delzenne, N. M. (1998). Dietary fructans. *Annual Review of Nutrition*, 18(1), 117-143.

Ruxton, C. H. S., Reed, S. C., Simpson, M. J. A., & Millington, K. J. (2004). The health benefits of omega-3 polyunsaturated fatty acids: a review of the evidence. *Journal of Human Nutrition and Dietetics*, 17(5), 449-459.

Sagar, N. A., Pareek, S., Sharma, S., Yahia, E. M., & Lobo, M. G. (2018). Fruit and vegetable waste: Bioactive compounds, their extraction, and possible utilization. *Comprehensive Reviews in Food Science and Food Safety*, 17(3), 512-531.

Sarriá, B., Martínez-López, S., Sierra-Cinos, J. L., Garcia-Diz, L., Goya, L., Mateos, R., & Bravo, L. (2015). Effects of bioactive constituents in functional cocoa products on cardiovascular health in humans. *Food Chemistry*, 174, 214-218.

Sawicki, C., Livingston, K., Obin, M., Roberts, S., Chung, M., & McKeown, N. (2017). Dietary fiber and the human gut microbiota: Application of evidence mapping methodology. *Nutrients*, 9(2), 125.

Sawicki, C., Livingston, K., Obin, M., Roberts, S., Chung, M., & McKeown, N. (2017). Dietary fiber and the human gut microbiota: Application of evidence mapping methodology. *Nutrients*, 9(2), 125.

Shahidi, F. (2009). Nutraceuticals and functional foods: whole versus processed foods. *Trends in Food Science & Technology*, 20(9), 376-387.

Shihabudeen, H.M.S., Priscilla, D.H., & Thirumurugan, K. (2011). Cinnamon extract inhibits α-glucosidase activity and dampens postprandial glucose excursion in diabetic rats. *Nutrition & Metabolism*, 8(1), 46.

Sofi, F., Macchi, C., Abbate, R., Gensini, G. F., & Casini, A. (2010). Effectiveness of the Mediterranean diet: can it help delay or prevent Alzheimer's disease?. *Journal of Alzheimer's Disease*, 20(3), 795-801.

Solanki, I., Parihar, P., & Parihar, M. S. (2016). Neurodegenerative diseases: from available treatments to prospective herbal therapy. *Neurochemistry International*, 95, 100-108.

Ting, Y., Jiang, Y., Ho, C. T., & Huang, Q. (2014). Common delivery systems for enhancing in vivo bioavailability and biological efficacy of nutraceuticals. *Journal of Functional Foods*, 7, 112-128.

Tome-Carneiro, J., & Visioli, F. (2016). Polyphenol-based nutraceuticals for the prevention and treatment of cardiovascular disease: Review of human evidence. *Phytomedicine*, 23(11), 1145-1174.

Tur, J. A., & Bibiloni, M. M. (2016). Functional foods. In: *Encyclopedia of Food and Health*. first ed. Elsevier Ltd.https://doi.org/10.1016/B978-0-12-384947-2.00340-8

Tur, J. A., Bibiloni, M. M., Sureda, A., & Pons, A. (2012). Dietary sources of omega 3 fatty acids: public health risks and benefits. *British Journal of Nutrition*, 107(S2), S23-S52.

Van Duijn, C. M., van der Lee, S. J., Ikram, M. A., Hofman, A., Hankemeier, T., Amin, N., & Demirkan, A. (2016). Metabolites associated with cognitive function in the Rotterdam study and Erasmus rucphen family study. *Alzheimer's & Dementia: The Journal of the Alzheimer's Association*, 12(7), P165.

Van Loo, J., Coussement, P., De Leenheer, L., Hoebregs, H., & Smits, G. (1995). On the presence of inulin and oligofructose as natural ingredients in the western diet. *Critical Reviews in Food Science & Nutrition*, 35(6), 525-552.

Vauzour, D., Vafeiadou, K., Rodriguez-Mateos, A., Rendeiro, C., & Spencer, J. P. (2008). The neuroprotective potential of flavonoids: a multiplicity of effects. *Genes & Nutrition*, 3(3), 115.

Williams, D. J., Edwards, D., Hamernig, I., Jian, L., James, A. P., Johnson, S. K., & Tapsell, L. C. (2013). Vegetables containing phytochemicals with potential anti-obesity properties: A review. *Food Research International*, 52(1), 323-333.

Zhou, Y., Zheng, J., Li, Y., Xu, D. P., Li, S., Chen, Y. M., & Li, H. B. (2016). Natural polyphenols for prevention and treatment of cancer. *Nutrients*, 8(8), 515.

Ziegenfuss, T. N., Hofheins, J. E., Mendel, R. W., Landis, J., & Anderson, R. A. (2006). Effects of a water-soluble cinnamon extract on body composition and features of the metabolic syndrome in pre-diabetic men and women. *Journal of the International Society of Sports Nutrition*, 3(2), 45.

6

Advancement in the Processing of Condiments and Spices

Himani Singh and *Murlidhar Meghwal*

Department of Food Science and Technology, National Institute of Food Technology Entrepreneurship and Management, Kundli Sonipat, Haryana-131 028, India

Drying of the Spices

Drying is an important mode of preservation and it ensures microbial safety of the spices.The main purpose of drying is to increase the shelf storage and make the seasonal spices available round the year with minute losses of nutrients, flavor, taste and color (Maroulis, 2003). During drying, there is a continuous transfer of heat and mass due to the heat application which eventually leads to moisture removal from the product (Akpinar, 2006) (Hashim, 2014). Since, the sensory, nutritional and functional attributes of spices are partially or totally affected by the drying process, the adequate choice of drying technique becomes requisite. There exists a number of conventional drying methods including solar drying (El-sebaii, 2012), freeze drying (Ciurzy nska 2011), vacuum drying (Nadi, 2012), osmotic dehydration,hot air drying and fluidized bed drying (Onwude, 2016). However, these drying techniques lead to poor quality of the spices because of the longer duration of treatment and high amount of energy. To combat with the issues of poor quality, there has been a significant development of the novel techniques, which includes- microwave, ohmic, infrared, pulse electric field, radio- frequency, ultraviolet, ultrasound, supercritical and heat pump heating, in the drying of the agricultural crops. These novel techniques are found to produce higher quality products as compared to the conventional drying techniques (Moses J. N., 2014). The microwave and radio frequency heating technologies are developed in such a manner that they are capable of partial or complete removal of the conventional methods of heating/ drying (Onwude, 2016). The use of these novel techniques has gained a keen attention for the drying purpose of spices. Some of the researchers have also proved that combination of conventional and novel drying techniques leads to better product output.

Microwave Drying

Microwave drying has found an important place in the processing of spices. It has been proved to be a productive dehydration technique. Apart from drying, this technique is also been used for the blanching and sterilization of spices. Though drying of spices not only damages the growth of micro-organisms and prevents the harmful biochemical changes, but it also leads to changes in the appearance and loss of volatiles due to dehydration, esterification, oxidation, rearrangement etc. The use of microwave drying has increased widely because of requirement of short drying time, high drying rates, rapid and voluminous heating, lower energy requirements and good quality dried products Kubra et al., 2016.

This technique has been combined with many conventional drying techniques in order to remove all the lacunas and get the best product. The most important factors of the Microwave drying are drying temperatures and power of the microwave. The drying parameters including: drying speed, drying time, drying efficiency, drying curve and the quality of the product are greatly affected by the drying temperature and power of the microwave (Li Z. R., 2010).

Mechanism: Like other food materials, spices are also dielectric in nature. The Microwaves are electromagnetic waves with frequencies ranging from 300-30000 MHz. They are a combination of magnetic and electrical fields. The water molecules being bipolar in nature rotate with the rapidly changing electromagnetic field. This rapid rotation of the water molecules generates friction between the molecules and heat is evolved. The microwave heating is volumetric because these waves penetrate directly into the material. Thus, providing uniform and fast heating throughout the product. Since, the water molecules absorb the energy quickly, it leads to rapid evaporation of water and higher drying rates are achieved. An outward flux of the rapidly escaping vapor is created. This outward flux further helps in improving the drying rate and prevents the tissues of the material from shrinking, which is generally seen in other convective methods (Kubra et al., 2016). Table-1 gives the data of numerous studies which have been conducted to study the effect of various combinations of microwave power levels and drying temperatures, in order to improve the drying process using microwave.

Table 1: Optimized parameters for Microwave Drying

Spice	Optimum condition	Quality parameters	References
Ginger (Zingiber Officinale)	Microwave drying (800W) (25 min)	Yield and composition of essential oil; total phenolic content and antioxidant activity	(Kubra I. &., 2012)
Parsley (Petroselinum crispum Mill.)	Microwave oven drying 900 W (3.5 min)	Color and drying rates	(Soysal Y. , 2004)
Rosemary (Rosmarinus officinalis L.)	Pre- drying using Convective drying followed by MD 480 W, 46 ^{0}C (84 min)	Drying kinetics, Volatile compounds, sensory data	(Szumny, 2010)
Garlic	Microwave- vacuum drying @ 376.1 W (3 min)	Retention of high allicin content	(Yu, 2007)
Oregano (Origanum vulgare)	Vacuum-microwave drying @360 W, 4–6 kPa (24 min)	Drying kinetics, Volatile compounds, sensory data	(Figiel, 2010)
Red bell pepper (Capsicum annuum L.)	Intermittent Microwave connective drying @ 597.2W, 35^{0}C at Pulse rate of 3	Drying time kinetics, physical (color and texture) improvements and sensory attributes	(Soysal Y. A., 2009)
Cumin (Cuminum cyminum)	Microwave drying @ 730 W (10min)	Yield and composition of volatile oils	(Behera, 2004)
Fenugreek (Trigonella foenumgraecum L.)	Microwave heating @ 850W, 130^{0}C (8 min)	Inactivation of trypsin and α-chymotrypsin	(Salah, 2004)
Mint	Microwave vacuum drying @11.2W/g	Drying kinetics, color changes	(Therdthai, 2009)
Peppermint	Microwave oven drying @ 700W, 50 ^{0}C	Total phenolic content retention	(Arslan, 2010)
Coriander leaves (Coriandrum sativum L.)	Microwave drying 180 W (14 min)	Drying rates, Color, Rehydration Capacity	(Sarimeseli, 2011)
Sweet basil (Ocimun basilicum L.)	Convective pre-drying and VM finish-drying (CPD–VMFD) 360 W, 40_C (»250 min)	Drying kinetics, Volatile compounds, sensory data	(Sanchez, 2012)

Radio-frequency drying

The radio-frequency drying uses electromagnetic energies with frequencies ranging between 1-300MHz to heat a dielectric material. In the conventional drying processes the heat was transferred into the food system from outside through conduction or convention, whereas, RF generates heat from within the food material. The generated heat is distributed throughout the food system because of the friction which is developed due to continuous and rapid rotational movement of the molecules. Thus, food materials which have very low thermal conductivities (mainly the solid, eg- spices and semi-solid materials) are heated rapidly (Casals et al., 2010). Since, RF has longer wave lengths, deep penetration depth and lower frequencies (13.56, 27.12 and 40.68 MHz) as compared to Microwaves (at 915 or 2450 MHz), it offers uniform heating and thereby effective drying of the spices. But radio frequencies are not completely homogenous (Kim et al., 2012). The heating and subsequently drying rates of food material are greatly affected by the heterogeneous dielectric properties of the food material. These dielectric properties are greatly influenced by the temperature, moisture content, frequency, density and salt content. Also, sometimes different spices are mixed together in order to achieve a desired flavor, the dielectric property of this mixture is different to that of the individual spices, which again is a barrier to effective drying of the spice mixture (Ozturk et al., 2018). Table-2 shows the studies conducted to investigate the efficiency of the RF drying of spices.

Infrared drying

The Infrared drying (IRD) is done using the Electromagnetic radiations with wavelengths ranging from 0.78 – 1000 μm. The spectrum of the IR radiations exists in three categories- Near IR (0.78- 1.40 μm), Medium IR (1.40- 3.00 μm) and Far IR (3- 1000 μm). The material to be dried is exposed to these radiations, which absorbs the thermal energy generated through these wavelengths. To obtain higher drying rates the IR radiations are adjusted between NIR and FIR radiations for different spices. When the IR radiations strike to the surface of the spice, it creates charged molecules and atoms, which have different rotational, vibrational and electronic energies without increasing the temperature of the surrounding air (Moses J. N., 2014)(Rastogi, 2012). The electrical energy is converted into heat more efficiently with the IR technology. It increases the uniformity of heating throughout the product, takes less heating time, the product quality attributes are retained to a large extent and less energy is consumed. The electrical energy is converted to heat more efficiently with the IR technology. It increases the heating uniformity, takes less time to heat, product quality attributes are retained and energy requirement is less (El-mesery, 2015).

The combined IR and Hot air drying is the most common application of the IR drying. The combined drying from IR and HAD provides a synergistic effect, which reduces the drying time, energy consumption and increases the efficiency of heat and mass transfer. Due to the combined IR and HAD, energy loss is minimized as all the energy emitted is directly transferred to the surface of the material to be dried without heating the surrounding air. This mechanism increases the molecular vibration in the internal layers of the food material, thereby, increases the moisture movement from the internal to the external surface. The convective air then removes the water vapors from the upper surface of the food material. The convective air also reduces the temperature of the food and hence a good quality dried product is achieved (Praveen Kumar, 2006). Table-2 shows significant findings of IR drying and its combination modes.

Hybrid Solar dryers

The hot air dryers mainly used energy from the fossil fuels. The ever increasing demand of the fossil fuels, higher costs, resource insecurity and the environmental concerns has compelled the scientists to think over the utilization of the renewable energy, like- solar power as a potential alternative to fossil fuels. Thus, several investigations are going on the effective utilization of these renewable energies. Mostly, the thermal and photovoltaic solar power systems are been investigated. More recently, the hybrid photovoltaic-thermal (PVT) solar power system has been developed. This system is known to convert the solar energy into thermal and electrical power simultaneously. The results of various studies carried out on this system prove that the hybrid system is more efficient than the individual thermal and photovoltaic systems (Nayak, 2011). The heat pump systems reduces the energy consumption and are an efficient and environment friendly technology because it recovers the wasted heat. The combination of hot air dryers and heat pump system results in greater efficiencies rather the conventional hot air dryers. Since, the heat pump dryers can work at low temperatures they are suitable for drying the heat sensitive materials. Recently, many scientists have studied the combined effect of heat pump systems and the PVT solar system. In this combined system, the PV unit reduces the panel temperature and cools it by removing the heat absorbed by the system from the solar radiations. The heat is basically condensed in the PV panel and is utilized for water or space heating. This process not only enhances the electrical efficiency of the PV panel but it also increases the performance coefficient of the whole heat pump unit. Though, the heat pump-assisted PVT solar dryers have been investigated a number of times for their efficiencies. The hybrid PVT solar dryers which are equipped with heat pumps are a novel technique. (Mortezapour, 2012) studied the performance of the heat pump assisted photovoltaic thermal solar dryer for drying the saffron. The purpose of the study was to reduce the consumption of fossil fuels and obtain a high quality product.

Table 2: Significant findings on Radio frequency, IR and IR- Hot air drying

Spice	Optimized condition	Quality parameter	Reference
Black pepper and Red pepper	RD at 27.12 MHz (with 50 s and 40 s treatments), maximum power 9Kw and sample temperature to achieve 60°C	Temperature measurements, microbial enumerations and color measurements.	(Kim, 2012)
Cumin, curry and garlic powder	RD at 27.12 MHz, maximum power 6Kw and sample temperature to achieve 70°C	Dielectric properties, heating rates and heating uniformity in various spice mixes.	(Ozturk, 2018)
Chilli (Pickino)	IR-HAD: λ- 2.4-3.0 μm, Temp. 50-70°C	IR-HAD consumes 33.5% less power than HAD	(Mihindukulasuriya, 2015)
Green pepper (Capsicum annuum L.)	IR-HAD: T=75°C; IRP < 240 W depending on temp and MC	Lowest value of VC retention (16.86%) due to high temp. of 75°C as compared to 46.54% when using HAD.	(Lechtanska, 2015)
Garlic	IR thin layer drying: IR intensities- 0.075, 0.15, 0.225 and 0.3Wcm^{-2}	The drying rate increases and time decreases with increasing radiation intensity.	(Younis, 2018)

Novel thermal and Hot -air oven combined drying

Diffusion and capillary actions are important mechanisms which controls the drying in porous regions of the agricultural products (Erbay, 2010). This can take longer drying times because of the internal resistance offered by the product. This issue can be foreseen by combined use of novel and conventional drying techniques. Many researchers have studied the beneficial effects of combined usage of novel and conventional drying techniques. Some of them are listed below here:

Microwave and Hot- air combined drying (MW- HAD)

During the microwave drying the product comes in contact with the microwave radiations. These radiations enter deep inside the product and cause volumetric heating because of friction that generates due to the ion movement and rotating dipoles in the drying product (Miura, 2003)(Sadeghi, 2013). This process leads to mass transfer due to the generation of vapour inside the product. The water vapors are forced towards the surface of the product from where the moisture is removed easily. In the combined application of MW and HAD, the moisture at the surface of the product is removed rapidly to the atmosphere because of the convective air flow using minimum energy (Kaur, 2014)(Amiri Chayjan, 2014). This method increases the rate of drying, efficiency and also reduces the time of drying. Moreover, the MW radiations penetrates deep inside the product during combine MW and HAD which not only leads to controlled and precise heating of the product but also improves the quality of spices with appropriate drying rate and time. (Soysal Y. A., 2009) Investigated IMW and HAD approach on the drying of oregano. In the IMW and HAD approach, the MW energy is applied after the constant rate initial period. The surface of the product is dry in this case and all the moisture is present in the center of the product. During this stage the heat is generated inside and the water vapours are forced to move to the surface, where they are removed by the convective hot air.

Infrared and hot- air combined drying

The IR drying technology provides improved efficiencies for increased surface heating uniformity, increased retention of product quality attributes, conversion of electrical energy into heat, significant reduction in energy demand, decreased net heating time and simple equipment set-up (El- mesery, 2015)(Sadin, 2014).

A synergistic effect is observed while using combination of IR and HAD. During the application of this combined approach the energy from the heated element is directly transferred to the surface of the product without the surrounded air being getting heated and the loss of energy is minimized. The mechanism of operation of IR and HAD involves the vibration of the inner layers of the product by the IR, which leads to an increase in the movement of the moisture from the inner layers to the surface from where the convective hot air removes moisture rapidly, thereby keeping the temperature of the product low and yielding a good quality dried product (Praveen kumar, 2006). (Mihindukulasuriya, 2015) investigated the drying of chilli using combine approach of infrared and hot air rotary dryer and found that drying with the IR and HAD requires less power than HAD alone. Also, it yielded good quality dried chilli.

Particle size reduction

Particle size reduction is an age old process. Apart from few spices mostly all spices cannot be used in food preparations as such, hence there is a need for grinding of spices. Grinding uses application of mechanical forces for the size reduction and the energy required to fracture a material increases with the reducing size of the particle required. It is well known that the quality of the spices is graded upon the extrinsic and the intrinsic factors; where retention of volatile oil, oleoresins, antioxidant property comprises the intrinsic factors and color, appearance, shape, texture are the extrinsic or physical factors (Balasubramanian, 2012).

The grinding process generates heat which is detrimental to the flavor and quality of the spice. During grinding the temperature of the product raises from 40 to 95 ^{0}C, due to which the spices loose a significant fraction of their volatile oil, impacting the aroma and flavor compounds of the spice (Balasubramanian, 2012). Therefore, the need for pre-cooling and maintaining low temperature during grinding in the mill came up. This precooling technology helped not only in reducing the loss of volatile oils and enhanced the flavor strength but also significantly helped in improving the overall quality of the spices (Murthy, 2007).

Cryogenic Grinding

Cryogenic grinding is also known as cryo-milling or freezer grinding, freezer milling, meaning the material is grinded near to its glass transition temperature or the embrittlement temperature. The cryogenic grinding consumes less energy, gives higher throughput, there is no mill clogging due to the oil content of the spices, no browning of spices, minimum losses of volatiles, effective control on particle size as compared to that of ambient or conventional grinding systems (Murthy, 2007)(Goswami, 2010).

Cryogenic grinding technique significantly reduces the loss of volatile oil. The refrigeration temperature to precool the spices is provided by the cryogens mostly used is the liquid nitrogen having a B.P of -195.6^{0}C. The liquid nitrogen maintains the desired low temperature during the grinding operation by absorbing the generated heat. The vaporization of the liquid nitrogen to the gaseous state maintains low temperature and also creates a dry and inert atmosphere for extra protection of quality of the spice. Therefore, the losses of volatile oils and moisture from the spices are reduced by pre-cooling the spice and maintaining the low temperature throughout grinding operation. Hence, the flavor strength of the spice is retained. The spices become embrittled due to extremely low temperatures in the grinder which solidifies the oils. This embrittlement permits easy crumbling and finer and more even sized size reduction of the spices.

Thus, under cryogenic processing very small and fine particle size can be obtained. This helps in equal or uniform spread of the flavor of the spice in the food in which they are used, and the large specs which occur in food products can be controlled (Singh, 1999).

Table-3 shows the extensive research done by scientists on the cryogenic processing of spices under the heads such as – volatile oil retention, effects on color, energy efficiency, particle size, overall product quality, etc.

Table 3: Effect of Cryogenic grinding on quality aspects of spices

Spice	Processing condition	Quality parameters	Reference
Ajwain (*Trachyspermum ammi* L.)	Cyo-grinding and ambient grinding	Increased yield, more retention of polyphenols, flavonoids and antioxidants	(Sharma L. A., 2015)
Black pepper and turmeric	Cryo-grinding	Maximum retention of flavoring components in terms of volatile oil	(Barai, 2015)
Coriander (*Coriandrum sativum* L.)	Cryo-grinding	Inc. Volatile oil (0.14- 0.39%), Oleoresin (13.8- 19.58%), TPC (32.44- 92.99mg GAE/g crude seed), TFC (15.28- 20.85mg QE/g),higher DPPH scavenging activity	(Saxena, 2015)
Pepper (*Piper nigrum* L.)	Cryogenic grinding and hammer milling- flavor retention studies	High concentration of aroma constituents retained, better odorant potency, color, sensory attributes. More monoterpenes retained in cryo-ground pepper	(Liu, 2013)
Cumin (*Cuminum cyminum* L.)	Cryo-grinding	Inc. in yield by 29.9%, Volatile oil (33.9- 43.5 %), Cuminaldehyde (inc. from 48.2- 56.1%) and inc. in fatty acid methyl esters (FAME)	(Sharma L. A., 2016)
Cinnamon	Cryo and ambient grinding	Ave. particle size (dec. from 0.454mm to 0.356mm), lower energy consumption, good color retention	(Barnwal, 2014)
Fenugreek	Cryo and ambient grinding at varied moisture content	Cryo processed- finer particle size, grinding time and energy consumption inc. with inc. moisture content but was less for the cryo-grinding	(Meghwal, 2013)

Essential oils

Essential oils are a bundle of tremendous biological properties, due to which they have made a crucial place in enhancing the functionality of various products such as drinks, foods, pharmaceuticals, cosmetics, perfumes and green pesticides (Dima C. &., 2015) .

Essential oils are basically the volatile oils which constitute strong aromatic compounds having complex chemical compositions. The major components of these essential oils are hydrocarbons, aldehydes, alcohols, phenols, terpenoids, ketones, esters, etc. (Rassem, 2016). The demand of essential oil consumption as a flavoring agent in food and beverage industry, in pharmaceutical and cosmetics is increasing on a fast pace. Hence, with the ever growing demand, there is a need for newer and more efficient production techniques which would not only cater to the needs of better yield but also good quality of essential oils extracted. For that matter scientists have been working on development of various newer and greener technologies which are proving to be a boon for extraction and development of essential oils.

Figure: 1. represents the various conventional and novel/ greener techniques of extraction of essential oils. The essential oils are thermo-labile and almost all the conventional extraction techniques are based on the principles of high temperatures, due to which the components of the essential oils undergo chemical alterations (including- oxidation, hydrolysis, isomerization) thereby disturbing the natural essence of the oils. Hence, compromise with the quality of the essential oils especially with longer extraction time. Moreover, the conventional extraction techniques consumes high energy, takes longer time periods for extraction, simultaneous extraction of other polar compounds such as plant pigments, causes environmental pollution and degradation of the chemical composition of the oils (El Asbahani, 2015)(Palvic, 2015). Novel techniques of extraction takes good care of the eco- friendliness, sustainability, economy, competitiveness, higher efficiency and good quality of the essential oils. The novel extraction techniques have emerged with the concept and principles of green extraction. These new techniques save energy, time, solvent and carbon dioxide emissions. The novel methods for the extraction of essential oils have been discussed. Table-4 shows some the studies carried out to optimize and to study the effect of novel extraction techniques on the essential oil and bioactive compounds of spices.

Fig. 1: Essential oils extraction techniques

Table 4: Extraction of Bioactive compounds and Essential oils using Novel techniques of extraction

Spices	Bio-active Components/ Oil	Method	Optimum condition	Reference
Allium cepa	Polyphenols/flavonoids	Solvent free microwave hydro-diffusion and gravity (MHG)	500W, 1W/g, 23 min	(Zill-e-Human, 2011)
Curcuma longa dried rhizomes	Curcumin	MAE	140W, 5min, acetone	(Wakte, 2011)
Tunisian cumin (Cuminum cyminum L.)	Essential Oil	MHG	203.3W, 16min, improved EO yield	(Benmoussaa, 2018)
Rosemary	Essential oil	SFME	100^{0}C, 30 min	(Filly, 2014)
Black pepper	Essential oil	UMAE	Microwave and ultrasonic powers as 500 W & 50W, 100^{0}C, 7 min	(Wang, 2018)
Fennel	Essential oil	SFEAP	200 bar, 40^{0}C, 40N	(Hatami, 2018)
Thyme	Essential oil	Ohmic assisted hydrodistillation	Device operated at 220V, 50Hz, extraction time-24 min to 1hr.	(Gayahian, 2012)
Oregano	Essential oil	Ohmic assisted hydrodistillation	Device operated at 0-300V voltage and 0-16 A current	(Hashemi, 2017)

Supercritical Fluid Extraction (SFE)

Supercritical extraction is efficient over the conventional extraction processes. This extraction is performed at low temperatures and hence is suitable for the thermo-labile components of the essential oils. It has high selectivity, as the solvation power (interaction of the solute with the solvent) of the fluid with respect to temperature and pressure can be controlled. Though, maintaining high pressures can lead to simultaneous extraction of undesirable higher molecular weight compounds such as waxes during essential oil extraction. As compared to the liquid solvents the supercritical fluids enters into the porous solid samples more effectively due to their lower viscosities and higher diffusivity, leading to high rates of mass transfer and rapid extraction. The SFE uses non-toxic fluids like carbon dioxide and is hence environmental friendly (Ch Stratakos, 2016).The most commonly used liquid in supercritical fluid extraction id Carbon dioxide because of its unique properties, including its inertness, lower toxicity, decreased values of its critical parameters (T_{cr} = 31.1 8C and P_{cr} = 7.4 MPa), lower price and easy availability. In its supercritical phase the carbon dioxide exhibits properties of a non- polar liquid having high diffusing property which results in the extraction of non-polar compounds of the material (Dima C. I., 2015).

Supercritical Fluid Extraction assisted by Cold pressing (SFEAP)

The Supercritical fluid extraction assisted by the Cold pressing (SFEAP) is a novel approach for the extraction of volatile oil from the spices, developed recently by (Johner, 2018). The performance of the integrated approach of SFE and Cold pressing has already been evaluated by extraction- from the pulp of pequi (Caryocar brasiliense) (Johner, 2018) and from the powder of fennel (Hatami, 2018). Hatami et al. (2018) confirmed that SFEAP is more effective than simply SFE and increased the overall extraction yield by 24.5%. Though the combined extraction from SFE and cold pressing yields better results, it still needs more in-depth investigations on its economic feasibility as well as superiority over other green extraction techniques.

Subcritical Liquid Extraction

Subcritical liquid extraction is one of the novel methods for the extraction of essential oils. The subcritical liquid extraction is also called as subcritical water extraction (SWE), pressurized hot water extraction (PHWE) and pressurized low-polarity water extraction (PLPWE) (Pavlic, 2015). In this extraction technique, the water is superheated at temperatures ranging from 100 – 375°C (critical temperatures) at high pressures (> 20 bar). There is a decrease in the dielectric constant of water which decreases the polarity of water, ensuring that the non-polar components get extracted from the spice sample (Palvic, 2015)(El Asbahani, 2015). The SWE saves energy, uses less time, yields larger quantities, maintains the quality of the components of oil and is environmental friendly.

Ultrasonic-Assisted Extraction (UAE)

Ultrasonic assisted extraction (UAE) facilitates component extraction and is hence used to extract essential oils from spices. The UAE have minimum damaging effects on the extraction compounds, it can work well in the absence of organic solvents and requires less time for completing the extraction process (Ch Stratakos, 2016). The ultrasound works on the principle of cavitation, where production and breakdown of microscopic bubbles helps in the extraction process. The microscopic bubbles increase in the size and then they collapse violently, inducing mechanical forces which damages the cell membranes and results in high extraction yields and faster extractions (Cameron, 2009). There are numerous studies which illustrates the advantages of UAE in terms of the extraction of volatile oil. (Assami, 2012) reported that the combined treatment of ultrasound and hydrodistillation on Carum carvi seeds resulted in higher extraction yield of volatile oil in short duration of time without causing any harm to the composition of the oil. The pretreatment with ultrasound has been combined with many of the newer extraction techniques and have been found to yield better and good quality results (CHAP, et al., 2016). Though, the capital cost is expensive, but ultrasound is a viable alternative to the conventional as well as novel and is a greener technique of volatile oil extraction providing lesser extraction time and higher yields.

Microwave Assisted Extraction (MAE)

Microwave assisted extraction (MAE) involves heating the solvent which is in contact with the spice sample using the microwave energy. The hydrogen bonds are broken down because of the dipole rotation of the molecules and migration of the ions. This rotation and migration increases the penetration of the solvent into the matrix of the spice sample, dissolving the components of interest and carry out the extraction (Mandal, 2008). The main characteristic feature of the MAE is that it enhances the dissolution of the compounds to be extracted with the simple application of heat which occurs due to the microwave field applied to the sample. Similar to other noval techniques for the extraction of volatile oils, the MAE not only conducts extraction in short durations, uses less solvent and lead higher recoveries but also requires less sample and the whole process is reproducible. MAE uses microwave radiation as a source of energy for heating solvents efficiently and quickly.

Solvent-free Microwave-assisted extraction (SFME)

Solvent free microwave assisted extraction technique is one of the modified version of Microwave assisted extraction. The SFME takes just 20-30 min for the extraction of essential oils as compared to the conventional methods which took hours for the extraction. Also, it is not too expensive of a technology and even the operating cost is low (Chan, 2011). The volatile oils extracted via SFME contains more amount of oxygenated compounds which means the oil has more of aromatic compounds, and hence the oil extracted from SFME is considered to be more valuable in terms of quality (CHAP, et al., 2016). SFME is based on a simple principle of microwave assisted dry distillation of the material without any addition of water or any organic solvent. Here the selective heating of the in-situ water content of spices takes place, which make the tissues swell and burst, making the essential oil freely available. This free essential oil then gets evaporated by the azeotropic distillation with the water which is present in the spice sample (Li, et al., 2013). The excess water is then refluxed and store back into the material as the original water of the plant material. Microwaves have strong interactions with the salt and nutrient containing physiological water, therefore more efficient extracts can be obtained from the materials which tend show a higher dielectric loss. Thus, upon heating the tissues undergo excessive swelling and ruptures, making the essential oil to flow towards the layer of the water. But this mechanism is dependent upon the solubility of the essential oil in water. This is the reason that solubilization becomes the limiting step in the SFME and solubility one of the essential parameters. The other mechanism of action depends upon the dipolar moments of the compounds of the essential oils. Essential oils are consists of organic compounds which absorbs microwave energy on a large scale. The proportion of extraction of these compounds by microwave extraction depends upon their high or low dipolar moments. Compounds with higher dipolar moments tend to interact vigorously along the microwaves and are extracted easily in comparison to the compounds which have low dipolar moments (Filly, 2014).

Microwave hydro-diffusion and gravity

The microwave assisted extraction has minimized the emission of carbon dioxide into the atmosphere compared to hydro-distillation, a conventional method of extraction (Farhat, 2017). The microwave hydro-diffusion and gravity have expanded the areas of the microwave assisted extraction technique by producing good quality extracts supported by innovative, efficient, fast and environmental friendly process (Lopez-Hortas, 2016)(Prez, 2014). The MHG extraction is a combination of the microwave irradiation and the earth's gravity at atmospheric pressure without the water or solvent (Lopez-Hortas, 2016)(Binello, 2014). Benmoussaa, 2018 stated that, the effect of MHG on the quality and quantity of the cumin essential oil. Their study showed an improvement in the extraction yield, in shorter extraction time, less consumption of electrical energy, lower emissions of carbon dioxide and less water wastage.

Ultrasonic- microwave assisted extraction

The green ultrasonic and microwave assisted extraction techniques are known best for their advantages. Though the MAE is known for non-homogeneous heating effects and UAE lacks in a strong thermal effect. Therefore, the combination of both MAE and UAE compliments the extraction process with their advantages. The defect of uneven heating by the MAE process can be compensated by the mechanical oscillations and continuous stirring action of the ultrasonic waves and the MAE process can provide with the sufficient thermal effect while compensating for the defect of insufficient heat production by ultrasound. This newer technique is also known for its greener nature i.e. it consumes less time and energy, it is fast and ensures safe and purified product (Chemat, 2017).

Wang, 2018 performed a comparative study on the efficiency of the essential oils of white and black peppers extracted by MAE, UAE and UMAE. They found higher oil yields with UMAE. Also, the essential oils extracted with UMAE obtained more of monoterpenes and sesquiterpenes than the oils extracted from MAE and UAE. Moreover, the UMAE required less solvent, time and energy and is proved to be eco-friendly.

Ohmic heating assisted extraction

Ohmic assisted extraction is an electro- technology where electric current is involved. The electric current is in direct contact with the food material or spices. The ohmic processing is referred to as mild processing which not only preserves the nutritional, functional, structural but also the sensory properties of the spices. It is an environmental friendly technology which reduces the burden on the environment by reducing the use of non-renewable resources and is an economic technology. It reduces the monitory expenditure and also improves the quality of the product. The ohmic heating assisted extraction works on the principle of electroporation. The cellular tissues get electrophoresed due to the presence of electric field. This electroporation results in better extraction of the bioactive compounds (Pereira, 2016) . In order to achieve better results this greener and newer technique for extraction of bio-active compounds is combined with conventional extraction techniques such as hydro-distillation.

Encapsulation

Essential oils from spices are known for their various health benefits because of the pharmaceutical properties they exhibit, also they are famous for their antimicrobial activities. Plus these volatile oils are known for their aroma and flavor in foods. Though the essential oils are a bundle of benefiting components, these components are highly unstable, which makes the essential oil sensitive towards various physico-chemical factors. These factors include light, oxygen, pH and temperature. The EO in the presence of light and oxygen undergo oxidation process of their unsaturated compounds thereby producing free radicals. The EO tends to lose their aromatic compounds when treated or stored at high temperatures (Dima, et al., 2015).

Due to their sensitivity to light, temperature, oxygen and pH, EO needs protection during storage, transportation and processing. This protection can be done in the form of encapsulation. Encapsulation protects the EO from being getting destroyed by the physicochemical factors and also maintains their flavor and biological value. Through encapsulation their odor and taste is masked and they are also transformed into water soluble powders (Augustin, 2012). These encapsulated components of the essential oils are released slowly in order to preserve their flavor and extend their shelf life. The encapsulation of the EO into micro-particles and nanoparticles should be done through integrated processes along interlinked stages/ phases. The encapsulation process should be selected such that it increases the functionality of the foodstuff to which the encapsulated EO is added. The foodstuff should exhibit safety, high nutritional value, health benefits, and acceptable sensory attributes and of course should be affordable (Donsi, 2011). The material of the encapsulant and the encapsulating technique should be friendly to the nature of the food system containing the EO. The EOs to be added to liquid food systems should be first converted into nano-emulsions and microemulsions or to liquid colloidal dispersions (Piorkowski, 2014)(Salvia-Trujillo, 2015) or the EO should be mixed with the water- soluble molecular systems like- cyclodextrins (Pinho, 2014).

Encapsulation is a process in which the bioactive oil/ essential oil droplets are packaged in a homogeneous or heterogeneous matrix, or are surrounded by a coating to form capsules which have many useful properties. As discussed earlier encapsulation interests the pharmaceutical, food and cosmetic industries because of the convenience system it provides for the delivery of these bioactive compounds at right time and right place. Other industries which show their potential interest in the encapsulated bioactive oils include: cultural products, personal care, industrial chemicals, veterinary medicine, biomedical, biotechnology and sensor industries. The encapsulation retards the thermo-oxidation of the essential oils (Soltani, 2015) . Encapsulation is efficient and feasible approaches which modulates the easy release of bioactive oils, enhances their physical stability, retards oxidation reactions, enhance bioactivity, decreases volatility and reduces toxicity. The choice of an encapsulating system is dependent upon the intended use of the final formulation, which may further depend upon its size, shape and nature.

Microencapsulation

The microencapsulation methods are majorly characterized by their particle size which ranges from 1-1000μm (Singh et al., 2014). The microencapsulation technique has been used successfully by the pharmaceutical, cosmetics and food industries. The microencapsulation is been used to protect and deliver the sensitive compounds of food, like- oils and fats, vitamins, minerals, enzymes, aromatic compounds and colorants (Chen et al., 2013). It limits the deleterious effects of environment (oxygen, heat, light, humidity, etc.) and thereby prevents the oil from being getting oxidized. It also prevents the oxidation of unsaturated fatty acids. Thus, increases the stability and shelf life of the oils. The encapsulation leads to easier handling of the sensitive substances by controlling their hygroscopicity, promotes dispersibility and flowability and increases the solubility. It also reduces the evaporation of volatile compounds from spices and provides control release of active products. Also, it masks the odor and taste of the volatile substances and allows their slow release when it is necessary in small amounts or when the surrounding environment is supportive (Aghbashlo et al., 2013).

Spray drying (Schmitz-schug et al., 2016), fluidized bed (Ivanova et al., 2005), coacervation (Yang et al., 2014), spray chilling (Dutra et al., 2016), extrusion (Rijo et al., 2014), interfacial polymerization (Gibbs et al., 1999), etc. are the different methods used to produce the microparticles. But all of these technologies are not used for encapsulating oils. Many studies have been conducted to investigate the effect of microencapsulation technologies on encapsulation of bioactive oils, as shown in table-5. Prior to encapsulation, emulsions are formed and therefore emulsion preparation is a key to good encapsulation. The emulsions consist of a disperse phase and a dispersing medium. The dispersed phase which is in the form of small droplets is generally the bioactive/ essential oil. The dispersed phase also known as core or the encapsulated material. The continuous phase is the wall material or the encapsulating medium. These emulsions formed can also use emulsifiers in order to prevent the coalescence, which depends upon the ratio of the lipophilic to aqueous phase and the wall materials. The emulsifier to be selected is also dependent upon the size of the particle size of the emulsion. The encapsulation efficiency and the stability of the microparticles are greatly influenced by the particle size. During the process of emulsification interfacial forces are formed. These forces prevent the interactions between the active compounds in the aqueous phase and the core material. This barrier property is exhibited by the interfacial membrane lying between the two phases. Thus, prevents the oxidation or other deteriorative processes of the active compounds by the continuous phase (Anandharamakrishnan & Ishwarya, 2015). The core material influences the morphology of the beads and the process of deposition/ coating of the shell/ membrane/wall material. There are three types of capsules which can be clearly distinguished from each other- Mononuclear capsules having a core surrounded by the shell, Polynuclear microcapsules having many cores surrounded by a single layer of shell and Matrix microcapsules where the core material is homogeneously distributed within the shell/ wall material.

Table 5: Microencapsulation methods and application of Bioactive oils

Bioactive Oil	Technology	Applications	Encapsulating Materials	Shape of the capsule	References
Rosemary essential oil	Spray dryer	Food industry	Maltodextrin and modified starch	Spherical with some irregularities	Fernandes et al., 2014
Cinnamon oil	Simple coacervation	Antimicrobial effect	β-Cyclodextrin and chitosan	Irregular shape rough surface and aggregates	Xing et al., 2011
Thyme oil	In-situ polymerization	Insect repellent	Melamine–formaldehyde 23–78% 1–10 μm	Spherical shape and smooth surface	Chung et al., 2013
Oregano oil	Spray drying	Food	IN, 5%, 15%, & 25%		Beirao da Costa, et al., 2013
Clove oil	Complex coacervation	Pharmaceutical	Gelatin with sodium carboxymethyl guar gum		Thimma & Tammishetti, 2003
Oregano oil	Supercritical solvent impregnation	High antioxidant activity	Rice starch, 20% &Gelatin, 1%		Almeida et al., 2013
Oregano oil	Spray-drying & Freeze-drying	Food	Rice starch porous spheres, IN, & gelatin/sucrose capsules		Beirao da Costa et al., 2012
Ginger Oil	Spray drying	Antimicrobial and antioxidant potential	Cashew gum and inulin		(Fernandes, 2016)

Nanoencapsulation

Nanocapsules are the nano-vesicular systems, which have a typical core and shell structure in which the bioactive compounds are encapsulated by a reservoir surrounded by a polymer membrane. The active compound is encapsulated in either of the forms- liquid, solid or as dispersion. The region where the active compound is received can be hydrophilic or lipophilic depending upon the components and methods employed. The size of the nanocapsules ranges from 1 to 100 nm (Wilczewska et al., 2012). Research has shown that particles smaller than 200nm decrease their clearance, hence they reside for longer time in the organisms.

Nanospheres and nanocapsules are the two types of nanostructures. Nanocapsules as discussed earlier contain a polymeric membrane which surrounds a liquid nucleus that can be either hydrophilic or lipophilic. Whereas, nanospheres have a polymer matrix which can be either porous or solid (Santos, Lopes et al., 2014; Santos, Lorenzoni et al., 2014). To reduce the immunological interactions and molecular interactions in the particle surface of various chemical groups, the nanocapsules are coated with non-ionic surfaces. The nanocapsules are absorbed more in the body than the other encapsulated systems. Depending upon the method of processing, the bioactive substance can either adher or remain inside of the polymeric membrane. When the Nanocapsule reaches the target tissue the bioactive oil is released from the system by various mechanisms, such as- desorption, diffusion or erosion. There are many methods for the preparation of nanocapsules as listed down in Table-6.

Table 6: Nanoencapsulation methods and application of bioactive oils

Bioactive Oil	Technology	Applications	Encapsulating Materials	Shape of the capsule	References
Peppermint	Solvent-precipitation		Zein/gum arabic	Spherical	Cheng & Zhong, 2015
Curcumin	Spray-drying	Anti-inflammatory agent	Chitosan		Sowasod et al., 2013
Oregano essential oil	Emulsification and ionic gelation	Antioxidative and antimicrobial activity	Chitosan	Spherical	Hosseini et al., 2013
Oregano, thymol oil	Liquid–liquid dispersion (solvent-precipitation)	Antioxidant properties	Zein	Spherical	Wu et al., 2012
Turmeric oil	Emulsification–gelification	Antibacterial, antifungal, antiplatelet, insect repellent, antioxidant, antimutagenic and anticarcinogenic properties	Chitosan–alginate	Spherical	Lertsuthiwong et al., 2009
Curcumin	Solvent free nanoencapsulation	Therapeutic properties	Sodium caseinate (0.1-10% conc.)		(Rao, 2016)

References

Akpinar, E. (2006). Mathematical modelling of thin layer drying process under open sun of some arometic plants. *Journal of Food Engineering*, 77, 864-870.

Amiri Chayjan, R. K. (2014). Modeling drying characteristics of hawthom fruit under microwave-convective conditions. *Journal of Food Processing and Preservation.*, 39; 239- 253.

Arslan, D. O. (2010). Evaluation of drying methods with respect to drying parameters, some nutritional and color characteristics of peppermint (*Mentha* x *piperita* L.). *Energy Convers Manage*, 51, 2769-2775.

Assami, K. P. (2012). Ultrasound induced intensification and selective extraction of essential oil from Carum carvi L. seeds. *Chemical Engineering Process, Process Intensif.*, 62, 99-105.

Augustin, M. A., & Sanguansri, L. (2012). Challenges in developing delivery systems for food additives, nutraceuticals and dietary supplements. In Encapsulation technologies and delivery systems for food ingredients and nutraceuticals (pp. 19-48). Woodhead Publishing.

Balasubramanian, A. G. (2012). Cryogenics and its application with Reference to Spice Grinding: A review. *Critical reviews in Food Science and Nutrition*, 52: 781-794.

Barai, R. (2015). Effect of Cryogenic and ambient grinding on flavoring componenets of black pepper and turmeric. International Journal of Emerging Trends in Engineering and Basic Sciences, 2(3), 49-54.Barnwal, P. M. (2014). Effect of cryogenic and ambient grinding on grinding characteristics of cinnamona nd turmeric. *International Joutnal of Seed Spices*, 4(2), 26-31.

Behera, S. N. (2004). Microwave heating and conventional roasting of cumin seeds (Cuminum cyminum L.) and effect on chemical composition of volatiles. *Food Chemistry.*, 87, 25- 29.

Benmoussaa, H. E. (2018). Microwave hydrodiffusion and gravity for rapid extraction of essential oil from Tunisian cumin (*Cuminum cyminum* L.) seeds: Optimization by response surface methodology. *Industrial Crops and Products.*, 633-642.

Bera, M. S. (2001). Development of cold grinding process, packaging and storage of cumin powder. *Journal of Food science and Technology*, 257-259.

Binello, A. O. (2014). Effect of microwaves on the *in situ* hydrodistillation of four different Lamiaceae. *C.R. Chim.*, 17, 181-186.

Buckenhuskes, H. &. (2004). Hygeinic problems of phytogenic raw materials for food production with special emphasis to harbs and spices. *Food Science and Biotechnology*, 262-268.

Cameron, M. M. (2009). Impact of ultrasound on dairy spoilage microbes and milk components. *Dairy Science Technology*, 89, 83-98.

Casals, C. V. (2010). Application of radio frequency heating to control brown rot on peaches and nectarines. *Postharvest Biology and Technology*, 58, 218-224.

CHAP, Stratakos, A. &. Koidis, T.(2016). Methods of Extracting Essential Oils. In Essential Oils in Food Presevation, Flavor and Safety (pp. 31-38). Belfast, United Kingdom: Elsevier Inc.

Chan, C. Y. (2011). Microwave assisted extractions of active ingredients from plants. *Journal of Chromatography*, 37, 6213-6225.

Chemat, F., Rombaut, N., Sicaire, A. G., Meullemiestre, A., Fabiano-Tixier, A. S., & Abert-Vian, M. (2017). Ultrasound assisted extraction of food and natural products. Mechanisms, techniques, combinations, protocols and applications. A review. *Ultrasonics sonochemistry*, 34, 540-560.

Ciurzyńska, A., & Lenart, A. (2011). Freeze-drying-application in food processing and biotechnology-a review. *Polish Journal of Food and Nutrition Sciences*, 61(3), 165-171.

Dima, C., & Dima, S. (2015). Essential oils in foods: extraction, stabilization, and toxicity. *Current Opinion in Food Science*, 5, 29-35.

Dima, C., Ifrim, G. A., Coman, G., Alexe, P., & Dima, Ş. (2016). Supercritical CO2 Extraction and Characterization of *C oriandrum Sativum* L. Essential Oil. *Journal of Food Process Engineering*, 39(2), 204-211.

Dima, Ş., Dima, C., & Iordăchescu, G. (2015). Encapsulation of functional lipophilic food and drug biocomponents. *Food Engineering Reviews*, 7(4), 417-438.

Donsi, F. A. (2011). Nanoencapsulation of essential oils to enhance their antimicrobial activity in foods. *Food Science and Technology.*, 44, 1908-1914.

Eikani, M. G. (2007). Subcritical water extraction of essential oils from Coriander seeds (*Coriandrum sativum* L.) . *Journal of Food Engineering*, 80, 735-740.

El Asbahani, A. M. (2015). Essential Oils: from extraction to encapsulation. *International Journal of Pharmaceuticals*, 483; 220-243.

El-Mesery, H. S., & Mwithiga, G. (2015). Performance of a convective, infrared and combined infrared-convective heated conveyor-belt dryer. *Journal of Food Science and Technology*, 52(5), 2721-2730.

El-Sebaii, A. A., & Shalaby, S. M. (2012). Solar drying of agricultural products: A review. *Renewable and Sustainable Energy Reviews*, 16(1), 37-43.

Erbay, Z., & Icier, F. (2010). A review of thin layer drying of foods: theory, modeling, and experimental results. *Critical Reviews in Food Science and Nutrition*, 50(5), 441-464.

Farhat, A. B. (2017). Efficiency of the optimized microwave assisted extractions on the yield, chemical composition and biological activities of *Tunisian Rosmarinus Officinalis* L. essential oil. *Food Bioprod. Process*, 105, 224-233.

de Barros Fernandes, R. V., Botrel, D. A., Silva, E. K., Borges, S. V., de Oliveira, C. R., Yoshida, M. I., ... & de Paula, R. C. M. (2016). Cashew gum and inulin: New alternative for ginger essential oil microencapsulation. *Carbohydrate Polymers*, 153, 133-142

Figiel, A., Szumny, A., Gutiérrez-Ortíz, A., & Carbonell-Barrachina, Á. A. (2010). Composition of oregano essential oil (*Origanum vulgare*) as affected by drying method. *Journal of Food Engineering*, 98(2), 240-247.

Filly, A. F. (2014). Solvent-free microwave extraction of essential oil from aromatic herbs: From laboratory to pilot and industrial scale. *Food Chemistry*, 193-198.

Gayahian, M. F. (2012). Coparison of ohmic-assisted hydrodistillation with traditional hydrodistillation for the extraction of essential oils from Thymus vulgaris L. *Innovative Food Science and Emerging Technologies*, 14, 85-91.

Goswami, T. (2010). Role of cryogenics in food processing and preservation. *International Journal of Food Processing*, 6(1):2.

Hashemi, S. N. (2017). Efficiency of ohmic assisted hydrodistillation fo rthe extraction of essential oil from oregano (Oreganum vulgare subsp. viride) spices. *Innovative Food Science and Emerging Technologies*, 41, 172-178.

Hashim, N. O. (2014). A prelimminary study: kinetic models of drying process of pumpkins (*Curcurbita Moschata*) in a convective hot air dryer. *Agriculture and Agriculatural Science Procedia*, 2(2), 345- 352.

Hatami, T. J. (2018). Exraction and Fractionation of fennel using supercritical fluid extraction assisted by cold pressing. *Industrial Crops and Products*, 123, 661-666.

Jarén-Galán, M., & Mínguez-Mosquera, M. I. (1999). Effect of pepper lipoxygenase activity and its linked reactions on pigments of the pepper fruit. *Journal of Agricultural and Food Chemistry*, 47(11), 4532-4536.

Johner, J. H. (2018). Developing a supercritical fluid extraction method assisted by cold pressing for extractiono f pequi(Caryocar brasilense). *Journal of Supercritical Fluids*, 137, 34-39.

Kaur, K., & Singh, A. K. (2014). Drying kinetics and quality characteristics of beetroot slices under hot air followed by microwave finish drying. *African Journal of Agricultural Research*, 9(12), 1036-1044.

Kim, S. Y., Sagong, H. G., Choi, S. H., Ryu, S., & Kang, D. H. (2012). Radio-frequency heating to inactivate Salmonella *Typhimurium* and *Escherichia coli* O157: H7 on black and red pepper spice. *International Journal of Food Microbiology*, 153(1-2), 171-175.

Klieber, A., & Bagnato, A. (1999). Colour stability of paprika and chilli powder. *Food Australia*, 51(12), 592-596.

Kubra, I. R., & Jagan Mohan Rao, L. (2012). Microwave drying of ginger (*Z ingiber officinale R oscoe*) and its effects on polyphenolic content and antioxidant activity. *International Journal of Food Science & Technology*, 47(11), 2311-2317.

Kubra, I. K. (2016). Emerging trends in microwave processing of spices and herbs. *Critical reviews in Food Science and Nutrition*, 2160-2173.

Lechtanska, J. S. (2015). Microwave and infrared assisted convective drying of green pepper: Quality and energy considerations. *Chemical Engineering and Processing: Process Intensification.*, 98, 155-164.

Li, Y., Fabiano-Tixier, A. S., Vian, M. A., & Chemat, F. (2013). Solvent-free microwave extraction of bioactive compounds provides a tool for green analytical chemistry. *TrAC Trends in Analytical Chemistry*, 47, 1-11.

Li, Z. R. (2010). Temperature and Power control in microwave drying. *Journal of Food Engineering*, 97, 478-483.

Liu, H. Z. (2013). The effect of cryogenic grinding and hammer milling on the flavor quality of ground pepper (*Piper nigrum* L.). *Food Chemistry*, 3402-3408.

Lopez-Hortas, L. C. (2016). Flowers of Ulex europaeus L.- comparing two extraction techniques (MHG and distillation). *C.R. Chim* , 19, 718-725.

Mandal, V. M. (2008). Microwave assisted extraction of curcumin by sample-solvent dual heating mechanism using Taguchi L9 orthogonal design. *J. Pharma Biomed. Anal.*, 46, 322-327.

Maroulis, Z. B., & Saravacos, G. D. (2003). *Food Process Design* (Vol. 126). CRC Press.

Maskan, M. (2001). Drying, shrinkage and rehydration characteristics of kiwi fruits during hot air and microwave drying. *Journal of Food Engineering*, 48: 177-182.

Meghwal, M., & Goswami, T. K. (2013). Evaluation of size reduction and power requirement in ambient and cryogenically ground fenugreek powder. *Advanced Powder Technology*, 24(1), 427-435.

Mihindukulasuriya, S. D., & Jayasuriya, H. P. (2015). Drying of chilli in a combined infrared and hot air rotary dryer. *Journal of Food Science and Technology*, 52(8), 4895-4904.

Miura, N. Y. (2003). Microwave dielectric properties of solid and liquid foods investigated by time- domain reflectometry. *Journal of Food Science.*, 68(4) 1396- 1403.

Mortezapour, H. G. (2012). Saffron drying with a heat pump-assisted hybrid photovoltaic thermal solar dryer. *Drying Technology*, 30, 560-566.

Moses, J. N. (2014). Novel drying techniques for food Industry. *Food Engineering Reviews*, 6(3), 43-55.

Murthy, C. T., & Bhattacharya, S. (2008). Cryogenic grinding of black pepper. *Journal of Food Engineering*, 85(1), 18-28.

Nadi, F., Rahimi, G. H., Younsi, R., Tavakoli, T., & Hamidi-Esfahani, Z. (2012). Numerical simulation of vacuum drying by Luikov's equations. *Drying Technology*, 30(2), 197-206.

Naidu, M. K. (2012). Effect of Drying methods on the quality characteristics of Fenugreek (Trigonella foenum-graceum) greens. *Drying Techology*, 30, 808-816.

Nayak, S. K. (2011). Drying and testing of mint (Mentha piperita) by a hybrid photovoltaic-thermal (PVT)-based greenhouse dryer. *Drying Technology*, 29(9), 1002-1109.

Onwude, D. H. (2016). Recent advances of novel thermal combined hot air drying of agricultiural crops. *Trends in Food Science and Technology*, 57, 132- 145.

Ozturk, S. K. (2018). Dielectric properties, heating rate, and heating uniformity of various seasoning spices and their mixtures with radio frequency heating. *Journal of Food Engineering*, 228, 128-141.

Palvic, B. V. (2015). Isolation of coriander (*Coriandrum sativum* KL.) essential oil by green extractions versus traditional techniques. *Journal of Supercritical Fluid.*, 99; 23-28.

Pavlic, B. V. (2015). Isolation of coriander (*Coriandrum sativum* L.) essential oil by green extractions versus traditional techniques. *Journal of Supercritical Fluids*, 99, 23-28.

Pereira, R. R. (2016). Effects of ohmic heating on extraction of food-grade phytochemicals from coloredp otato. *LWT-Food Science and Technology*, 74, 493-503.

Pinho, E. G. (2014). Cyclodextrins as encapsulation agents for plant bioactive compounds. *Carbohydrate Polymers*, 101, 121-135.

Piorkowski, D. T., & McClements, D. J. (2014). Beverage emulsions: Recent developments in formulation, production, and applications. *Food Hydrocolloids*, 42, 5-41.

Praveen Kumar, D. H. (2006). Suitability of thin layer models for infrared and hot air drying of onion slices. *LWT-Food science and Technology*, 39(6), 700-705.

Prez, L. C. (2014). Microwave hydrodiffusion and gravity processing of Sargassum muticum. *Process Biochem.*, 49, 981-988.

Rao, P. J., & Khanum, H. (2016). A green chemistry approach for nanoencapsulation of bioactive compound–Curcumin. *LWT-Food Science and Technology*, 65, 695-702.

Rassem, H. N. (2016). Techniques for Extraction of Essential OIls from Plants: A review. *Journal of Basic and Applied Sciences*, 10 (16); 117-127.

Rastogi, N. (2012). Recent trends and developments in ifrared heating in food processing. *Critical Reviews in Food Science and Nutrition*, 52(9), 737-760.

Sadeghi, M. M. (2013). Mass transfer characteristics during convective, microwave and combined microwave- convective drying of lemon slices. . *Journal of the Science of Food and Agriculture.*, 93(3), 471- 478.

Sadin, R. C. (2014). Development and performance evaluation of a combined infrared and hot air dryer. *Journal of Biological and Environemental Sciences*, 8(22), 11-18.

Salah, A. (2004). Inactivation of trypsin and chymotrypsin inhibitors in fenugreek (Trignella foenumgraecum L.) defatted seed flour. *Food Science and Agricultural Research Center*(130), 5-17.

Salvia-Trujillo, L., Rojas-Graü, A., Soliva-Fortuny, R., & Martín-Belloso, O. (2015). Physicochemical characterization and antimicrobial activity of food-grade emulsions and nanoemulsions incorporating essential oils. *Food Hydrocolloids*, 43, 547-556.

Calín-Sánchez, Á., Lech, K., Szumny, A., Figiel, A., & Carbonell-Barrachina, Á. A. (2012). Volatile composition of sweet basil essential oil (*Ocimum basilicum* L.) as affected by drying method. *Food Research International*, 48(1), 217-225.

Sarimeseli, A. (2011). Microwave drying characteristics of coriander (*Coriandrum sativum* L.) leaves. *Energy Convers Manage.*, 52, 1449-1453.

Saxena, S. S. (2015). Effect of Cryogenic grinding on volatile oil, oleoresin content and antioxidant propertie of coriander (*Coriandrum sativum* L.) genotypes. *Journal of Food Science and Technology*, 52(1), 568-573.

Schweiggert, U. C. (2007). Conventional and alternative processes for spice production- a review. *Trends in Food science and Technology*, 260-268.

Sharma, L. A. (2015). Effect of cryogenic grinding on oil yield, phenolics and antioxidant properties of ajwain (*Trachyspermum ammi* L.). *International Journal of Seed spices.*, 5(2), 82-85.

Sharma, L. A. (2016). Effect of cryogenic grinding on volatile and fatty oil constituents of cumin (*Cuminum cyminum* L.) genotypes. *Journal of Food Science and Technology*, 53(6), 2827-2834.

Singh, K. K., & Goswami, T. K. (1999). Design of a cryogenic grinding system for spices. *Journal of Food Engineering*, 39(4), 359-368.

Soltani, S., & Madadlou, A. (2015). Gelation characteristics of the sugar beet pectin solution charged with fish oil-loaded zein nanoparticles. *Food Hydrocolloids*, 43, 664-669.

Soysal, Y. (2004). Mic rowave drying characteristics of parsley. *Biosystems.*, 89, 167- 173.

Soysal, Y. A. (2009). Intermittent microwave conevctive air drying of oregano. *Food Science and Technology International.*, 15 (4), 397- 406.

Soysal, Y. A. (2009). Intermittent microwave- convective drying of red pepper: Drying kinetics, physical (color and texture) and sensory quality. *Biosystem Engineering.*, 103, 455- 463.

Szumny, A. F. (2010). Composition of rosemary essential oil (*Rosmarinus officinales*) as affected by drying method. *Journal of Food Engineering.*, 97, 253- 260.

Therdthai, N., & Zhou, W. (2009). Characterization of microwave vacuum drying and hot air drying of mint leaves (*Mentha cordifolia* Opiz ex Fresen). *Journal of Food Engineering*, 91(3), 482-489.

Vega-Mercado, H., Góngora-Nieto, M. M., & Barbosa-Cánovas, G. V. (2001). Advances in dehydration of foods. *Journal of food Engineering*, 49(4), 271-289.

Wakte, P. S. (2011). Optimization of microwave, ultra-sonic and supercrtitical carbon dioxide assisted extraction techniques for curcumin from Curcuma longa. *Sep. Pur. Technol*, 79, 50-55.

Wang, Y., Li, R., Jiang, Z. T., Tan, J., Tang, S. H., Li, T. T., & Zhang, X. C. (2018). Green and solvent-free simultaneous ultrasonic-microwave assisted extraction of essential oil from white and black peppers. *Industrial Crops and Products,* 114, 164-172.

Younis, M., Abdelkarim, D., & El-Abdein, A. Z. (2018). Kinetics and mathematical modeling of infrared thin-layer drying of garlic slices. *Saudi Journal of Biological Sciences*, 25(2), 332-338.

Yu, L. X. (2007). Preparation of garlic powder with high allicin content using combined microwave- vacuum and vacuum drying as as microencapsulation. *Journal of Food Engineering*, 83, 76- 83.

Zill-e-Human, V. M. (2011). A remarkable influence of microwave extraction: Enhancement of antioxidant activity of extracted onion varieties. *Food Chemistry*, 127, 1472-1480.

7

Past, Present and Future Perspective of Army Operational Rations (AOR'S)

***Kumar Ranganathan*[1*], *Shreelaxmi*[2], *Vijayalakshmi Subramanian*[3] *Shanmugam Nadanasabapathi*[4] and *Anil Dutt Semwal*[5]**

[1]*Head, Food Engineering and Packaging Technology Division, Defence Food Research Laboratory, DRDO, Mysuru, Karnataka, India*
[2]*Junior Research Fellow, Food Engineering and Packaging Technology Division Defence Food Research Laboratory, DRDO, Mysuru, Karnataka,India*
[3]*Senior Research Fellow, Food Engineering and Packaging Technology Division Defence Food Research Laboratory, DRDO, Mysurur, Karnataka,India*
[4]*Former Head, Food Engineering and Packaging Technology Division, Defence Food Research Laboratory, DRDO, Mysuru, Karnataka,India*
[5]*Director, Defence Food Research Laboratory, DRDO, Mysuru Karnataka India*

Introduction

Military personnel's are deployed for operation in adverse climatic conditions like desert, cold (arctic) and high altitude regions. Stressful and challenging living conditions have affected the outcome of many battles, campaigns and wars. Acclimatization to the environmental condition is a very essential stage of deployment. The time for acclimatization is restricted/limited for the effective operation towards the purpose of deployment. The performance of the military personnel in various deployed regions depends on the nutrition provided by the food consumed. The operational rations are designed to provide the essential nutrition to enhance their performance during training and/or operational situations.

Nutrition requirement of military personnel is different from that of the civilians. The nutrition requirement also varies with the kind (sedentary or continuous work) and region (plain, desert or high altitude) of deployment. The environmental stress caused by extreme heat, cold, etc. also influences the nutrient requirement/ intake. The insufficient dietary nutrient intake affects the immune function, pace of recovery from illness and injury and also the physical performance (Tharion et al., 2005).

The operational ration is also designed in such a way that it provides to the military personnel the required nutrition from the build in nutrition reserves during certain operational situations. Considering the vital criteria for nutritional requirement, the menu for the military personnel is designed. The chapter discusses the past, present and future of the army operational ration (AOR) in various countries of the world and the influence of advances in food processing methods in AOR menu.

Military Rations

The menu of the deployed personnel is known as military ration. Military ration are of four types: group feeding, individual packaged ration, restricted ration and specialty ration. The type of ration provided to military personnel depends on the mission of the unit, location, tactical scenario, and availability of food service equipment and personnel. Table 1 list out the classification of military rations in US based on the type of operation and food products served (Thomas et al., 1993). The work nature of deployment determines the design and development of military ration. The technological advancement in food processing and preservation helps in the enhancement of shelf stability of the food to be part of ration, minimizing the need for preparation and ensuring the nutritional quality and safety of food. These advances reduce the hardship in menu design and development according to the requirement (Neil Hill et al., 2011).

Table 1: Types of military ration in US (Source: Adapted from Thomas et al., 1993)

Group feeding	Individual ration	Restricted ration	Speciality ration
Garrison Feeding: Semi-perishable and perishable food items that require cold storage, food service equipment and personnel	Meal-Ready-to-Eat (MRE) - Basic operational: Heat processed food items that are packed in flexible pouches and does not require any preparation	For special missions upto 30 days: Dehydrated foods high in energy but restricted in calorie content	For operations in extremely cold conditions: Freeze-dried, cooked entrees. Meal portions are provided for a supply of 24hrs
Field Feeding: Canned & dehydrated foods	Back-up ration: For war and emergency situations. Shelf stable products packaged in a menu bag	For initial assault & special operations of 10 days or less: Pre-cooked freeze- dehydrated food	Cold weather: includes a supplement module providing an additional energy (1020 cal/ meal)- comprises of bread, soup, extra beverage, cookies and candy
Field ration: Ready-to- heat and serve products that are packaged in metal cans		Used for <5 days: Survival packets of six compressed bars	

Ration design and development

Food plays a vital role for sustained performance and morale in the field of deployment. Ration developed should provide adequate nutrition for effective and efficient performance at the field. The main purpose of ration design and development is to provide adequate nutrition as it is essential for enduring the following traits in the military personnel:

- Fighting fit condition
- Vigour/ vitality
- Zeal and enthusiasm
- Resistant to diseases
- Speedy recovery from injury and illness

The provision of inadequate or inappropriate military ration results in key issues like: weight loss (both voluntary and involuntary) due to inadequate ration consumption, dehydration, boredom due to monotony with rations and gastrointestinal complaints (Thomas et al., 1993). The various factors determining the operational ration design and development are (Meiselman,1996)

Food

- Portion size
- Acceptances, Liking, Preferences
- Perceptions (appearance, labels, packaging, origins)
- Presentation (temperature, utensils, dishes)
- Variety, monotony, sensory specific satiety

Individual

- Age, sex, cultural influences
- Expectations (marketing, education influence)
- Attitudes (neophobia, involvement, dietary restraint)
- Commensality (non-obese vs. obese)
- Food and Mood/Emotion

Environment

- Location
- Time of Day (meal appropriateness, snacks, presentation)
- Choice
- Ambiance (comfort)
- Convenience & access (effort, time)
- Price, value

The menu of operational ration especially depends on the purpose of deployment, i.e. general operation and special purpose operation. Based on the purpose of the ration, the nutrition composition, type of food item and portion size of the food varies. For instance, Table-1 represents the type of ration developed for various operational purposes.

During the selection of food items for ration, various factors such as behavioural, psychological impact on military personnel and environment of deployment need to be investigated before inclusion (RTO Technical Report 2010). The ration designing should also focus on easy logistics, i.e. it should reduce the issues involved in supplying the ration to remote areas of deployment. The advances in food processing and packaging comes handy in reducing the logistic issues.

The recommended dietary allowance, environment of deployment, portion size and acceptance by the military personnel are the vital factors considered while designing and development of military operational ration.

Required dietary allowance

The basic nutritional requirement of comprises energy, macro- and micro- nutrients such as carbohydrate, fat, protein, minerals and vitamins. The energy requirement varies with level of physical work and place of deployment. In general the energy requirement ranges from 3200- 4500 Kcal per day for military personnel. The operational rations are also classified based as normal and special purpose. The energy expenditure required of normal operation and special purpose is determined to be 3600 Kcal/day and 4900 Kcal per day respectively. In general, total daily energy requirement is contributed by the dietary protein, carbohydrate and fat consumed. The key nutrients provided by Combat ration of various countries is as listed in the Table-2. Table-3 represents the nutritional requirement of a special purpose combat ration. The recommended dietary intake of macronutrients such as carbohydrate, protein and fat should be 404-584g, 135- 225g and 80-140g respectively for a menu of 3600 Kcal per day and 552 -797g, 184 – 307g and 109 – 191g respectively for 4900 Kcal per day menu. The nutritional requirement varies with the region of deployment i.e. the recommended dietary intake for personnel deployed in cold region is different from that of the need of personnel deployed in high altitude. Table 4 depicts the nutritional requirement in Cold weather operations.

The high altitude ration is designed to help to maintain body weight, nutrition status and metal and physical alertness of the deployed personnel. High altitude ration should provide high energy and carbohydrate. It should also provide adequate protein to protect form muscle loss. Iron content should also be considered based on the requirement of the military personnel to enhance the oxygen supply at the high altitudes (Singh et al., 1999 and Anusha et al., 2017). The nutrient content for diet of active individuals at high altitudes of 9000 to 12000 feet is as listed in Table-5. Thus military/ army operational ration is designed considering the recommended dietary requirement of the personnel deployed in various regions.

Table 2: Key nutrients of the combat ration in various countries (Source: adapted from RTO Technical Report, 2010)

	AUS	BEL	CAN	CZE	FRA	DEU	ITA	NLD	NOR	SVN	UK	USA
Energy (Kcal)	3700	3200	4395	3551	3200	3524	3650	3682	3762	3537	4294	3995
Carbohydrate (g)	593	440	681	414	440	501	521	540	649	518	618	528
Protein (g)	108	104	141	111	104	96	100	108	90	95	107	126
Total fat (g)	116	114	123	133	114	126	129	126	89	110	155	157
Calcium (mg)	968	>800	1016	746	>800	900	1079	718	NP	NP	1444	1705
Iron (mg)	32	20	26	21	20	19	33	33	NP	NP	21	26
Sodium (mg)	5860	NP	9381	2458	NP	8300	5250	8061	NP	NP	8293	6850

NP- Not Provided, AUS- Australia; BEL- Belgium ; CAN- Canada; CZE- Czech Republic;
FRA- France; DEU- Germany; ITA- Italy; NLD- Netherland; NOR- Norway; SVN- Solvenia;
UK- United Kingdom; USA- United States of America

Table 3: Recommended Dietary intake for special purpose combat ration

Nutrients	Value
Energy (kcal)	4900
Carbohydrate (g)	550-800
Protein (g)	158-185
Total fat (g)	110-190
Total fiber (g)	30
Vitamin A (μg)	900
Thiamin (mg)	1.2
Riboflavin (mg)	2.5
Niacin (mg)	16
Vitamin B6 (mg)	2.6
Vitamin B12 (μg)	2.4
Folate (μg)	400
Pantothenic acid (mg)	6
Biotin (μg)	30
Vitamin C (mg)	45
Vitamin D (μg)	5
Vitamin E (mg)	10
Vitamin K (μg)	70
Choline (mg)	550
Calcium (mg)	1000
Phosphorus (mg)	1000
Zinc (mg)	15
Iron (mg)	14
Magnesium (mg)	410
Iodine (μg)	150
Selenium (μg)	70
Molybdenum (μg)	45
Copper (mg)	1.8
Chromium (μg)	35
Manganese (mg)	5.5
Fluoride (mg)	4
Sodium (mg)	2300-12000
Potassium (mg)	3800

(*Source*: adapted from RTO Technical Report, 2010)

Table 4: Nutritional content of operation rations used in Cold weather

Nutrients	Military RDA Values	
	Men	Women
Energy (Kcal)	4500	3500
Protein (g)	100	80
Carbohydrate (g)	580- 680	580 - 680
Fat (g)	140 - 220	140- 220
Vitamin A (IU)	5000	4000
Vitamin E (mg)	10	8
Vitamin C (mg)	60	60
Thiamin (mg)	1.6	1.2
Riboflavin (mg)	1.9	1.4
Niacin (mg)	21	16
Vitamin B6 (mg)	2.2	2
Folacin (μg)	400	400
Vitamin B12 (μg)	3	3
Calcium (mg)	800 - 1200	800- 1200
Phosphorus (mg)	800 - 1200	800 – 1200
Iron (mg)	10 - 18	18
Sodium (mg)	4800- 7800	4800 - 7800
Potassium (mg)	5500	4100
Magnesium (mg)	350- 400	300
Zinc (mg)	15	15
Cholesterol (mg)	180- 490	180 - 490

(*Source*: Adapted from Marriott and Carslon, 1996)

Table 5: Nutritional requirement of Active individuals at high altitudes (9000-12000 ft)

Nutrients	Quantity
Energy (Kcal)	4829
Carbohydrates (g)	746.8
Proteins (g)	144.0
Animal protein (g)	40.0
Fat (g)	147.9
Vitamin A (IU)	6279
Thiamine (mg)	4.5
Riboflavin (mg)	3.8
Nicotinic acid (mg)	37.5
Ascorbic acid (mg)	247.6
Iron (mg)	91.5
Calcium (g)	1.55

(*Source*: Adapted from Singh et al., 1999)

Environmental factors

The military personnel are exposed to various environment conditions. The diversified environments of operation are classified based on the temperature and altitude of region. The military personnel are exposed to the following extreme environments (Thomas et al., 1993, RTO Technical Report 2010, Anusha et al., 2017):

- Hot environment- deserts
- Humid forests
- Hot humid coastal regions
- Cold environment- Arctic and Antartic regions
- High altitude
- Sea level
- Space

The exposures to various extreme environmental conditions are both natural and man-made due to activities (Table-6). The activities such as flying, swimming/ diving also has its own impact on the personnel and these factors need to counted while designing the operational ration.

The nutritional requirement varies with the deployment at extreme conditions (Edwards et al., 1992 and Edwards et al., 1995). The extreme environmental condition has its own effect on the work performance of the military personnel employed (Figure 1) (Selvamurthy and Singh, 2003). The extreme environment exposure results in condition like increased energy and water requirement, hypophagia, hyodipsia, impaired thermo regulation, muscle depletion, motion sickness when travelling in various environment, etc. According to Anusha et al., (2017), at high altitudes the deployed personnel suffer from hypoxia, cold, high solar radiation as physical stresses apart from other psychological stresses. Similarly other environmental condition also has its physical and psychological impacts that affect the performance of the personnel. The operational ration needs to be designed and developed considering these environmental factors and the corresponding nutritional requirements.

Energy requirement of the military personnel depends on both the level of physical activity and deployment environment (Table 7). The energy requirement at temperate condition varies form 32- 63 Kcal/ kg bodyweight depending on the level of physical activity. Similarly for hot and cold environments it ranges form 40 – 75 Kcal/ kg body weight and 35- 68 Kcal/ kg body weight respectively. The energy requirement at varying environment conditions proves that the nutritional requirement varies with environment conditions. Hence, this plays a major role in the operational ration design and development.

Table 6: Exposure of military personnel to various extreme environmental conditions (Source: Adapted from Selvamurthy and Singh, 2003)

Primary natural	Primary man made	Condition	Environment
X		Low temperature	Arctic / Antarctic / Altitude
X		High temperature	Tropics
X	X	Reduced pressure	Altitude/ flight
X		Increased pressure	Diving
X		Reduced gravity	Space
	X	Increased gravity	Flight
X		Decreased oxygen availability	Altitude
	X	Increased oxygen	Diving
	X	Change in inspired air composition	Diving
X		Lack of water	Desert
X	X	Lack of food	Anywhere
X		Increased radiation	Space / Altitude
X	X	Isolation	Arctic/ Antarctic / Space

Table 7: Energy requirement of various levels of physical activity at temperate, cold and hot environment (Source: Selvamurthy and Singh, 2003)

Physical activity	Environment		
	Temperate	Cold (>0°C)	Hot (> 30°C)
Light	32-44	35-46	40-54
Moderate	45-52	47-55	55-61
Heavy	53-63	56-68	62-75

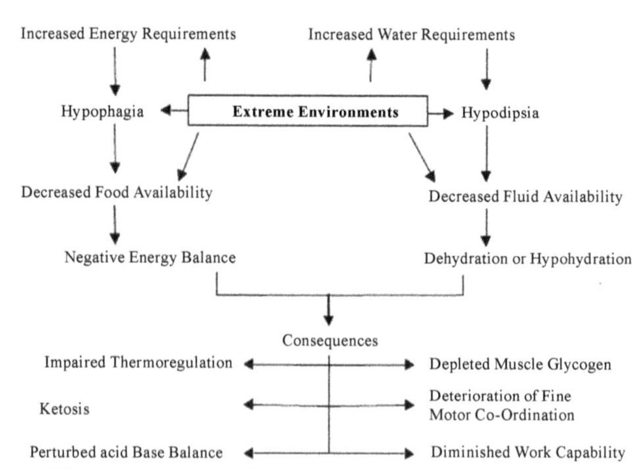

Fig. 1: Effect of extreme environmental condition on work performance (*Source*: Adapted from Selvamurthy and Singh, 2003)

Optimization considerations

The various factor need to be considered while optimizing the operational ration depends on the purpose, target group, environment of deployment and food processing method. Six major factors must be taken into consideration in optimizing combat ration design (Kumar et al., 2015).

1. Performance enhancement is the key factor and was the focus of the workshop on which this volume is based.
2. Caloric densification is extremely important for situations in which a soldier must have highly compact rations that are also easy to handle. These rations have many calories packed into a very small volume. Some of the components have a caloric density of about 7kcal/mL.
3. Component preservation is the basis for providing shelf-stable operational rations in the field. The components must be microbiologically stable, which can be achieved either by destroying all the pathogenic and spoilage microorganisms or by inhibiting their growth. Moreover, even after accomplishing that, one must ensure that the food remains biochemically, physically, and chemically stable. It should have the same structure and appearance after 3 years of storage at a nominal temperature of 27°C as it had when it was first produced.
4. Heating and cooling of such rations is an important consideration, because hot food and appropriately cool beverages affect morale and proper food and water intake.
5. Unless there is consumer acceptance of the food, it will not be consumed and nutrition could be compromised. Meeting all of the nutritional requirements is predicated upon consumption of the entire ration. Satisfaction with the ration contributes as well to the social well-being of the soldier.
6. Lastly, quality monitoring by using time-temperature integrators is also important. It makes it possible to ensure that, despite high temperature stresses over long storage periods, the food has the intended attributes and nutrients at the time of consumption.

Technical Concerns for Ration Development:

- Liquid versus Solids—stability differences, sterilization required
- Long-term storage requirements
- Minerals and electrolytes—choice of salts
- Vitamins—choice of form
- Requirements for additives
- Stabilizers-antioxidants
- pH control buffers and acidulents for liquids

- Packaging the product to maximize stability
- Tastc and acceptability—as a function of temperature

Acceptance of ration

The ration designed based on the nutritional requirement and other factors need to be accepted by the military personnel who are the ultimate consumers. The rations are generally designed based on the palatability and requirement of the military personnel. The field trials are conducted at the target region and group for analysing the acceptance and its effect on the personnel's health thereof. In general, the food item developed is provided to the selected target group for a designated period of time. The subjects of the targeted group rate the food product developed according to their liking using a 9- point hedonic scale. By the end of the field trial, the subjects are subjected to biochemical analysis in order to check the effect on their nutritional reserve and physical activity. The prolonged consumption of a particular food item also has its impact on the military personnel, where they retrieve from consumption of monotonous ration. This ultimate affect the physical and psychological ability of the personnel. Hence, the feedbacks from the subjects need to be considered with utmost care and given importance during ration designing. Thereby, the acceptance of the developed food item needs to be analysed for a designated period of time before inclusion into the operational ration.

Operational ration size

The operational ration size depends on the purpose of deployment and the group size. The combat rations are designed to provide troops with self- sustained feeding. The individual combat rations can be of 24hr menu size and involving products that require minimum preparation and no catering skills. The rations can also be of 10 Man pack rations etc. based on the size of group. The group feeding rations can be in the pallets containing individual combat rations. The operational rations are generally packed in portion sizes that are consumed once opened. The operational ration should also be designed in such a way that no special facilities are needed for preservation and preparation. The packaging of operational rations plays an important role as its directly impacts the logistics and disposal issues. Hence, the size of the operational ration plays a vital role in ration design and development.

Food processing: Shelf life extension

The common or widely employed food processing technologies include thermal processing. The conventional thermal processing techniques involved in the different types of operational products are known to be pasteurization, canning, retort processing, microwave processing, dehydration, sun drying, etc.

Thermal processing is the most widely used process technology in the food industry which ensures microbiological safety of the products. These methods rely essentially on the generation of heat outside the product to be heated, by combustion of fuels or by an electric resistive heater, and its transference into the product through conduction and convection mechanisms. Microbial control processes at temperatures in the 65-95°C range is called pasteurization and treatment at temperatures from 100-150°C is known as sterilization. Thermal processing of vegetables undergoes physicochemical, textural and morphological changes. One of the negative side effects of heat sterilization of vegetables and fruits is the extreme softening of the product, resulting in an undesirable texture. Blanching and freezing of beef and other meat products are important techniques used in developing frozen food products. Canning of fruits with syrup and vegetables and meat products with brine are the most commonly employed in operational preparation. Retort processing and aseptic processing techniques of ready-to-eat meals are the recent advancements in thermal processing of food, flexible and carton boxes packaging respectively. Microwave processing, a common thermal processing technique, has been used not only to bake food products but also to prepare ready-to-cook products, reheat – to -eat, ready-to-eat foods etc.

Conventional thermal operations have the tendency to induce permanent changes to the nutritional and sensory attributes of foods. Therefore, recent developments in food processing operations have aimed at technologies that have the potential to substantially reduce damage to nutrients and sensory components by way of reduced heating times and optimized heating temperatures. This had lead to the invention and improvisation of non thermal technologies.

Non-thermal methods are generally performed at a temperature lower than that of the thermal processing techniques or at ambient temperature, thereby minimizes the changes in flavor, essential nutrients and vitamins, etc., the various types of non-thermal food processing techniques are: irradiation, high hydrostatic pressure, ultrasound, pulsed electric filed, pulsed light, oscillating magnetic field, etc. Each of these techniques can be used either alone or in combination to optimize product quality, processing time and bacterial and enzyme inactivation. These non-thermal processing techniques have been more effective when combined with a mild heat treatment. The application of irradiation (ultra violet and gamma), high pressure processing (HPP), ultra sonication, pulsed electric field (PEF) in combination with mild heat treatment has been established to effective in the processing of operational ration products. Gamma irradiation of dosage 1-6kGy has been proved to be effect in processing ready-to-eat meat curries, fruit juice powders, etc. HPP processing has also been established for its efficacy to enhance the shelf life of fruit and vegetables juices, purees, meat products, etc. PEF is widely employed in the processing of liquid food products like milk. Fruit juices, soups, etc. Ultrasonication has been effective individually and in combination with thermal or HPP of food products especially liquid foods. Likewise all the other non-thermal processing techniques have been studied for the successful application on the preparation of HMR products.

Food packaging plays a major role in the HMR foods processing and preservation. The successful application of both thermal and non-thermal processing techniques depends on the type of package used to contain the treated food product. Food products tend to change with time in composition and quality, be it a raw material or processed product. The need of packaging techniques propped up to enable storage of the food material, by retaining its nature, minimizing deterioration and extending its life, making it available around the world in all seasons. The main concept of packaging is to curb the interaction between the product and the environment surrounding, to prevent the initiation of deterioration. The packaging used plays a major role in the retention of the quality attributes of the HMR foods till consumed by the consumer. The multilayer packages has been used widely for the packaging of retort processed food, fried and baked products, milk, etc. The recent advancements like intelligent packaging using quality indicators, absorbers, emitters etc., help in the retention of the quality of the packed food products. The biodegradable or ecofriendly packages are the more recent and widely appreciated advancement in food packaging, to reduce waste, etc. The retort processed products, which are generally packed in multi-layered pouches (poly propylene/aluminium/nylon/cast poly propylene) in recent times, where earlier, packed in lacquered tin or aluminium cans. The meat products are placed on a plastic or aluminium tray and covered by a see through film and stored under cold condition. MAP is another important food packaging technique used in the preservation of food products. Nano-packaging techniques are the most recent packaging techniques wherein the nanoparticles present in the film or package contributes to the antibiotic activity, physical and mechanical strength. Thereby the food packaging ensures the retention of the quality attributes and extending the shelf life of the operational rations.

Food processing advancements: nutrition

The operational rations such as survival or emergency Ration was designed for military, disaster relief, marine, aviation, outdoor adventure and extreme conditions applications. The survival ration comes in handy for oneself, for hurricane survival, tornado survival, earthquake survival and unforeseen non-natural disasters seen every day around the world.

The survival ration comprises of food and water for the endurance at extreme conditions. The major criteria to be met by the survival ration were it should possess an extended shelf life with minimal change in quality; provide the required nutrition on consumption, etc. The food items which enter into the survival ration were basically classified into three categories: Freeze dried, Meal Ready-to- Eat (MRE) and Food bars.

Meals ready-to-eat (MRE) are products which are basically retort processed products. The meals are prepared, packed and retort processed, where the final product has an extended shelf life and is microbiologically stable. The retort processed products need to be just reheated and consumed. The major part of the survival ration includes the retort processed ready-to-eat meals. This food processing is basically performed on the basis of the F_0 value of the index microorganism. Initially canned products were treated and used for the purpose. The transportation and storage of canned foods was a major problem logistically and in operation. Later on, with the advancement in packaging materials, flexible multilayered packages with the desirable barrier properties were adopted for the retort processing which was a major breakthrough. The reheating of these packages was also easy when compared to the conventional cans and bottles.

Dehydration/drying have been one of the conventional methods of extending the life of the food products. Freeze drying and spray drying techniques has gained importance in the survival ration development. Freeze-drying, also known as lyophilization, lyophilization, or cryodesiccation, is a dehydration process typically used to preserve a perishable material or make the material more convenient for transport. The lyophilization process sublimates the water within the desired product, leads to concentration or densification of the product and also makes it to rehydrate readily or easily available for absorption. On consumption the product was claimed to give an aromatic, nutritional and flavourful components of the product, in an unaltered, intensely flavourful state. Freeze drying increases caloric density by weight, but by itself contributes nothing to ration space reduction. To also maximize caloric density by volume, airspace in the food must be reduced or eliminated. Such densification can be accomplished by physical compression of the dehydrated product. Compression is carried out by subjecting the product to pressure in a die of appropriate size and dimension-i.e., square, rectangle, or other geometric shape- in order to eliminate void space and to produce a compact and dense food bar.

Spray drying converts directly, the fluid materials into solid or semi-solid particles. Spray drying is a unit operation by which a liquid product is atomized in a hot gas (mostly air or inert gas such as nitrogen) current to instantaneously obtain a powder. The initial liquid feeding can be a solution, an emulsion or a suspension. It is suitable for both heat-resistant and heat sensitive products. The spray drying process can produce a good quality final product with low water activity and reduce the weight, resulting in easy storage and transportation. Spray drying is preferred, since it results in a product with a relatively uniform particle size distribution. Microencapsulation is the principle behind spray drying process, where the micronutrients and bioactive compounds are encapsulated in an active form inside a carrier agent. Spray-drying of cooked, homogenized pulses produces an easy to use ingredient with strong nutritional profile, has the process produces an easily rehydratable powder, which is highly convenient and nutritious. The spray dried powders are also compressed into bars like freeze dried products, to reduce the void space for easy transportation.

The most recent method of producing energy, protein or nutritionally dense products is vacuum infusion or impregnation. The introduction of compounds can be achieved through classical infusion, through the immersion of the products in hypertonic solutions of the respective compound, or through a new technology, vacuum infusion or impregnation.One of the main directions in the alimentary industry is focused upon the preservation of the existing natural compounds either through the minimum processing of the raw materials or through the strengthening of the foods with multiple physiologic active compounds such as prebiotics, probiotics, vitamins, fiber, mineral salts etc.Vacuum infusion is also performed: to soften the tissues using enzymes such as pectin methylesterase; nutrient enrichment by infusing vitamins, minerals, etc.; to maintain texture of the product; to enhance mass transfer rate, etc.

Food bars find the first and foremost position in the survival ration. The food bars are generally energy or protein dense products, which are in the compressed and concentrated form. These bars provide the required energy and nutrients as per the dietary allowance during the time of emergency. The food bars comprises of compressed powder (freeze or spray dried) or thermally treated products such as chikki, etc. These condensed or dense food bars satiate the hunger and also provides the necessary amount of calories, proteins and other nutrients. The most recent advancement in the food bars is the incorporation or enrichment of the bars with encapsulated or by infusing with the nutraceutical compounds, energy providing compound, etc.

Food fortification

Fortification for nutrient enrichment has been in practice for many years. According to Thomas et al., (1993) the food products such as cookies, energy bars, beverages, etc. were fortified with vitamins and minerals to combat various nutritional deficiencies among the military personnel. Criteria for the optimal fortification of supplementation that may well be applied to any performance-enhancing supplements.

- The supplementation regimen should be based on knowledge of the nutrient content of an individual's basic diet. This knowledge may be simpler for combat rations than for normal diets, but many advances in analytic techniques may be required.
- Supplements should be in a form that is acceptable and consumable by troops under battlefield conditions. Carriers must be highly preferred items.
- Fortified items should be designed to maximize nutrient retention in the form that they are consumed by users. Both storage stability and nutrient stability when reconstituted for use need to be optimized.
- The interactive effects among different components must be considered and controlled. Reactive micronutrients will be separated by encapsulation or by use of separate carriers.
- The levels and specific forms of individual fortifying nutrients should be chosen to give maximum bioavailability, with little or no chance of negative effects from either under- or over supplementation.

- Supplementation should be based on knowledge of the metabolic roles of the -supplement's components and the basic dietary components.
- Food components or food derivatives that might have ergogenic effects are generally classified as either a nutritional or a pharmacological ergogenic aid.

Such aids enhance performance by:

1. Acting as a central or peripheral nervous system stimulant,
2. Increasing the stored amount or availability of a limiting substrate,
3. Acting as a supplemental fuel source or reducing reliance on a limiting substrate during prolonged physical exertion,
4. Reducing or neutralizing metabolic by-products that interfere with energy-producing reactions or muscle contraction, and
5. Enhancing recovery.

The technological advancements with micro-encapsulation, nano-encapsulation and vacuum infusion/impregnation, helps in effective fortification of nutrient supplements and bioactive compounds in operational rations. These adaptation of these innovations helps in providing the military personnel with a highly nutritive and quality desired product.

Revolution of operational ration in India

Defence Food Research Laboratory (DFRL) is the prime institution working on the operational ration requirement of the military personnel from the time of inception. Table 8 list the development of various pack rations by DFRL since the time of inception. The operational ration designed for personnel deployed in plains and high altitude are analysed for the nutrient supplied and the values are as mentioned in Table 9 (Babusha et al., 2008). The menu of the operational ration formulated for Indian army forces are based on their food habitat and nutritional requirement. The menu comprises of products made of wheat and rice, biscuits, curries (such as dal fry, vegetable curries etc.), biscuits, energy bars, sweets (eg. sooji halwa, kheer, etc.) and candies, milk powder and beverages, etc. The menu comprises of retort processed products, freeze dried and dehydrated products that require minimal/ no cooking. The menu also comprises of functional products in the form of candies and beverages. The operational menu composition also varies with the environment of deployment for operation. The recent advances such as non-thermal treatment, various fortification techniques and food packaging has enabled DFRL to develop energy and protein rich food products and other functional products to combat motion sickness and other environment related issues on deployment.

Table 8: Packed ration developed since inception by DFRL (1967-2011)

S.No.	Type of Ration
1.	5 Man Compo Pack (Arid zone ration)
2.	Survival Ration
3.	Action Messing Meals for Navy
4.	Meal for Tank Crew
5.	2,4,6,10 men pack ration for submarine crew
6.	Arid Zone Ration
7.	One Man Compo Pack Ration
8.	Compo ration for Army
9.	MRE ration
10.	Emergency Flying Ration for Aircrew
11.	MBT Ration
12.	Six types of MRE ration and Modified MRE
13.	Modified compo pack ration
14.	Prototype in-flight ration
15.	Composite ration for mechanised infantry
16.	Submarine crew ration for marine commando force
17.	RTE meals for submarine crew
18.	Theatre terrain weapon platform specific MREs

Table 9: Nutrient requirement of Indian soldiers deployed at plain and high altitudes (*Source*: Adapted from Babusha et al., 2008)

Nutrients	Intake values (per day)	
	Plain	High altitude
Energy (Kcal)	3632	4180
Protein (g)	124	104
Carbohydrate (g)	565.6	630
Fat (g)	98.8	138
Vitamin A (μg)	34.3	961
Vitamin C (mg)	15.0	117
Thiamin (mg)	2.03	1.5
Riboflavin (mg)	1.08	1.5
Niacin (mg)	26.8	25.5
Iron (mg)	34.7	24.07
Calcium (mg)	1474	1303
Phosphorus (mg)	3190	2360
Sodium (mg)	8189	9876
Potassium (mg)	2679	10391
Zinc (mg)	13.58	21.32
Copper (mg)	3.31	8.89
Total dietary fiber (g)	80.9	77.2

Conclusion

As long as there have been conflicts among human populations, the ability to provide sustenance to combat forces has been a requisite for battlefield success. The operational rations are formulated to provide the military personnel with required nutrition and also to serve a specific purpose. The operational ration designed plays a major role in maintaining the physical and psychological valour of the army personnel. The nutritional composition of ration depends on the environmental condition of region and country in which the personnel to be deployed for operation. The past, present and future of the operational ration goes in hand with the evolution of various food processing and packaging methods. The ultimate purpose of application of technological advancements in operational ration production is to provide a highly nutritive, shelf stable, quality and safe product.

References

Anusha, M.B., Shivanna, N. and Anilakumar, K.R. (2017). Nutritional requirement at high-altitude with special emphasis to behaviour of gastro-intestinal tract and hormonal changes. *Defence Life Science Journal.* 2(2): 120-127.

Babusha, S.T., Singh, V.K, Shukla, V., Singh, S.N and Prasad, N.N. (2008). Assessment of ration scales of the armed forces personnel in meeting the nutritional needs at plains and high altitudes-I. *Defence Science Journal.* 58(6):734-744.

Edwards, S.A, Roberts, D.E and Mutter, S.H. (1992). Rations for use in a cold environment. *Journal of Wilderness Medicine* 3, 27-47

Kumar, R. Vijayalakshmi, S., Reddy, K. R., George, J. and Nadanasabapathy, S. (2015). Calorie densification technologies for the development of light weight foods- an overview. *Indian Food Industry.* 34(3):36-47

Marriott, B.M., and Carlson, J.S. (1996). Nutritional needs in cold and in High altitude environments: National Academy Press, Washington DC. 100- 250.

Meiselman, H.L. (1996). The contextual basis for food acceptance, food choice nad food intake: the food, the situation and the individual. In: Food choice, acceptance and consumption. Meiselman, H.L. and MacFie, H. (Eds) Chapman & Hall Publications. 239-263.

Neil Hill, Joanne Fallowfield, Susan Price and Duncan Wilson. (2011). Military nutrition: maintaining health and rebuilding injured tissue. Phil. Trans. R. Soc. B (2011) 366, 231–240

RTO Technical Report. (2010). Nutrition science and food standards for military operations. North Atlantic Treaty Organization. AC/323 (HFM-154)TP/291. www.rto.nato.int

Selvamurthy, W. and Singh, S.N. (2003). Nutritional Requirement for Human adaptation in extreme environments. *Proceeding of the Indian National Science Academy.* B69(4), 485-506.

Singh, S.N., Sridharan, K and Selvamuthy, W. 1999. Nutrition in high altitudes. *Bulletin of Nutrition Foundation of India* 20(3), 1-3

Tharion, W. J., Lieberman, H.R., Montain, S.J., Young, A.J., Baker-Fulco, C.J., DeLany, J.P and Hoyt, R.W. Energy requirements of military personnel. *Appetite*, 44, 47-65.

Thomas, C.D., Baker-Fulco, C.J., Jones, T.E., King, N., Jezior, D.A., Fairbrother, B.N., Speckman, K.L. and Askew, E.W. (1993). Nutritional Guidance for Military Field Operations in Temperate and Extreme Environments. US Army Research Institute of Environmental Medicine. Natick, MA. 1-79.

8

Developments in the Technology of Oils and Refineries

Sandeep Singh

P.G. Department of Food Science and Technology, Khalsa College Amritsar-143002, Punjab,India

List of abbreviations

%	:	Percentage
&	:	And
<	:	Less than
°C	:	Degree Celsius
cm	:	Centimetre
e.g.	:	Example
gm	:	Gram
i.e.	:	That is
kg	:	Kilogram
kPa	:	Kilopascal
L	:	Litre
mg	:	Milligram
min	:	Minute
ml	:	Milliliter
mm	:	Millimeter
ppm	:	Parts per million
ppb	:	Parts per billion
sec	:	Second
wt	:	Weight
μm	:	Micrometer

Introduction

Oils and fats are utilized both for their food applications as well as for industrial purpose. As a food product, oils and fats are consumed in a variety of forms such as shortening, ghee, butter, margarine, cooking oil, salad oil, etc. Other uses include their usage in the form of lubricants, greases, soaps, cosmetics, paints as well as animal feed and biofuels. Among the Asian countries, Indonesia leads as the prime producer with more than fifty percent of the total palm oil production in the world. As a second-best producer of oils and fats, China was the major manufacturer of rapeseed as well as soybean oil in year of 2018. In general, Asian countries contributed to more than fifty percent of the total fat and oil manufacturing in the world in 2018 (Kushairi et al., 2018). Even though fats and oils are primarily utilized for food applications, there has been an increased interest in their industrial utility for the manufacture of biofuels. Fats manufactured from various animal sources like tallow or lard also contribute to a considerable quantity and compete with vegetable oils for the production of biofuels.

The resources for producing oils and fats are diverse depending on their end use, yet they are obtained from two major sources: animal sources and plant sources. Oils and fats are used as starting materials for the preparation of shortenings, margarines, and other tailored products that further act as functional ingredients in food products prepared by food processors, restaurants, and in home. They have been used for food and other variety of applications since the prehistoric times, due to their easy availability. For example, fatty tissues from animals liberate free-floating fats on being boiled, whereas oil can be pressed out from oil fruits and oilseeds.

The role of dietary oils and fats in human nutrition is also one of the important areas of interest. Over the time, oil and fat industry has provided significant contributions toward nutritional improvement of various products. With the growing concern linked between cardiovascular diseases and consumption of oil and fat products, low-calorie variants of different foods, free or low in cholesterol are also available now. Excessive fat consumption among the populations of developed countries and consequences of obesity has led to the development of reduced-fat and very-low-fat variants of foods.

Oils and fats find wide utility because of their properties that contribute to flavour, lubricity, texture, and satiety to foods. The sources and the properties associated with the oils and fats as ingredients form a key role in the performance and nutritional quality of the prepared food products. Increasing consumption of oils and fat products for either general use or specific applications could be attributed to the combined efforts of oil and food processors. Although many seeds and fruits contain oil, only a few of these sources are of economic importance. In the area of oil and fat extraction, existing research has focused on improving the quality, paying particular consideration to optimizing the efficiency of extraction and refining, thereby reducing the duration of various steps and processes. Most of the developments in the processing techniques have been focused to improve the efficiency of processes to achieve higher outputs during their manufacture and to promote consumption.

Sources of oils and fats

Oils and fats for food purpose are derived either from oilseeds or from animal sources. According to global production, the requirement is increasingly dominated by palm oil and soybean oil. Palm oil is primarily used for its lower economic cost, edible properties, and easy availability. Soybean oil is the second extensively manufactured and utilized oil in the world. Expansion in the palm oil as well as soybean oil production may also be due to their use as a source material for manufacturing biofuels. Some of the oils produced worldwide in majority are palm oil, soybean oil, rapeseed or canola oil, and sunflower oil (Verhé, 2004). Minor oils produced commonly are cottonseed, groundnut, olive, and corn oil, palm kernel oil and coconut oil. Sesame oil, linseed oil, and castor oil are few of the oils that are produced at low levels. Animal fats such as tallow or lard, and marine oils like fish oil are generally heat rendered or boiled in water to separate them from proteins and the tissues. Rendering may be accomplished using steam or by dry heat.

Vegetable oils and fats are obtained through different means of extraction from their source materials. Traditionally, hydraulic presses were used for mechanical expression of oil, however they are lesser efficient as some portion of oil is left in the cake or meal after extraction. Solvent extraction has largely replaced these methods as it provides improved oil recovery. Using a suitable solvent (hexane), the oil is extracted from the pre-processed oilseeds or flakes followed by solvent striping. The recovered solvent is reused for further oil extraction, while the finished oil shows no solvent residues after processing.

Most of the vegetable oil sources are not only cultivated as for their oil, but also for providing protein-rich foods. Oil from olives or coconut is extracted from the fruit pulp rather than the seed of the fruit. Palm also has seeds, which provide palm kernel oil. Most of the oil-bearing tree fruits and kernels provide the highest oil yields. Olive oil production is almost entirely from Mediterranean countries. It has a high reputation as healthy oil based on its fatty acid composition, its minor components, and its method of extraction. Soybeans are grown more than any other oilseed, specifically crushed to meet the demands for meal. Corn oil and cottonseed oil are the by-products of corn milling and cotton industry, respectively. Rice is also a valuable crop and widely consumed in Asian countries like Japan, India, Korea, China, and Indonesia (Ghosh, 2007). Rice bran is a by-product of rice milling industry and is processed to produce rice bran oil. Rice bran contains high proportion of unsaturated fatty acids (oleic acid and linoleic acid) along with multiple dietary phytochemicals such as tocopherols, squalene, phytosterols, polyphenols, and γ-oryzanol (Patel & Naik, 2004).

Marine oils are usually by-products of fish that are procured for their processing as a protein source. However, some fish are caught with the purpose of oil production only. Production of animal fats is dependent on the meat consumption. Similarly, milk fat production such as butter is dependent on the milk production. A shift from animal fats to vegetable oils began early primarily due to animal fat supply shortages and new process developments that helped gain popularity of vegetable oils. In view of the high demand for oils and fats at present for food and non-food purposes, the surge in production of these sources is obvious. This may be achieved and further improved by seed breeding to give higher yields, by developing seeds, which are more drought-resistant and give better yields. Rapeseed or mustard oil is characterised by high levels of erucic acid. For food purposes high-erucic oils has been replaced by low-erucic acid varieties and designated as "canola oil". Olive oil is extracted from the fleshy mesocarp of olives by means of pressing. The first pressings are of highest quality and are designated as virgin oils. This extraction procedure avoids high temperature and requires no solvent for extraction. Since the oil is not subjected to refining, it retains important minor components that add to its nutritional value.

Technologies of oil extraction

Preparation or pre-processing of raw materials

The production method of oil depends on the raw material. The only raw materials that can be stored without quality deterioration for a long period are oilseeds. Oilseeds can be stored and transported easily. Oilseed pre-treatment prior to oil extraction normally affects oil yield and quality (Olaniyan, 2010). As a pre-treatment, oilseeds are cleaned, possibly dehulled, reduced in size, and heat-treated. The oil from oil fruits can be press-extracted, while oilseeds can be mechanically pressed, solvent extracted, or by the combination of both. The press and extraction meal from seed oil production is usually used as fodder. The situation for oil fruit delivering pulp oils is very different. As soon as the oil fruit is harvested enzymatic degradation reactions start, and the oil quality deteriorates substantially and quite rapidly. Therefore, as a pre-treatment, palm fruit is cooked and sterilized, and the fruits from the bunch stalk are separated. Similarly, milk fat needs to be processed immediately after milk production, as milk is highly susceptible to spoilage. Oil extraction from oilseed starts only after the seeds have been subjected to pre-treatments, or seed preparation, depending upon the type of oil seed.

Cleaning, dehulling, and size reduction

During harvestation, collection, and transportation of oilseeds, several unwanted materials may be added unintentionally with the seeds as impurities. These unwanted materials include stems, leaves, stones, sand, dirt, and weed seeds. The seed preparation process starts with the cleaning of the incoming oilseeds by removal of all such unwanted materials. Hulls, coverings, and shells are abrasive and add to the bulk entering the extraction processes. Removing hulls reduces wear and tear of the pressing equipments and increases the oil yield capacity. Bar or disk hullers may be used for medium-sized seeds such as sunflower or cottonseed, while hot aspiration systems are used for soybeans. Cracking machines are used for larger oilseeds. Seeds enveloped in thick, hard shells are dehulled by various devices that crack the hulls and then separate the hulls by screening or aspirations. The objective of the above procedures is to remove as much hull as possible with minimal loss of meats.

Oil extraction from oil seeds is facilitated by reduction of the seeds to smaller particles. During this operation, the oil-bearing material is broken into smaller pieces manually or by mechanical grinding or milling. Grinding or crushing of oilseeds prior to extraction is to ensure that oil-bearing minute cells embedded in fibrous structures are broken or ruptured to release the oil (Akpan et al., 2006; Tayde et al., 2011). Hammer mill or attrition mill may be used for initial reduction of large oil seeds such as copra or palm kernel; however, milling rolls are used as concluding size reduction step.

Flaking, cooking, and drying

Oilseed meats are flattened into thin flakes for easier extraction of oil during various extraction processes. Flaking helps to reduce power consumption of a screw press and ensures a uniform heating or cooking of all oilseed particles. Flaking can be achieved by using heavy steel rolls that revolve in opposite directions so that the rolls pull the meats between them. Speed differential makes one roll to revolve faster than the other roll, which helps the smearing action of the rolls. The meats are flattened into thin flakes as they pass through the rolls.

Seed cooking or conditioning is important to achieve high oil yields. This pre-treatment becomes necessary for most of the oilseeds, particularly for the seeds that are too soft to endure high pressures generated within a screw press. Heat treatment causes coagulation of proteins, resulting in break-up of inter-cellular emulsion, which in turn increases the oil extractability. Live steam is added into the heating chamber to elevate the oilseed moisture and provide a humid atmosphere within the cooker. The oilseeds are then heated to a temperature around 98-100°C for about 15-20 minutes, in a sealed vessel or chamber that prevents any loss of moisture or drop in the temperature. This heating process reduces the oil viscosity for the quick release of oil and allows the oil to flow more readily out of the screw press. At the same time, high temperature serves to inactivate and destroy most of the enzymes, microorganisms, and toxins that may otherwise adversely affect the oil or meal quality. Cooking enables to bind gossypol to the cottonseed protein (Dunning and Turstage, 1952), ensuring light-coloured oil. Hence, cooking, when done properly, serves to sterilize the oilseeds by destroying the harmful factors present in the oilseeds.

In an alternate method of conditioning, a high-temperature short-time (HTST) treatment is given (Eggers, 2004) to the oilseeds prior to pressing. The oilseeds are treated with live steam for ~20 sec, which raises the temperatures between 120°C and 145°C. The treatment is carried out with the aim of enriching the oil with valuable phenolic compounds along with deactivation of enzymes. Different valuable compounds are released from the cell structure by the hydrolysis effect of steam pretreatment, which enriches the oil during pressing (Eggers, 2004).

The oilseeds are finally dried to lower the moisture content depending on the suitability for pressing. The moisture content is reduced to around 2-3% for more efficient pressing. Cooking and drying may be carried out in the same vessel or in separate vessels.

Development of extrusion as a means of material preparation

Based on the technology of a screw press, Zies and Baer (1954) developed equipment called expanding extruder (Zies, 1963). An expanding extruder consists of a worm shaft, rotating rapidly within a cylindrical barrel. Steam is injected into the barrel, which raises the moisture and temperature and softens the material. The oilseed, at high moisture and temperature, is subjected to high-pressure (1379-4137 kPa) for few seconds before discharged from the expander. When the crushed oilseed emerges from the expander, some absorbed moisture flashes, which inflates the particles giving a porous and sponge-like internal structure. Oilseeds enter one end of the barrel, which are forced out through a die plate at the discharge end.

Extrusion cooking has been found to be highly effective for the inactivation of enzymes in the raw materials. It has many merits over the previous methods used for the preparation of oilseeds. Cooking is achieved within a very short span of time (~20 seconds) and activity of troublesome enzymes such as lipase in rice bran and urease in soybean can be checked. Lusas and Wathius (1988) found that extrusion is also suitable for inactivating aflatoxins in peanut meal. The first oil-preparation application for extrusion was for rice bran (Williams and Baer, 1965; Baer et al., 1966). Rice bran is obtained as a fine powder, which makes it difficult to handle for solvent extraction plant. The powdered rice bran also contains an enzyme lipase, which splits triglycerides into free fatty acids. As soon as bran is obtained during rice milling, lipase enzyme is activated. The cooking conditions within an expander inactivate lipase and convert the bran into porous, sponge-like particles called 'collets' that permit rapid flow of solvent. Similarly, canola contains enzymes that release phosphorus compounds into the oil. During traditional cooking, phosphatides are already released into the oil before the temperature reaches high enough to inactivate the enzymes. On the contrary, an expander brings the canola to full temperature in short span of time, leading to considerable reduction of phosphorus compounds, reduced chlorophyll, free fatty acid, peroxide value, and colour pigments in the oil.

Williams et al., (1977) used extrusion as a process for cooking and puffing of cereal grains for the pet food industry. Later, it was used for the preparation of oilseeds to improve extraction of fats for the oil industry (Williams, 1993). In 1970, this expander was used to extrude flaked soybean and other oilseeds into porous 'collets' (Williams, 1995).

The same procedures also transform flaked soybean and flaked cottonseed into porous collets that handle better in an extractor. Extrusion is beneficial because it converts poor quality flakes into easily extracted collets. Even compared to good flakes, extruded collets offer great advantages because collets are larger, heavier, stronger, and more porous than flakes. This permits faster solvent flow (because of larger particles), greater extractor capacity (because the heavier, more porous collets occupy less space), and better drainage.

Some oilseeds are mechanically crushed and then solvent-extracted. Expanding extruder is also used to prepare oilseeds for mechanical crushing. High shear crushes the oilseeds and rupture the oil cells. Heat generated by the friction raises the temperature, which causes drying by rapid moisture reduction as the material flows out of the extruder. The partially deoiled "cake" from the screw press is sent to a solvent extractor. Simultaneously, expanding extruder converts the press cake into porous collets.

Agglomeration (The ALCON process)

This process was the first commercially applied rapid cooking method for inactivating phospholipases in soybean flakes. It is a unique heat treatment of soybean between the flaking and solvent extraction step. In the treatment (Fig. 1), steam is injected to raise the water content of soybean flakes and simultaneously rapidly heated to inactivate the enzymes. The wetted flakes agglomerate under stirring, and the agglomerates are dried and cooled to extraction temperature. Intensive agitation provides rapid heating of the flakes and the required temperature is achieved uniformly in a few minutes. The heated flakes are discharged into a cooling chamber before the hexane extraction. The oil obtained after Alcon treatment is free of non-hydratable phosphatides. The method provides number of advantages such as higher bulk density compared to conventional flakes; higher percolation rate; reduction of hexane retention in the meal; and increased proportion of hydratable lecithin, leading to higher lecithin yield (Penk 1980). The refining costs are greatly reduced, yet the disadvantage is lower yield of meal.

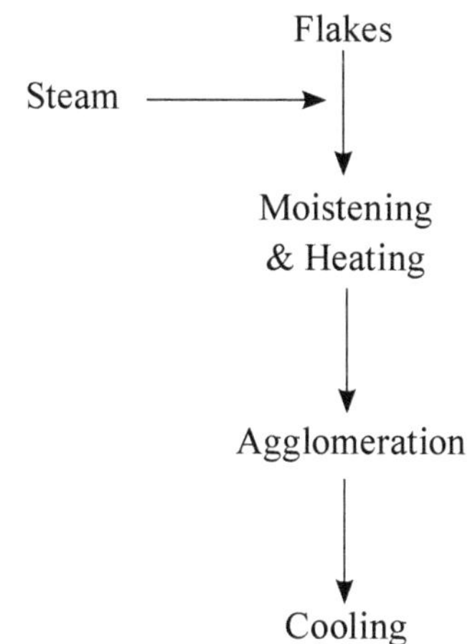

Fig. 1. Schematic presentation of Alcon Process

Oil extraction

Oil extraction is a process of obtaining oil from the prepared source material such as oilseeds (cottonseed, sesame seed, soybeans, groundnuts, etc.), fleshy fruits (olive and palm), or animal by-products. Oils or fats can be extracted by three general methods: mechanical pressing, for oil-bearing seeds and nuts; extraction using volatile chemical solvents, for a more complete extraction than possible with pressing; and rendering, used with oleaginous fruits and animal products. The extraction processes can be reviewed as five basic operations that evolved with time and later developed further with the advancement in technology. They are hot water floatation, ghanis, manual pressing, powered expellers, and solvent extraction. There are number of other methods, which can be used for the extraction of oils; however, these methods are still not as popular or widely successful as the conventional methods. They include, supercritical fluid extraction, distillation using water or steam (or a combination of both), use of microwaves, etc.

Traditional methods

Many traditional methods are still used to extract oil from oil-bearing materials. Such systems include the use of heavy stones, wedges, levers, and twisted ropes to apply pressure to the material to squeeze out the oil. However, they are inefficient, perform at low capacity, and labour intensive. Hence, their utility has lowered down due to the availability of more efficient and economically viable methods. Hot water floatation is probably the simplest method and may be used in many rural areas of economically poor countries. The ground material is placed in boiling water and heated for several hours. On removing from the fire and cooling, the oil floats to the surface and is skimmed off. The oil is then heated in a shallow pan to drive off the last traces of water. This improves the stability of oil as water contributes to the development of rancidity in oils. The extraction efficiency is generally low and problems often occur with the formation of oil–water emulsions, which make the final separation difficult. In some cases, salt is used to break such emulsions.

In many parts of India, especially in the rural areas, oil is extracted from the oilseeds using a large rotating pestle and mortar system, called 'ghani' which is powered by an animal or a motor, although sometimes human power may also be used. The oilseeds as well the oil extracted out are held in the middle of a circular mortar, made of wood or stone. The mortar is firmly fixed in the ground and as the pestle rotates, oil is released by friction and pressure and runs out of a small aperture at the base of the mortar. A ghani has a life of 4-5 years, after which the pit becomes worn out. A typical bullock driven ghani can process ~40 kg of the material/day. In the case of motor driven or power-ghani, either the pestle or mortar is fixed, while the other is rotating. Power ghani is usually operated in pairs and has a capacity of processing ~100 kg of oilseeds per day (Achaya, 1993).

Conventional methods

The conventional methods are well-known and widely practiced methods of oil extraction namely, mechanical extraction and solvent extraction. Many oils are extracted by any of the two methods, or a combination of the two.

Mechanical extraction

Mechanical expression involves the application of pressure (using hydraulic or screw presses) to force oil out of an oil-bearing material (Arisanu, 2013). In a pressing unit, the oilseeds are subjected to extremely high pressure and temperature so that oil is mechanically forced out from the oil cells. Although, a large number of variations may occur in the pressing operation, however a typical pressing operation may consist of cooking, pressing, cake cooling and finishing, and oil filtration. Oil yield showed a linear increase with the increase in pressure, duration of pressing and increase in the bulk temperature in a ground soybean sample with the oil yield showing a peak value at about 75 °C, (Mwithiga and Moriasi, 2007).

The pressing methods are of two types: Batch pressing and continuous pressing. Batch pressing is an old process for extraction of oil, in which pressure is applied to small batches of oil-bearing materials confined in bags, clothes or cages. It is also termed as hydraulic pressing. Continuous expellers or screw press are now used to the almost complete exclusion of hydraulic presses for the mechanical extraction of most of the oilseeds like soybean, cottonseed, flaxseed and groundnut. Screw presses serve two applications for oilseeds: full pressing and pre-pressing. Full pressing generates very high pressure (~96,500 kPa) to squeeze out as much oil as possible to produce meal with very low amounts residual oil (~3-5%). Similarly, animal materials can be full-pressed to reduce to 7-10% residual oil or fat (Gunstone and Padley, 1997). Pre-pressing, in turn is more preferred and used to de-oil the seeds partially for further solvent extraction. A prepress has more advantages than a full-press as it generates lesser pressure than a full press, consumes less energy per kilogram of oilseed, and runs at higher capacity. However, higher level of oil is retained in the solids, which is 15-30%, as compared to 3-5% in full pressing. This is not a demerit of the process as it is used in conjunction with solvent extraction.

Mechanical presses (manual or powered) meant for small scale oil extraction are simple, safer and containing fewer steps compared to solvent extraction of vegetable oils (Oyinlola *et al.*, 2004).With regard to oil yield, screw presses have an advantage over hydraulic presses for churning out slightly higher yields, in addition to their continuous mode of operation (Arisanu, 2013). On the industrial scale, industrial machines or expellers are used for extracting vegetable oils mechanically. Mechanical pressing is frequently used for extraction of vegetable oil from oilseeds having oil content higher than 20% (Sinha *et al.*, 2015). Two types of mechanical pressing methods are, cold-pressing and hot-pressing method. Cold-press or scarification method is carried out at low temperature (below 50°C) and pressure, whereas the hot-press method is carried out at elevated temperature and pressure. Hot-press method increases the yield of oil due to decreased seed oil viscosity at high temperatures. This enhances oil flow during extraction. Thus, high temperature increases the efficiency of the extraction process and yields of up to 80% of available oil in seed are possible (Patel *et al.*, 2016), but they may also engender oil degradation, with attendant deterioration of oil quality.

Seed oils produced though cold-pressing tend to be better than those from hot-pressing as adverse effects caused by high temperatures are avoided in the former. Some of the likely adverse effects are decreased oxidative stability, degradation of valuable oil components and reduced oil keeping quality. In cold-pressed oils, the purity and natural characteristics of the oil remain preserved (Azadmard- Damirchi *et al.*, 2011; Bhatol, 2013).

Generally, these methods have the advantage of low operation cost, and of producing high-quality, light-coloured oil with low concentration of free fatty acids (Carr, 1976; Kirkothmer, 1979). However, the yields are low compared to solvent extraction and are therefore comparatively lesser efficient, often due to a larger portion of oil left in the cake or meal after extraction (Buenrostro and Lopez-Munguia, 1986; Anderson, 1996). In addition, pressing is time consuming and labour intensive (Bhuiya *et al.*, 2015). In castor oil extraction, mechanical pressing removes only about 45% of the oil, while the remaining oil is obtained by solvent extraction from the meal (Ogunniyi, 2006).

Except for olive oil or similar materials, pressing has lost its significance. It is applied only as a pre-pressing step prior to solvent extraction and has been completely replaced by direct solvent extraction. Oil obtained by pressing is considered more natural and superior to solvent-extracted oil but most commodity vegetable oils, other than palm and olive oil, are solvent extracted.

Solvent extraction

Solvent extraction method is a conventional extraction method commonly applied to oilseeds with oil content lesser than 20%. There are three main steps in solvent extraction: seed preparation (flaking or extrusion into collets), oil extraction, and finally desolventization of the oil and meal. The prepared flakes or collets are extracted in an extractor using a solvent. The oil and solvent mixture or the 'miscella' is collected and subjected to desolventization. The oil is collected separately while the solvent is condensed and stored for reuse. Finally, desolventization of the meal is carried out. There are different types of extractors, including rotary or deep-bed, horizontal belt, and continuous loop extractors. The solids may be immersed in the solvent (immersion extraction) or the solvent may percolate through the solids (percolation extraction), or a combination of both immersion and percolation may also be used.

Extraction of oil from oil seeds by using an appropriate solvent is an effective method of oil extraction. This method is one of the most efficient methods in vegetable oil extraction, with lesser amount of residual oil left in the cake or meal (Buenrostro and Lopez-Munguia, 1986; Tayde et al., 2011). The choice of solvent is based mainly on the maximum leaching characteristics of the desired solute substrate (Dutta et al., 2015). Other considerations are high solvent-solute ratio, relative volatility of solvent to oil; oil viscosity, and cost and market availability (Muzenda et al., 2012; Takadas and Doker, 2017). Solvents commonly used are hexane, ethanol, diethyl ether and petroleum ether. Hexane remains to be the preferred solvent as it is a clear hydrocarbon, fully miscible with oil and immiscible with water, and does not impart any objectionable odour to the oil or meal. However, hexane is inflammable and presents a safety hazard.

Solvent extraction may also be carried out to obtain fat free residue or residue in which proteins are not heat denatured. Solvent extracted meal is preferred for the manufacture of protein concentrates and isolates, adhesive, fibres, or plastics. Solvent extraction method offers improved advantages due to higher yields and consistent performance. Other advantages are its repeatability and reproducibility. Although highly efficient, this method has disadvantages such as high investment, higher energy requirements, higher costs of production, and long extraction time than mechanical pressing methods (Buenrostro and Lopez-Munguia, 1986; Del Valle and Aguilera, 1999). In addition, there are concerns about high solvent consumption, plant security problems, emission of volatile organic compounds into the atmosphere, relatively high number of processing steps, etc., (Dawidowicz et al., 2008; Takadas and Doker, 2017).

Innovative or alternate extraction techniques

Environmental concerns and the possible short supply of petroleum-based solvents may have spurred the researchers for alternate techniques of extraction processes.

Conventional methods for oil extraction are time consuming, in addition of being thermal and environment safety hazards. These shortcomings can be overcome by the use of alternative or innovative methods commonly termed as 'Green Techniques' and are preferred over the conventional methods. Some of the commonly called methods are such as supercritical fluid extraction, microwave-assisted extraction, ultrasonic-assisted extraction, etc. (Bampouli et al., 2014).

Supercritical and subcritical fluid extraction

Supercritical fluid extraction or Supercritical carbon dioxide extraction is based on a relatively simple processing idea with however a technically sophisticated procedure. This extraction process takes advantage of the principle that if gases are compressed and become supercritical, their properties change and some of their parameters become similar to those of liquids. Advantages of such a process are the mild temperature treatment, free from oxidation during drying, solvent free residues, and zero flammability risks. Supercritical carbon dioxide at temperatures of ~40°C and at a pressure of 300-700 bar shows solvent characteristics like those of liquid solvent as it dissolves neutral lipids but leaves behind the polar lipids (Azmir et al., 2013). Extractions are performed by circulating supercritical carbon dioxide through a pressurized cell, filled with the oilseed. The viscosity of supercritical fluids is much lower than that of ordinary fluids, offering great advantages. Oil extraction from seeds can be conveniently carried out by this process as the solvent removal is much easier and residual carbon dioxide is a natural component in air and in foods. However, the disadvantages of the process are high investment costs, difficulty in continuous application and high-energy requirements.

Higher quality soybean oil can be obtained using supercritical carbon dioxide extraction as compared to the oil obtained using hexane (Friedrich et al, 1984). Other oilseeds studied for supercritical fluids extraction are cottonseed (List, 1984a), wet-milled and dry-milled com germ (List, 1984b), and rice bran (Zhao, 1987). In general, extraction with supercritical carbon dioxide provides oil extraction without carryover of undesirable lipids and other components. Usually, such oils have a light colour and do not need refining. These advantages might compensate for the lower yield compared with hexane extraction.

Supercritical carbon dioxide extraction has also proven useful in the development of reduced cholesterol foods or cholesterol-free foods, in which majority or a part of the cholesterol is separated by treatment using cholesterol-degrading enzymes, differential extraction, or supercritical fluid extraction. This method of extraction has been useful in various food products like eggs, fish, beef, and milk fat (Sieber, 1993). Presently, supercritical fluid extraction is used for high value materials such in decaffeination of coffee and tea, debittering of hops and in the production of natural flavour extracts.

In contrast to supercritical carbon dioxide extraction, which extracts only non-polar compounds, subcritical carbon dioxide extraction (SCO_2) involves extraction of plant oils below the critical pressure (7290 kPa) and critical temperature (31.1°C) of carbon dioxide. The carbon dioxide is maintained in a liquid state during subcritical conditions, which allows the non-polar and polar components to be extracted out of the seed or plant tissues (Tunna et al., 2018). Since, adoption of supercritical carbon dioxide extraction in the edible oil industry is limited to few products, probably due to high operating pressures, the relatively lower extraction temperature and pressure in SCO_2 may prove to be more economical alternative than the former.

Microwave-assisted extraction

Microwave assisted extraction is another novel technique owing to its unique heating mechanism, reasonable cost, and good performance under atmospheric conditions. It is based on the irradiation of the material using microwaves that results in rapid internal heating of polar solvent molecules (Kappe, 2006). Extractions are conducted in closed or open vessels where the sample and solvent are combined and then exposed to microwave energy. The extraction time is significantly reduced because with microwaves the sample/solvent mixture is directly heated. Microwaves are electromagnetic and non-ionizing radiations having a frequency range between 300 and 300,000 MHz that causes molecular motion. The magnetic field produces the direct action of waves on the material, which is able to absorb a component of the electromagnetic energy and to transform it into heat. Microwave assisted extraction is based on heating the solvent through absorption of microwave energy by polar molecules, which increases the solvent infiltration into the sample (Duarte, 2014). Microwave assisted extraction has been helpful for extracting organic compounds as it saves solvent and is quick and capable in terms of energy use. Microwave assisted extraction technique has been utilized for oil extraction from a wide variety of oilseeds including soybean, castor, peanut, canola, olive, sunflower, hazelnuts, rapeseed etc. (Mgudu et al., 2012). Its field of use for extraction of vegetable oils from oilseeds is narrow, however has been applied at smaller scale for the extraction of flavours and essential oils (Ramanadhan, 2005; Rassem et al., 2016). The technique allows for better retention and availability of desirable nutraceuticals such as phytosterols, tocopherols, and phenolic compounds in the extracted oil. It therefore represents a new step forward for the production of nutritional vegetable oils with improved shelf life because of high antioxidant content and properties.

Hence, microwave assisted extraction provides several advantages such as improved stability towards oxidation of the product due to increase in phenolic antioxidants, better extraction rates, high oil yields and good oil quality, reduced extraction time and reduced energy consumption costs, and reduced solvent consumption. The disadvantage of microwave assisted extraction is that it may not always be suitable for plants, since high microwave energy disrupts plant structure Additionally, microwaves would degrade the polyunsaturated fatty acids in vegetable oil, resulting in unrepresentative fatty acid profile (Uquiche et al., 2008).

Ultrasonic-assisted extraction

Ultrasonic-assisted extraction has gained interest in the food processing industry owing to its ability to facilitate extraction of components like oils, proteins and polysaccharides. Ultrasound is a form of mechanical energy that creates compression and expansion cycles when passing through the liquid medium. This innovative technique makes use of ultrasonic sound waves to increase vibration and heat, resulting in the destruction of rigid plant cell walls, thereby enhancing contact between the solvent and plant materials (Takadas and Doker, 2017). It is an innovative method of increasing extracted oil yield by making plant cell walls thinner, and thus enhancing the interaction of the solvent. The major advantages of the application of ultrasound are the minimum effects on extractable compounds, reduction or avoidance of organic solvents, and a reduction in extraction time (Vilkhu et al., 2008). Ultrasonic-assisted extraction has the capability for its utility in oil extraction processes to improve efficiency and reduce the process time, which may have a significant impact on edible oil industry. Ultrasonic waves provide a greater solvent penetration into the sample matrix, increasing the contact between the sample and the solvent (or reagent) and improving the mass transfer rates. In addition, this technique is also useful for extracting compounds from living organisms since it promotes the breaking of biological cell walls. This technique serves many advantages like- use of low quantities of solvent, reduction of working times, and increase in yield and quality of extract. Moreover, it is also inexpensive, fast, and versatile compared to traditional techniques, since it can use several solvents of different polarities. However, ultrasonic-assisted extraction has some drawbacks, including difficulties in combination with other instruments and automation (Pico, 2013).

The good performance of these innovative techniques has encouraged researchers to explore the prospects of combining some of them, with the aim of synergizing oil extraction. Ultrasound-assisted aqueous extraction is a step further than the simplicity of ultrasonic extraction, which can be carried out using an ultrasonic transducer and an ultrasonic water bath. The mechanical effects of ultrasound create a greater penetration of solvent into the cellular structure of plant and enhance mass transfer. This in turn can facilitate polar and non-polar components leaching from the cellular structure (Azmir et al., 2013). Advantages of the ultrasonic-assisted extraction process are reduction in extraction time (Stanisavljevic et al., 2007), increased extraction yield (Takadas and Doker, 2017), lower energy consumption and being eco-friendly (Tian et al., 2013; Hashemi et al., 2015). Ultrasonic waves can thus provide a viable alternative to many conventional and costlier commercial oil production processes.

Enzyme-assisted extraction

Enzyme assisted aqueous extraction or aqueous enzyme extraction involves water as a medium to extract oil from oilseeds using additional enzymes, which hydrolyze the vegetable cell walls and release the oil into the aqueous medium. The enzymes commonly used for oil extraction, are cellulase, α-amylase, pectinase and protease. Plant cell walls consist of interlinked polymer network, comprising proteins and carbohydrates such as pectin, cellulose, hemicellulose, and starch. Commercial exogenous or food-grade enzymes can be utilized to break down the cell walls and to liberate oil from the oil-bearing cells. As the lipid molecules are amphipathic in nature, the water-soluble components diffuse into water and end up forming an emulsion (Li et al, 2014). The oil in water emulsion can be broken by altering the temperature or using enzymes. Hence, in the process of aqueous enzyme extraction, enzymes are involved which separate the desired extracted constituents. Therefore, this technique has revealed very promising applications in the coming future and its contribution to green extraction technologies.

In the aqueous enzyme extraction, the basic material is initially subjected to a pre-processing that consists of dilution, grinding, and heating for inactivation of the endogenous enzymes. The mix is shifted to a reactor, the exogenous enzyme is added to the substrate and maintained under optimum incubation conditions followed by centrifugation and separation of the oil.

A number of factors affect the recovery of oil from oilseeds. Usage of enzymes either on their own or along with other techniques is achievable depending on the type and structure of raw material, experimental characteristics, composition and type of enzyme, etc. For example, heat-treated soybean flour when treated separately with different enzymes (cellulase, pectinase, hemicellulase or protease) resulted in higher yield with protease alone than rest of the enzymes (Wu et al, 2009). Similarly, rapeseed predominantly containing pectin in the cell walls when treated by pectinase showed 85.9% increase in the oil yield (Lamsal et al, 2006). Tano-Debrah & Ohta (1997) and McGlone et al. (1986) investigated aqueous enzyme extraction of coconut oil from fresh or dried coconut meat with added enzymes or combinations of enzymes to achieve higher oil yields

The main advantage of this method is that resulting oil is phospholipid-free and consequently has better clarity. Somehow, due to high cost of commercial enzymes, the industrial relevance of this process becomes limited (Karlovic et al., 1994; Shende and Sidhu, 2014). Still, there are many other advantages for industrial use of oil extraction with enzymes, such as low environmental impact, significant reduction of energy consumption, and quality improvement of the final product. Moreover, the process is environment friendly, safer, healthier and extraction of oil and proteins can be done simultaneously without compromising the quality. In addition, it is cost-effective as consumption of solvent is reduced and is effective in removal of anti-nutritional factors, toxins and avoids degumming process (Latif et al, 2011; Chabrand and Glatz, 2009; Yang et al, 2011). These several merits make aqueous enzyme extraction a promising green technique not only for oilseed processing but also to extract the desired compound.

Table 1: Merits and Demerits of Extraction methods

Method	Merits	Demerits
Traditional methods		
	Simple and economical.	Time consuming, low oil recovery and high residual oil content in press cake.
Conventional Methods		
Cold pressing	Relatively economical; pre-treatments not required; absence of heat; oil is more refreshing or flavoursome	Longer extraction time, poor oil recovery and high residual oil in press cake

Method	Merits	Demerits
Mechanical expression	Simplicity of the process and equipments, the low investment cost and the high quality of the products.	Lower extraction efficiency, high residual oil in press cake.
Expeller pressed extraction	Simplicity of the process and equipments, the low investment cost and the high quality of the products.	Poor product quality due to high processing temperatures; high residual oil quantity in the cake.
Solvent extraction	Production of high quality oil and defatted meal as well as high extraction efficiency, repeatability and reproducibility.	Long extraction times, high solvent requirement, time and energy consumption, environmental and safety issues, high cost of installation and maintenance
Innovative Techniques		
Microwave-assisted extraction	Moderate investment, no provision of hazardous fumes, small amount of solvent, fast extraction time	Energy can result in lipid oxidation.
Ultrasonic-assisted extraction	High quality product, lower energy requirement, fast extraction time, inexpensive equipment.	Higher initial capital cost, large volumes of solvent, labour intensive.
Enzyme-assisted extraction	Low environmental impact, low energy consumption, high quality end product, No need of a solvent, environment friendly, lower capital investment	High cost of enzymes, Dependence on several physico-chemical factors.
Supercritical fluid extraction (SFE)	Rapid extraction time, non toxic & non flammable solvent, no residual solvent, no hazardous waste, no filtration required.	Expensive equipment, Higher technical complexity, risk of system clogging.

Solvent recovery by membrane technology

Oil separation from the solvent is achievable using membrane technology immediately after solvent extraction, to improve recovery of solvent from crude oil. A combined process of conventional distillation and membranes could be used for improving solvent (hexane) recovery (Koseoglu and Engelgau, 1990). The method comprises filtration of the micelle, using a compatible nano-filtration membrane to filter into two fractions. One would be a hexane-rich fraction, to be recycled in the oil extractor, while the other would be an oil-rich fraction. Oil-rich fraction would be subjected to second filtration stage and processed by distillation to recover the remaining hexane to maximize oil recovery (Raman, Cheryan, and Rajagopalan, 1996).

Refining

Most of thc fats or oils immediately after extraction, are not suitable for human consumption with the few exceptions such as milk fats, cold pressed oils and virgin olive oil. Freshly extracted fats or oils from oilseeds, nuts, oil fruits, fish or animals, may contain varying amounts of naturally occurring non-glyceride materials that must be removed through a series of processing steps. They may also contain some components that are inappropriate either in terms of quality or in terms of their application.

The prime objective of refining is to remove the contaminants such as free fatty acids, phosphatides, colour components or pigments like chlorophyll and carotenoids, proteinaceous materials, waxes, flavouring substances, trace metals, antioxidants, carbohydrates, moisture, and dirt (Gunstone, 2005). These impurities if present adversely influence the oil quality, thereby reducing the shelf life and consumer acceptance. Some non-glyceride materials may not however be undesirable elements, for e.g. Tocopherols and other antioxidants, contribute to the function of protecting oils from oxidation and provide vitamin E. Hence, processing must be carried out carefully and in such a way so that retention of these substances can be managed.

The main quality control parameters for refining of crude oil are the free fatty acid and phospholipid content. Free fatty acids (FFAs) are formed due to the action of enzyme (lipase), heat and moisture resulting in cleavage of ester bonds in lipids (Akoh & Min, 2002). FFAs when present can behave as pro-oxidants, and lead to initiation of oxidation mechanism in lipids (Chaiyasit et al., 2007). FFA content is one of the most significant factors that influence quality of oil. Phospholipids are major constituents of cell membranes and therefore form a part of the vegetable oils during extraction. Oils with high proportion of phospholipids are prone to changes due to oxidative rancidity and darkening, especially in the presence of air or sunlight. These changes ultimately hinder deodorisation process of refining (Dijkstra, 2009). Hence, removal of phospholipids during refining becomes a very critical step. Along with these major contaminants, minor components, such as metal ions, volatile matter should also be removed effectively to ensure edible quality of oil.

Due to wide variation in the crude oil composition, the refining process also need to be modified appropriately, depending upon the raw plant/ oilseed sources, geographical location of the source and method of oil extraction. Meat fats may contain some free fatty acids, water, and proteins that have to be separated, while crude soybean oil may contain some proportion of proteins, free fatty acids, along with phosphatides, which must be removed through subsequent processing to produce the desired quality of oil.

Refining comprises of several processes that are undertaken to transform crude oils into edible oils. There are two types of refining operations, chemical and physical refining, with the former being most common method in the industry. Recently, extensive work has been carried out to develop refining technologies, using either conventional physical/chemical processes or several unconventional processes including biological and membrane processes.

Chemical refining

In a conventional process, chemical refining consists of four basic steps, degumming, neutralization, bleaching/ clarification, and deodorization (Ortega-Nieblas & Vázquez-Moreno, 1993). These discrete steps are carried out for the separation of gums or phosphatides, FFAs, pigments, and odoriferous compounds, respectively.

Degumming

Degumming is the initial process undertaken for refining of edible oil, which removes phospholipids and some portion of trace metals along with mucilaginous substances. In an extraction plant, this stage is usually considered as a pre-processing step (Erickson, 2002; Xu and Diosady, 2004). A common method of degumming consists of hydration with water to remove of hydratable phospholipids (Dijkstra & Opstal, 1987), followed by an acid treatment for non-hydratable phospholipids. Since, majority of the phosphatides are non-hydratable, addition of organic acids such as citric or phosphoric acid, hydrate the phosphatides, and permit their subsequent separation from the oil.

During the removal of phospholipids, metals such as iron and copper are also removed thereby improving stability of the oil against oxidation. Phospholipids if present in the oil, contribute to loss of neutral oil and darkening of the oil during neutralization. Their presence also causes formation of deposits in tanks and pumping systems, causing problems during storage and transportation of crude oil. The efficiency of a chemical process depends on the nature and quantity of phosphatides present in the oil (Dijkstra, 2013). Thus, elimination of phospholipids becomes vital for the stability of oils during storage.

Deacidification or neutralization

Neutralization, also known as caustic or alkaline refining, refers to the addition of sodium hydroxide solution or any appropriate alkali to the oil to cause neutralization of the free fatty acids, batch wise, semi continuously or through continuous operating systems. In all cases, the heated oil is brought into contact with the caustic soda solution, which reacts with free fatty acids to produce soap and water. The soap produced is insoluble in oil, making the subsequent separation much easier. The soap produced during the process also adsorbs natural pigments, gums, and mucilages that were not removed during the previous step of degumming. This soap must be separated from the oil prior to further processing, which is accomplished either by simply draining it off or by centrifugation. The separated oil is then washed once or twice with water to ensure complete elimination of all traces of soap. The washed oil is vacuum dried to a moisture content of approximately 0.1% prior to bleaching.

The viscous soap stock obtained after removal from the oil, is subjected to acidulation during which it is acidified with a concentrated mineral acid, such as sulfuric acid. The acid reacts with the soap to convert it back to a combination of free fatty acids and neutral oil called 'acid oil', which is sold as a by-product for the manufacture of feed for animals.

Bleaching

Bleaching is defined as a treatment carried out for partial or complete removal of colour components or pigments to generate a light coloured or colourless oil. These pigments if present, limit the use and marketability of the oil. Bleaching involves contact of the oil with surface-active substances that adsorb the undesired particles. The adsorbents along with the adsorbed particles are filtered off to obtain oil with the desired color.

Bleaching process uses bleaching clays or charcoal to remove many minor impurities like- colour compounds, oxidation products (peroxides), trace metals, phospholipids and soaps. Hence, the oxidative stability and sensory characteristics of the treated oil are largely improved (Garcia-Moreno et al. 2013). Bleaching earths or clays used for clarification can be of two types, naturally activated clays, and acid-activated clays (by treatment with sulphuric or hydrochloric acid). Preference for the type of clay depends on the quality and clarification requirements of the oil.

Activated carbon is mostly added together with the bleaching earth. If used for the removal of colour pigments, the bleaching earth–activated carbon ratio is typically 80: 20 or 90: 10. Usual dosage for removal of polycyclic aromatic hydrocarbons from coconut oil can be five kg/tonne or higher. Coconut oil, some fish oils, and some rapeseed oils may require activated carbon treatment to lower down the quantities of heavy polycyclic aromatic hydrocarbons (Kemeny et al., 2011). Apart from its higher cost compared to bleaching earth, the major drawbacks of activated carbon are its higher oil retention and poor filtration characteristics.

Deodorization

Deodorization is the finishing step in the refining of edible oils, to remove the odour causing components and trace elements, which ultimately contributes to extension in shelf life of oil. The main objectives of deodorization are: removal of oxidized compounds with undesirable taste and odour to produce oils with better sensorial or organoleptic properties, reduction of FFA content to minimum values, destruction of peroxides for greater oil stability, and improvement of colour by destruction of thermo-sensitive pigments, such as carotenoids. The final product of this process is high quality refined oil having light colour and no odour. Deodorization is a steam distillation, semi-continuous or fully continuous process, carried out at temperature around ~200 °C and at low pressure (2.5-9.0 mbar). It is however, preferable to use temperatures below 200 °C as it affects the oil quality (Camacho et al., 2001 and Fournier et al., 2006). In deodorization or distillation, the volatile compounds of oil like free fatty acids, odoriferous compounds, tocopherols, phytosterols, along with contaminants such as pesticides and polycyclic aromatic hydrocarbons also get removed (De Greyt and Kellens, 2005; Dijkstra, 2007a, b).

Physical refining

Physical refining is characterized by the complete absence of neutralization step and the separation of free fatty acids is carried out during the distillation stage, using highly efficient equipments (Gupta, 2008; O'Brien, 2008). The physical refining process is technically more costly, yet has the advantage of being environment friendly. The effluents are drastically reduced and the refining losses are lowered down as well. This process is beneficial for refining crude palm oil and coconut oil owing to their low non-hydratable phospholipids and high FFA contents. In addition, physical refining process is suitable for the oils that have processed through intensive degumming or "super degumming" and with a phospholipid content of less than 750 ppm (parts per million). However, the process is not suitable for oilseeds due to the higher levels of phospholipids in these oils. Soybean oil can be successfully refined by physical refining process only if the phospholipid content is reduced to less than 400 ppm (Phosphorus content <10 ppm) via ultra-degumming process (Gupta, 2008). In physical refining, difference in the volatility of different components results in the removal of FFA and odoriferous compounds in a single operation (Sampaio et al., 2011). The main advantages of physical refining are reduced demand for chemicals (such as phosphoric acid) or alkaline solutions and water, reduced waste production, and higher oil yield, all of which result in improved overall economic and environmental assessment of the process. However, this process is highly sophisticated and is not applicable to all types and qualities of crude oil. New developments in deodorisation technology are driven by the continuous need for more efficient processes (lower operating cost, higher refined oil yield, and better valorisation of side streams) and the increased attention paid to the (nutritional) quality of food oils and fats.

Alternate technologies of refining

Enzymatic degumming

Most of the methods of refining intend to reduce the phospholipid content of the oil, measured as phosphorus, to below 10 ppm, which is also achievable in the recent method called enzymatic degumming. Degumming through use of enzymes has gained much importance owing to its substrate specificity and ability to carry out the reactions under mild conditions than most of the similar processes using chemicals (Mei et al., 2013). The enzymes convert the phospholipids present in the oil into diacylglycerols and in comparison, to the conventional processes, it offers relatively better output during degumming (Dijkstra, 2010). The phospholipids are eliminated to improve the shelf life and organoleptic characteristics of the oil. The process involves conversion of non-hydratable phospholipids to hydratable-phospholipids, which are easy to eliminate with the water phase by centrifugation. Its applications can be for a wide variety of oils, and the number has grown since the method was first introduced. Advances in biotechnology have allowed the development of enzymatic degumming as a sustainable purification method and able to achieve the same, if not better, results than the traditional chemical methods while greatly increasing yields and reducing chemicals, wastes, energy, and overall costs. Interest in degumming through enzymes is on the increase because of commercial availability of several new, cost efficient and stable phospholipases with sufficiently high enzyme activity.

Membrane degumming and bleaching

Among different alternate methods, degumming by means of membranes has also gained importance for being energy efficient, less complex and non-chemical based method. The technique involves use of polymeric, ultra-filtration or nano-filtration membranes for the removal of phospholipids during degumming of crude oil (Pagliero et al, 2001). The membranes are highly efficient and retain both hydratable and non-hydratable phospholipids, resulting in removal of these compounds (Subramanian et al., 2001) A surfactant-aided membrane degumming has been useful in crude soybean oil, to lower down the phosphorus content between 20-58 ppm in the degummed oil (Subramanian, 1999). The quality of oil produced using membrane technology is quite superior, however higher processing cost, less throughput of membranes, and fouling of membranes restricts the industrial use of this method at present.

Membrane technology has been used to recover solvents after degumming the micelle to obtain solvent-free degummed oil. After degumming by ultrafiltration, the oil/solvent (hexane) micelle, could be fed into a nano-filtration process to increase the extent of solvent recovery (Snape and Nakajima, 1996).

Polymeric membranes have been tested in filtration process to remove phospholipids, pigments, and oxidation products during an experimental refining of crude oil from rapeseed and soybean (Subramanian, Nakajima, and Kawakatsu, 1998). In comparison to the original tocopherols content of the crude oils, the tocopherols content of the permeates increased by 12–26%. However, the process did not remove free fatty acids present in crude oils. Hence, single step processes using membranes were only adequate for the degumming and clarification stages, and not for deacidification.

Nano neutralization

In the search for better processing technologies for reducing the use of chemicals and oil losses in soap stock, the potential of Nano Reactor® technology has also been investigated. Neutralisation is a crucial step of refining to remove FFAs, which otherwise cause development of rancid flavours and speeds up oxidation process (Chaiyasit et al., 2007). Nano Reactors® are hydrodynamic cavitation reactors. Ultrasound cavitation (created by a cavitational effect) for edible oil degumming was studied by Moulton & Mounts (1990). Although the results were promising, the process was never industrially applied due to some inherent drawbacks: non-uniform cavitational effect, very high-energy requirement, and applicability only as a batch process. Hydrodynamic Nano Reactors® require less energy and are more suitable for large-scale oil processing especially in a continuous operation. For the industrial application, nano-neutralisation has been successfully introduced in the edible oil processing (Svenson and Willits, 2012).

Enzymatic bleaching

Recent developments in refining technology consists of enzymatic bleaching in which chlorophyllase enzymes, which can operate at pH 4.5-6.0 and at 55-65°C have been identified (Mikkelsen, 2011). When added during water degumming, these enzymes degrade the chlorophyll components in soybean or canola oil to very low residual levels (<50 ppb), eliminating the use of bleaching earths (Carlson et al., 2011). This process has been successfully tested at pilot scale; however, is yet to be ready for implementation on an industrial scale.

Dry condensing deodorization

Deodorization involves removal of FFAs along with other volatile components by bringing the oil in contact with steam at high temperature and low pressure. This process involves removal of high volumes of steam vapours, which in turn results in high consumption of steam (up to 75%) in a deodorizer. Utilization of steam can be lowered down by lowering the temperature of the condense water with a chiller. The total amount of steam used in a dry condensing system is quite low compared to that used in a conventional system. Hence, the amount of steam required for deodorization is much lower, with reduction in the overall effluents, and more cost-efficient conditions due to lower operating pressures (Kellens, 2012).

Supercritical CO_2 processing

Similar to the usage in the extraction processes, supercritical CO_2 processing has been tested successfully at laboratory scale for degumming in crude soybean oil (List et al., 1993), neutralization in crude rice bran oil (Dunford and King, 2001) and for the mild refining of palm oil (Ooi, 1996). However, it has never been applied at an industrial level, because of high investment and operating costs involved, making it viable only for high-value products.

Table 2: Merits and Demerits of Refining Methods

Methods	Merits	Demerits
Chemical refining	Efficient FFA removal to desired level, highly cost effective.	High loss of neutral oil, generates high volume of waste which is expensive to treat.
Physical refining	Simple method, reduces neutral oil loss, lesser steam, water and power consumption, improved oil yield, elimination of soapstock, minimal waste effluent quantity.	Stringent pretreatment requirements, physical deacidification reduces the tocopherols and carotenoids, requires high temperature and vacuum, not suitable for all oils.
Alternative methods		
Enzymatic degumming	Highly effective, provides a relatively good yield of oil,	Oxidative instability of the oil, High cost of enzymes
Membrane Technology	An efficient process, minimise energy consumption.	Choking of membranes, decrease in flow rate with time, poor deacidification, membrane fouling due to deposition.

Methods	Merits	Demerits
Nano Neutralization	Reduction in- degumming costs, alkali usage, soap content in oil, and oil losses in soap stock. Higher oil yield, energy efficient, low environmental impact, high oil quality.	Poor applications at industrial scale
Enzymatic Bleaching	Eliminating of bleaching earth	High cost of enzymes
Supercritical fluid extraction	Fast extraction time, small amount of solvent, non toxicity, non flammability, no hazardous waste, no filtration is required.	Costly, risk of system clogging.

Future trends

From a global perspective, the opportunities for fats and oil industry will be in the development of technologies that are more environment-friendly, and more health oriented. For the industry, it will mean processing with an attempt to safeguard the beneficial effects of natural antioxidants and alleviate the harmful effects of chemical processing. Although conventional methods of refining are cost effective and yield good outcome, destruction of some of the desirable components due to harsh chemicals and high temperatures warrant further improvement in the technology. There are numerous studies signifying the beneficial effects of phytosterols and phospholipids in the diet. They are thought to lower down cholesterol and risk factors contributing to cardiovascular diseases. Though necessary, these antioxidants are removed during refining processes like- degumming, neutralization, bleaching, and deodorization. Hence, the challenge lies in to enhance the presence of these compounds in fats and oils we consume, provided, their beneficial effects are confirmed. Health concerns related to composition of foods and their related benefits to human health i.e. presence of high proportion of saturated fatty acids or trans-fatty acids in our diet, are a major concern and an area guaranteed to receive more attention for research. Concerns to diminish the environmental implications of oil refining processes have to become more significant in the future.

Advent of green technologies has certainly contributed to incorporation of more environment friendly processes that will lower down the amounts of effluent discharges into the atmosphere and water. Waste generation and its mitigation is an important concern, which will demand our increasing attention. Together with an objective of reducing processing costs and/or improving oil quality, the manufacturers and processors have to be encouraged to improve the products or the concerned technologies. Eco-friendly techniques are effective, however are usually expensive, and are seldom practiced in the industry. The future prospects would be to design sustainable integrated refining processes with the consideration of health, environment, and the cost.

References

Achaya, K.T. (1993). Ghani: The traditional oil mill of India. Kemblesville, PA: Olearius Editions.

Akoh, C.C. & Min, D.B. (2002). Food Lipids: Chemistry, Nutrition and Biotechnology, 2nd edn. Florida, USA: Marcel Dekker.

Akpan, U.G., Jimoh, A., & Mohammed, A.D. (2006). Extraction, characterization and modification of castor seed oil. *Leonardo Journal of Sciences,* 8:43-52.

Anderson, D. (1996). A primer on oils processing technology, in: *Bailey's Industrial Oil and Fat Products,* Y.H. Hui Eds. John Wiley and Sons, pp 10- 17.

Arisanu, A.O. (2013). Mechanical continuous oil expression from oil seeds: oil yield and press capacity. *International Conference "Computational Mechanics and Virtual Engineering" COMEC 2013,* 24-25 October, 2013, Brasov, Romania.

Azadmard-Damirchi, S., Alirezalu, K., & Achachlousi, B.F. (2011). Microwave pretreatment of seeds to extract high quality vegetable oil. *International Journal of Nutrition and Food Engineering,* 5(9): 508-511.

Azmir, J., Zaidul, I.S.M., Rahman, M.M., Sharif, K.M., Mohamed, A., Sahena, F., Omar, A.K.M. (2013). Techniques for extraction of bioactive compounds from plant materials: A review. *Journal of Food Engineering,* 117(4), 426–436.

Baer, S., Williams, M.A. & Zies, C.W. (1966). Pre-treatment of oleaginous plant materials, U.S. Patent 3, 255, 220 (to International Basic Economy Corp.)

Bampouli, A., Kyriakopoulou, K., Papaefstathiou, G., Lauli, V., Krokida, M., & Magoulas, K. (2014). Comparison of different extraction methods of *Pistacia lentiscus* var. Chia leaves: yield, antioxidant activity and essential oil chemical composition. *Journal of Applied Research on Medicinal and Aromatic Plants,* 1(3):81-91.

Bhatol, K. (2013). Castor Oil Obtained by Cold Press Method. Shri Bhagwati Oil Mill (SBOM) manufacturer's Info. Banaskantha, Gujarat, India.

Bhuiya, M.M.K., Rasul, M.G., Khan, M.M.K., Ashwath, N., Azad, A. K., & Mofijur, M. (2015). Optimization of oil extraction process from Australian Native Beauty Leaf Seed. (*Calophyllum innophyllum*). 7th International Conference on Applied Energy- ICAE2015. *Energy Procedia,* 75:56-61.

Buenrostro, M.; Lopez-Munguia, A.C. (1986). Enzymatic Extraction of Avocado Oil. *Biotech. Letters,* 8(7), 505–506.

Camacho, M.L., Mendez, M.V.R., Constante, M.G. & Constante, E.G. (2001). Kinetics of the cis-trans isomerisation of linoleic acid in the deodorization and/or physical refining of edible fats. *European Journal of Lipid Science and Technology,* 103, 85–92.

Carlson K., Mikelsen R. & Soe J.B. (2011). New approaches for chlorophyll removal in oil processing. Paper presented at the 102nd AOCS Annual Meeting and Expo, Cincinnati, OH, USA.

Carr, R.A. (1976). Degumming and refining practices in the U.S. *Journal of American Oil Chemists Society,* 53, 347–352.

Chabrand, R.M., Glatz, C.E. (2009) Destabilization of the emulsion formed during the enzyme-assisted aqueous extraction of oil from soybean flour. *Enzyme and Microbial Technology,* 45:28–35

Chaiyasit, W., Elias, R.J., McClements, D.J. & Decker, E.A. (2007). Role of physical structures in bulk oils on lipid oxidation. *Critical Reviews in Food Science and Nutrition,* 47, 299–317.

Dawidowichz, A.L., Rado, E., Wianowska, D., Mardarowicz, M. & Gawdzik, J. (2008). Application of PLE for the determination of essential oil components from *Thymus vulgaris* L. Talanta, 76:878- 884.

De Greyt, W.F.J., Kellens, M.J. (2005). Deodorization. In: Shahidi, F. (Ed.), Bailey's Industrial Oil and Fat Products. John Wiley & Sons, New York, pp. 341–382.

Del Valle, J.M. & Aguilera, J.M. (1999). Extraction con CO2 a attar presion. Fundamentosy aplicaciomes el na industria de alimentos. *Food Science and Technology International,* 5:1-24.

Dijkstra, A.J. & Opstal, M.V. (1987). Process for producing degummed vegetable oils and gums of high phosphatidic acid content. US Patent 4698185.

Dijkstra, A.J. (2007a). Modification processes and food uses. In: Gunstone, F.D., Harwood, J.L., Dijkstra, A.J. (Eds.), The Lipid Handbook. CRC Press, New York, NY, pp. 263–353.

Dijkstra, A.J. (2007b). Vacuum stripping of oils and fats. In: Gunstone, F.D., Harwood, J.L., Dijkstra, A.J. (Eds.), The Lipid Handbook, third ed. Taylor & Francis Group LLC, Boca Raton, FL, pp. 235–253.

Dijkstra, A.J. (2009). Recent developments in edible oil processing. *European Journal of Lipid Science and Technology*, 111, 857–864.

Dijkstra, A.J. (2010). Enzymatic Degumming. *European Journal of Lipid Science and Technology*, 112, 1178–1188.

Dijkstra, A.J. (2013). The purification of edible oils and fats. Lipid Technology, 25, 271–273.

Duarte, K., Justino, C.I.L., Gomes, A.M., Rocha-Santos, T., & Duarte A.C. (2014). Green Analytical Methodologies for Preparation of Extracts and Analysis of Bioactive Compounds. *Comprehensive Analytical Chemistry*, 65, 59-78.

Dunford, N. & King, J. (2001). Thermal gradient deacidification of crude rice bran oil utilizing supercritical carbon dioxide. *Journal of American Oil Chemists Society*, 78, 121–125.

Dunning, J. W. & Terstage, R. J. (1952). Effect of overcooking of cottonseed meats on quality of meals, *Journal of American Oil Chemists Society*, 29: 153.

Dutta, R., Sarkar, U. & Mukherjee, A. (2015). Soxhlet extraction of *Crotalaria juncea* oil using cylindrical and annular packed beds. *International Journal of Chemical Engineering and Applications*, 6(2): 130-133.

Eggers, R. (2004). Processing of oilseeds with increase of minor components. Refining: handling by-products and minor components. March 17/18, 2014, Leipzig, Germany.

Erickson, M.C. (2002). Chemistry and function of phospholipids. In: Akoh, C.C., Min, D.B. (Eds.), Food Lipids Chemistry, Nutrition, and Biotechnology. Marcel Dekker, New York, NY, pp. 59–80.

Fournier, V., Destaillats, F., Juaneda, P. et al. (2006). Thermal degradation of long chain polyunsaturated fatty acids during deodorization of fish oil. *European Journal of Lipid Science and Technology*, 108, 33–42.

Friedrich, J.P. & Pryde, E.H. (1984). Supercritical CO_2 extraction of lipid-bearing materials and characteristics of the products, *Journal of American Oil Chemists Society*, 61: 223.

Garcia-Moreno, P.J., Guadix, A., Gomez-Robledo, L., Melgosa, M. & Guadix, E.M. (2013). Optimization of bleaching conditions for sardine oil. *Journal of Food Engineering*, 116, 606–612.

Ghosh, M. (2007). Review on recent trends in rice bran oil processing. *Journal of the American Oil Chemists Society, 84*(4), 315-324.

Gunstone, F.D. & Padley, F.B. (1997). Lipid technologies and applications. Marcel Dekker, Inc. New York.

Gunstone, F.D. (2005). Vegetable oils. In: Shahidi, F. (Ed.), Bailey's Industrial Oil and Fat Products. John Wiley & Sons, New Jersey, pp. 213–267.

Gupta, M.K. (2008). Practical Guide to Vegetable Oil Processing, second ed. American Oil Chemists' Society, AOCS Press – Academic Press, London.

Hashemi, S.M.B., Michiels, J., Yousafabad, S.H.A. & Hosseini, M. (2015). Kolkhoung (*Pistacia khinjuk*) kernel oil quality is affected by different parameters in pulsed ultrasound-assisted solvent extraction. *Industrial Crops and Products*, 70:28-33.

Kappe, C.O., Dallinger, D. (2006). *Nature Reviews. Drug Discovery* 5, 51–63.

Karlovic, D.J., Bocevska, M., Jakovlevic, J., Turkulov, J. (1994). Corn germ oil extraction by a new enzymatic process. *Acta Aliment*. 23 (4), 389–400.

Kellens, M. (2012). Dry Condensing Vacuum Systems for Deodorizers for Substantial Energy Savings, Chapter 10. In: Green Vegetable Oil Processing by Walter E. Farr and Andrew Proctor, Ed 1, AOCS Press, Urbana, Illinois.

Kemeny, Z., Hellner, G., Radnoti, A. Ergomoa, T. (2011) Polycyclic aromatic hydrocarbon removal from coconut oil. Paper presented at the 9th Euro Fed Lipid conference, Rotterdam, Netherlands.

Kirk-Othmer (1979). *Encyclopaedia of Chemical Technology,* 5:1-17.

Koseoglu, S.S., & Engelgau, D.E. (1990). Membrane applications and research in edible oil industry: An assessment. *Journal of American Oil Chemist's Society*, 67, 239–249.

Kushairi, A., Loh, S.K., Azman, I., Elina, H., Meilina, O.A., Zainal, B.M.N.I., Razmah, G., Shamala, S., Parveez, G.K.A. (2018). Oil palm economic performance in Malaysia and R&D progress in 2017. *Journal of Oil Palm Research*, 30, 163–195.

Lamsal, B.P., Murphy, P.A., Johnson, L.A. (2006). Flaking and extrusion as mechanical treatments for enzyme-assisted aqueous extraction of oil from soybeans. *Journal of American Oil Chemists Society*, 83(11):973–979

Latif, S., Anwar, F., Hussain, AI., Shahid, M. (2011) Aqueous enzymatic process for oil and protein extraction from *Moringa oleifera* seed. *European Journal of Lipid Science and Technology,* 113:1012–1018

Li, Y., Fine, F., Fabiano-Tixier, A., Abert-Vian, M., Carre, P., Pages, X., Chemat, F. (2014). Evaluation of alternative solvents for improvement of oil extraction from rapeseeds. *Comptes Rendus Chimie*, 17:242–251

List, G., King, J., Johnson, J., Warner, K. & Mounts, T. (1993) Supercritical CO_2 degumming and physical refining of soybean oil. *Journal of American Oil Chemists Society*, 70, 473ff.

List, G.R., Friedrich, J.P. & Christianson, D.D. (1984). Properties and processing of com oils obtained by extraction with supercritical carbon dioxide. *Journal of American Oil Chemists Society,* 61: 1849.

List, G.R., Friedrich, J.P. & Pominski, J. (1984). Characterization and processing of cottonseed oil obtained by extraction with supercritical carbon dioxide. *Journal of American Oil Chemists Society*, 61: 1847 (1984).

Lusas, E.W. & Wathuis, L.R. (1988). Oilseeds: Extrusion for Solvent Extraction. *Journal of American Oil Chemists Society,* 65, 1109-1114.

McGlone, O.C., Lopez-Munguia, A., Carter, J.V. (1986). Coconut oil extraction by a new enzymatic process. *Journal of Food Science*, 51, 695–697.

Mei, L., Wang, L., Li, Q., Yu, J. & Xu, X. (2013). Comparison of acid degumming and enzymatic degumming process for Silybum marianum seed oil. *Journal of the Science of Food and Agriculture*, 93, 2820–2828.

Mgudu, L., Muzenda, E., Kabuba, J. & Belaid, M. (2012). Microwave assisted extraction of castor oil. *International Conference on Nanotechnology and Chemical Engineering* (ICNCS 2012), December, 21- 22, Bangkok, Thailand.

Mikkelsen, R. (2011). Biotechnological approaches to remove chlorophyll components in plant oils. Paper presented at the 102nd AOCS Annual Meeting and Expo, Cincinnati, OH, USA.

Moulton, K.J. & Mounts, T.L. (1990) Continuous ultrasonic degumming of crude soybean oil. *Journal of American Oil Chemists Society*, 69, 443–446.

Muzenda, E., Kabuba, J., Mdletye, P. & Belaid, M. (2012). Optimization of process parameters for castor oil production. *Proceedings of the World Congress on Engineering 2012 Vol. III WCE 2012,* July, 4-6, 2012, London, U.K.

Mwithiga, G. & Moriasi, L. (2007). A study of yield characteristics during mechanical oil extraction of pretreated and ground soybeans. *Journal of Applied Sciences Research,* 3(10): 1146-1151

O'Brien, R.D. (2008). Fats and Oils – Formulating and Processing for Applications. CRC Press, New York, NY.

Ogunniyi, D.S. (2006). Castor Oil: a vital industrial raw material. *Bioresource Technology*, 97:1086 – 1091.

Olaniyan, A.M. (2010). Effect of extraction conditions on the yield and quality of oil from castor bean. *Journal of Cereals and Oilseeds*, 1:24-33.
Ooi, C. (1996) Continuous supercritical carbon dioxide processing of palm oil. *Journal of American Oil Chemists Society*, 73, 233–237.
Ortega-Nieblas, M. & Vázquez-Moreno, L. (1993). Caracterización fisicoquímica del aceite crudo y refinado de la semilla de Proboscidea parviflora (Uña de gato). *Grasasy Aceites*, 44, 30–34.
Oyinlola, A., Ojo, A., & Adokoya, L.O. (2004). Development of a laboratory model screw press for peanut oil expression. *Journal of Food Engineering*, 64:221-227.
Pagliero, C., Ochoa, N., Marchese, J. & Mattea, M. (2001). *Journal of American Oil Chemists Society*, 78, 793- 796.
Patel, M. & Naik, S. N. (2004). Gamma-Oryzanol from rice bran oil-A review. *Journal of Scientific & Industrial Research, 63*(7), 569–578
Patel, V.R., Durmancas, G.G., Viswanath, L.C.K., Maples, R. & Subong, B.J.J. (2016). Castor oil: properties, uses and optimization of processing parameters in commercial production. *Lipid Insights*, 9:1-12.
Penk 1980) Penk, G. (1980). Conditioning of Soybean Flakes. Process Technique and Economical Aspects, *DGF-Vortragstagung, Kiel*. p. 36
Pico, Y. (2013). Ultrasound-assisted extraction for food and environmental samples. *Trends Analytical Chemistry*, 43, 84-99.
Raman, L. P., Cheryan, M., & Rajagopalan, N. (1996). Solvent recovery and partial deacidification of vegetable oils by membrane technology. *Fett/Lipid*, 98, 10–14.
Ramanadhan, B. (2005). Microwave extraction of essential oils (from black pepper and coriander at 2.46 Ghz. *MSc. Thesis*, University of Saskatchewan, Canada.
Rassem, H.H.A., Nour, A.H. & Yunus, R.M. (2016). Techniques for extraction of essential oils from plants: a review: *Australian Journal of Basic and Applied Sciences*, 10(16): 117-127.
Sampaio, K.A., Ceriani, R., Silva, S.M., Tahama, T., Meirelles, A.J.A. (2011). Steam deacidification of palm oil. *Food and Bioproducts Processing*, 89 (4), 383–390.
Shende, D., & Sidhu, E.K. (2014). Methods used for extraction of maize (*Zea mays*, L.) germ oil – a review. *Indian Journal of Science and Technology*, 2, 48–54.
Sieber, R. (1993). Cholesterol removal from animal food—can it be justified? *Lebensmittel Wissenschaft & Technologie*, 26: 375-387.
Sinha, L.K., Haddar, S. & Majumdar, G.C. (2015). Effect of operating parameter on mechanical expression of solvent-soaked soybean grits. *Journal of Food Science and Technology*, 52(5): 2942-2949.
Snape, J. B., & Nakajima, M. (1996). Processing of agricultural fats and oils using membrane technology. *Journal of Food Engineering*, 30, 1–41.
Stanisalvjevic, I.T., Lazic, M.L. & Veljkovic, V.B. (2007). Ultrasonic extraction of oil from tobacco (*Nicotiana tabacum* L.) seeds. *Ultrasonics Sonochemistry*, 4(5): 646-652.
Subramanian, R., Ichikawa, S., Nakajima, M., Kimura, T., & Maekawa, T. (2001). Characterization of phospholipid reverse micelles in relation to membrane processing of vegetable oils. *European Journal Lipid Science Technology*, 103, 93-97.
Subramanian, R., Nakajima, M., & Kawakatsu, T. (1998). Processing of vegetable oils using polymeric composite membranes. *Journal of Food Engineering*, 38, 41–56.
Subramanian, S., Nakajima, M., Yasui, A., Nabetani, H., Kimura, T. & Maekawa, T. (1999). *Journal of American Oil Chemists Society*, 76, 1247–1253.
Svenson, E. & Willits, J. (2012) Nano-neutralization. In: W.E. Farr & A. Proctor (eds) *Green Vegetable Oil Processing*. Champaign, IL, USA: AOCS Press, pp. 157–168.
Takadas, F. & Doker, O. (2017). Extraction method and solvent effect on safflower seed oil production. *Chemical and Process Engineering Research*, 51:9-17.
Tano-Debrah, K., Ohta, Y. (1997). Aqueous extraction of coconut oil by an enzyme-assisted process. *Journal of Science of Food and Agriculture*, 74, 497–502.

Tayde, S., Patnaik, M., Bhagt, S.L. & Renge, V.C. (2011). Epoxidation of vegetable oils: A review. *International Journal of Advanced Engineering Technology,* II (IV): 491-501.
Tian, Y., Xu, Z., Zhang, B. & Lo, Y.M. (2013). Optimization of ultrasonic-assisted extraction of pomegranate (*Punicagranatum* L) seed oil. *Ultrasonics Somochemistry,* 20(1): 202-208.
Tunna, T.S., Sarker, M.Z.I., Ghafoor,K., Ferdosh,S., Jaffri, J.M., Al-Juhaimi, F.Y., Selamat, J. (2018). Enrichment, in vitro, and quantification study of antidiabetic compounds from neglected weed *Mimosa pudica* using supercritical CO_2 and CO_2- Soxhlet. *Separation Science and Technology*, 53(2), 243–260.
Uquiche, E., Jerez, M. & Ortiz, J. (2008). Effect of treatment with microwaves on mechanical extraction yield and quality of vegetable oil from Chilean hazelnuts (*Gevuina avellana* Mol). *Innovative Food Science and Emerging Technologies,* 9(4): 495 – 500.
Verhé R.G. (2004). Chapter 9. Industrial products from lipids and proteins, In: *Renewable Bioresources: Scope and Modification for Non-Food Applications* (Eds. Stevens C.V. and Verhé R.G.), John Wiley & Sons, Ltd, Chichester, pp. 208 – 250.
Vilkhu, K., Mawson, R., Simons, L., Bates, D. (2008). Applications and opportunities for ultrasound assisted extraction in the food industry - a review. *Innovative Food Science and Emerging Technology,* 9, 161–169.
Williams M.A. & Baer, S. (1965). The expansion and extraction of rice bran, *Journal of American Oil Chemists Society*, 42: 151 (1965).
Williams, M.A. (1993). Preparation of oilseeds to improve extraction of fats. *Extrusion Communication*, 6: 12
Williams, M.A. (1995). Extrusion preparation for oil extraction, *Inform* 6: 289.
Williams, M.A. (1997). Extraction of Lipids from Natural Sources Chapter 5: From Gunstone, F. D. and Padley, F. B. (1997). Lipid Technologies and Applications. Marcel and Decker Publications.
Williams, M.A., Horn, R.E. & Rugala, R.P. (1977). Extrusion, extrusion, extrusion. *Journal of Food Engineering,* 49: 99.
Wu, J., Johnson, L.A., Jung, S. (2009). Demulsification of oil-rich emulsion from enzyme-assisted aqueous extraction of extruded soybean flakes. *Bioresource Technology*, 100:527–533
Xu, L., Diosady, L.L. (2004). Nutritionally Enhanced Edible Oil Processing. AOCS Publishing, New York, NY.
Yang, L., Jiang, L., Sui, X., Wang, S. (2011). Optimization of the aqueous enzymatic extraction of pie kernel oil by response surface methodology. *Procedia Engineering,* 15:4641–4652
Zhao, W., Shishikura, A., Fujimoto, K. Arai, K. & Saito, S. (1987). Fractional extraction of rice bran oil with supercritical carbon dioxide. *Agriculture and Biological Chemistry,* 51: 1773.
Zies, C. W. (1963). Apparatus for the preparation of food compounds, *U.S. Patent* 3,108,530 (to International Basic Economy Corp).

9

Chronological Developments in the Technology of Weaning and Geriatric Foods

Sangeeta Pandey

Department of Nutrition & Dietetics, Mount Carmel College, 58, Palace Road Bengaluru, Karnataka

Weaning foods

Introduction

Breast milk is natural and best for the growth and development of newborn babies. Although breast milk is adequate to provide an optimum nutrient till the age of six months but after that it is not sufficient to sustain optimum growth and fulfill their nutritional requirements. Various nutrients are required to supplement milk until child is ready to eat adult meal. This additional food along with the breast milk is referred as "complementary Food". With adequate amount of complementary feeding breast-feeding continues up to two years of age and beyond so, referred as complementary feeding period.

Appropriate and adequate amount of complementary feeding helps in preventing malnutrition stunting and promoting growth. Introduction of this complementary food is called as weaning.

Weaning is the process of introducing adult food and beverage to the infant to support their growth and development. This is very crucial period in infant's life. This is a period for diet transition as the nutrition requirements for growth and brain development is quite high. Introduction of various taste and texture promotes biting and chewing skills (Guthrie, 1975). Most six-month age children start eating semisolid foods. Weaning homogenized infant foods play a major role in the nutrition of infants (Martinez, 2004).As infants grow, their nutritional requirements also increase. To keep up this growing demand WHO recommends that infants start eating solid, semi-solid or soft foods at the age of 6 months to ensure sufficient nutrient intake.

Importance of weaning foods

Weaning foods must bridge the gap between breast-feeding and family food. Weaning food is consumed by infants in the age group of 6-12 months whereas, weaning foods are also consumed by adult members in family .It is modified appropriately for infant's consumption by processing the ingredients to make it easily digestible. A nutritionally adequate weaning food is essential for achieving normal growth in the infancy period. Growth and development during the period of infancy influences both the well- being of the child and their long-term health as an adult.

Weaning foods are majorly prepared including a staple cereal or starchy tuber. Introduction of supplementary food prepared with locally available ingredients, which are easily available at low cost, is important to meet the nutritional requirements of the growing children (Saeeda, 2009). Appropriate consistency of weaning food is necessary, so that food can be swallowed and gulped easily. How ever, it is necessary to reduce the anti nutrient factors in the weaning foods so that nutrients can be absorbed without hindrance of anti nutritional factors.

Also, limited facilities for food preparation and storage allow contamination and rapid proliferation of micro-organisms which create risks of food-borne illness including diarrhoea. Several simple technologies are traditionally used for the processing of cereals, tubers and legumes to reduce their anti nutritional factors which maximize the nutrient absorption and reduce bulkiness in foods to make it easily digestible by the infants. These traditional technologies include roasting, germination and fermentation. Since these technologies have different effects on digestibility characteristics of the product, a combination of various technologies would therefore recommended to optimize product quality at a minimum cost (Nout, 1993). The characteristics of good quality weaning food is that it should meet protein and energy requirements and also must be high in nutrient density, low in viscosity and low in bulk density.

(i) Concept of calorie dense or dietary bulk foods

The total amount of calorie in a specific quantity and volume of food is termed as calorie density of the food. It is referred as good index for comparing the true value of various foods. Food consumption of most of the people is governed by the volume rather than weight. The calorie density is very important in preparation of weaning food to ensure appropriate amount of nutrients and also easy consumption (Malleshi , N.G.1988). The stomach capacity of infants is very limited so they cannot consume large quantity food in one feed.

Also, there may be much difficulty in feeding the baby frequently. Hence, it is important that weaning food should be calorie dense so the requirement of macro and micronutrients can be fulfilled. According to Obizoba, I.C et al., (1992) ,weaning food gruel must have energy density of 77 and 116 kcal /MI in order to meet nutritional requirements of infant receiving average breast milk intake.

The amount and type of protein used in the weaning food can be optimized by incorporating legumes and pulses on the basis of the amino acid (lysine) content.

When legumes are used, sufficient heat processing is included to destroy the heat-labile anti nutritional factors. Particular precautions must be taken with respect to phytic acid levels in such preparations. Optional ingredients include protein concentrates and amino acids, fruits, nutritive sweeteners, malt, milk or milk products, fats and oils, salt (including iodated salt), and spices. Vitamins and minerals may be added in accordance to the requirements and regulations of the country.

It is important to prepare weaning food with appropriate consistency for easy consumption by the infants (Balasubramaniam, 2014). The traditional weaning foods are prepared thick in consistency with high viscosity, which may cause choking and suffocation in infant during feeding. Initially when weaning food is introduced to infants it should be bland in taste and smooth in texture. Once infants start consuming the food with spoon, then variety of foods with different taste can be introduced. From six months and above is the right period to encourage new taste, as the infants are ready to accept new taste and variety, so the rejection rate is low. At this stage new foods should be introduced gradually to rule out food allergy.

The ideal weaning food characteristics are:

- It should be nutrient dense -rich in calories and providing adequate amount of protein, vitamins and minerals.
- It should be prepared with right consistency- semi solid and soft, which infant can swallow easily.
- The weaning food consists of low dietary bulk.
- It can be consumed with minimum processing like precooked and predigested, which minimizes the preparation time.
- Weaning food should be processed to reduce anti nutritional factors, microbial contamination and also indigestible fibre content.

(ii) Increase the calorie density of weaning foods

Usually introductory weaning food is prepared from staple cereals and starchy roots.

These preparations can be prepared using either water or milk. Starch molecules present in cereals or roots absorb water and gelatinized when it is cooked with water, which gives viscosity to food. Therefore, the food consumed by infants becomes inadequate to satisfy their nutritional requirements. Technology to process these cereals can solve these issues.

(iii) Commercial preparation of weaning foods

Simple traditional technologies like chapati making, popping, flaking and vermicelli extrusion were employed for formulating weaning foods from mix of cereals/millets and pulses. The protein efficiency ratio of weaning food ranging between 2.6 to 2.8 is acceptable to children. Viscosity of weaning foods can be reduced by adding 5% of barley malt powder in the formulations. Recommendation of these weaning food production to use at house hold community or factory level in rural and urban areas has to be given priority.

Table 1: Amounts of ingredients (grams) required to give the best possible protein value

Staples (g) / Supplement	Wheat	Rice	Sorghum, millet	Maize	Banana	Plantain
Legumes	80 / 10	65 / 25	75 / 10	55 / 35	105 / 55	85 / 55
Soy Beans	60 / 15	55 / 20	55 / 15	50 / 25	140 / 25	115 / 30
Dried Skim Milk	65 / 10	65 / 15	60 / 15	60 / 15	165 / 20	150 / 20
Chicken/Lean meat	65 / 40	65 / 25	65 / 25	65 / 35	185 / 40	160 / 45
Eggs	65 / 25	65 / 30	60 / 30	65 / 25	190 / 30	150 / 45
Fresh Fish	70 / 30	70 / 70	70 / 25	70 / 20	210 / 40	180 / 45

Source: Cameron, M., & Hofvander, Y. (1983). How to develop recipes for weaning foods.

Codex Committee has listed the ingredients used for weaning foods as cereals like wheat, rice, oats, sorghum, jowar, ragi, barley and other millets. Along with groundnut, sesame, soybean (defatted or low fat), and other legumes (Bender, 1973). Cereals have phytates an iron inhibitor and they also do not contain vitamin A and C and also deprived of good quality protein. Therefore, it need to be given along with pulses or animal foods. Oils and nuts help to increase energy density. Addition of sugar/ jaggery also increases energy content.

Technology development for weaning foods

Largely in the developed countries commercial weaning foods of excellent quality are either imported or locally produced. These are usually 10 to 15 times higher than the cost of the common staple foods due to processing, expensive packing, extensive promotion and solid profit margins (Bahlol, 2007).

Integrated Child Development Scheme (ICDS) and Food and Agriculture Organization (FAO) have suggested to develop supplementary foods based on locally available cereals, millets and legumes to combat malnutrition in children and mothers belonging to the low socio economic class (Imtiaz, 2011). However, the development of low-cost, high-protein food supplements for weaning infants is a constant prevailing problem in developing countries (UNICEF, 2012). Most technologies used in developing weaning foods are having their own drawbacks. For example drum drying and extrusion cooking are expensive and complicated so cannot be used in developing countries. Thus the practical method with low cost technology is required for processing. Dry roasting, steaming, popping, malting, boiling and flaking are some common processing methods, which can be used to prepare weaning foods.

Table 2: Bureau of Indian Standards specification for milk and cereal-based weaning foods

Ingredients	Amount
Moisture, % by mass, Max	5
Milk solid %	>20(min 5% by weight of mik fat)
Total protein, % by mass, min	12
Fat, % by mass, min	7.5
Total Carbohydrate, % by mass, min	55
Total ash, % by mass, max	5
Acid insoluble ash, % by mass, max	0.1
Crude fibre (on dry-weight basis), % by mass, max	1
Vitamin A (as retional), mg/100g min	350
Vitamin C, mg/100, min	25
Added Vitamin D, mg/100g	300-800
Thiamine (HCl), mg/100g, min	0.5
Riboflavin, mg/100, min	0.6
Nicotinic Acid, me/100g, min	5
I-Ascorbyl palmitate, mg/kg, max	200
Iron, mg/100g, min	5
Bacterial cout/g, max	50,000
Escherichia coli count/0.1g	Nil

Various food processing methods such as roasting, germination, milling, cooking, drying , fermentation and extrusion have lot of potential to improve nutrient bioavailability ,digestibility, nutrient density, food safety, storage stability and palatability of weaning foods. These processing techniques reduce high bulk by reducing viscosity to make ingredients suitable for weaning mixtures and also to promote nutrition repletion following diarrheal episodes (Rasane, 2014). Toasting, fermentation and germination of cereals and pulses are cost effective traditionally adopted processing methods. Fermentation process increases nutrients in food through the biosynthesis and bioavailability of micronutrients and essential amino acids by reducing anti nutrition factors. These process can also improve the quality of protein and digestibility of fibre (Hotz, et al.,2007).

Germination unlocks many nutrients that are in bound forms present in the food, thereby increasing their bioavailability, energy and sensory quality (Sangronis, and Machado, 2007). Toasting reduces anti- nutrients, improves the taste, texture and nutrients in the food and also reduces the moisture present in it, thereby increasing its shelf life (Parchure,1997). An integrated approach that combines the various traditional processing methods in preparation of weaning foods is the best way to ensure the nutrient content and bioavailability of micronutrients in plant based diets (Desikachar, 1982).

The combination techniques are more effective in removing anti nutritional factors in cereal, thereby producing nutrient dense weaning foods (King, 1985). Various Processing techniques and addition of additives have special potential to enhance the quality of weaning foods by enriching with micro nutrients, nutrient bioavailability, nutrient density, food safety, storage stability, palatability and convenience. Some of these processing methods are applicable for use at home, while others require the equipment and skills available in a small-or medium-scale food factory (Bressani, 1984). While the malting technique is simple and useful, germination requires space and time and demands sun-drying or mechanical drying facilities for thorough drying.

The roasted grains could also be flaked by pounding or mechanical flaking. Chapattis made by this could be dried in the sun or in a solar drier or partially on the pan and partially by aeration to a moisture level of less than 12 per cent, enabling them to be powdered. The vermicelli or ribbon like product can be sun-dried and later ground to powder. The first two processes above are both age-old household versions of the modern roller- drying or extrusion process for making weaning foods. The third technique is adopted for making puffed cereal grains, particularly from sorghum, maize, ragi, bengal gram, green gram, etc., and also as a first step for making flaked products (poha) from rice.

The high-temperature roasting helps in pre-drying the materials to a large extent and therefore is ideal for places where drying facilities are not available or the weather is rainy or damp. For using these techniques to produce weaning foods, trials were carried out with weaning foods formulated from different raw materials. Chapattis or rotis were made using mixtures of legume flours (Bengal gram and green gram) and specially prepared nutritious cereal flour from wheat, maize, or sorghum with the major portion of bran removed.

The vermicelli-type product was made from using various types of millet combination and roasted legume flours was made in dough with water, extruded through a hand-screw press, steam-cooked, dried in a current of air, and powdered. The slurry of the flour mix could also be cooked to a thick porridge or dumpling before extrusion.

Using the principle of flaking, preparation of rice-based formulations was prepared like rice flakes (poha, aval) and puffed green gram, bengal gram and soybeans. Similarly, a mixture consisting puffed chickpea (bengal gram) and puffed pearl millets was also made. Sensory evaluation of all the developed products was acceptable. Sweetening with about a spoon of sugar (5g) or jaggery (crude brown sugar) powder further increased the palatability of these products (Bressani, 1984). Below are various processing techniques used for the preparation of weaning foods:

(i) Cooking

Traditional cooking methods involved in preparation of weaning foods using mixture of cereals, oil seeds and pulses boiling them in 70– 90% of water to completely cook, tenderize and gelatinize the starch to form thick paste. Presence of amylase (sprouted grains) makes gruels nutrient dense as it hydrolyzes the starch to reduce the viscosity and increase the digestibility. Use of powdered beans in place of whole beans added to the boiling water can reduce cooking time by a factor of 10 (Nelson, & Steinberg, 1978) . Cooking processes hastens digestibility in a various ways. When starches are heated, the starch granules absorb water, a process known as gelatinization (Olkku, 1978).

Raw starch is resistant to enzymatic hydrolysis (Jenkins, 1986) and digestion before gelatinization occurs. However, cooking of sorghum in excess amount of water decreases digestibility. There are several factors present in some foods that also decrease the digestibility and/or absorption of specific nutrients that are denatured by heat during cooking. For example, protein digestibility is reduced by the protease inhibitors present in legumes (Aykroyd, 1982) unless eating inactivates the inhibitors.

The cooking or reheating of foods can destroy vegetative forms of entero pathogens. The temperature usually depends on the time of exposure, with short interval (less than one minute beating temperature) at greater than 75°C required. Even at these temperatures, however, heat-resistant spores survive. This is matter of concern when food is kept long between cooking and serving time. Foods held at temperatures between 20 and 50°C allow bacteria to multiply rapidly. Foods, which are cooked using high moisture amount, must be either sterilized or dried (moisture to be reduced) if has to remain stable for long period of time. Both packaging and sterilization are expensive procedure. Once the sterile package is opened the product becomes very susceptible to contamination and spoilage, then needs to be stored at controlled atmosphere. Both radiation and heat methods can be used for sterilization, but radiation is having limited applicability in developing countries.

(ii) Drying

Method of dehydrating is an age-old method for food preservation. Drying and adding the solutes like, NaCl or sucrose, to the food reduces the water activity (a_w) according to the nature of the solute added. Definition of water activity says the ratio of the water vapor pressure of the food as against of pure water. Water activity for multiplication of micro organisms (with bacteria being most sensitive) is also affected by other conditions like temperature or pH. Water activities of fresh foods above 0.98 are susceptible to spoilage and multiplication of entero pathogens. Most bacteria do not multiply at an a_w lower than 0.9, but some may grow in foods with an a_w as low as 0.75 (saturated NaCl). Values this low, however, is attainable only for dehydrated or dried foods, which can be stored for extended periods of time(Fig-1). Water must be available for spore germination, microbial growth, and toxin production

Drying is an effective preservation technique, although it is relatively inefficient as it requires large amount of heat and becomes expensive. Thermal efficiency increases with increasing drying temperature and dryer design improvements, but these enhancements require greater operator skill to minimize product damage. Sun drying is suggested cost effective method, but it is not suitable for drying pastes. Long drying times associated with solar drying can also lead to product deterioration as a result of bacterial growth before drying is complete. Cereal paste drying is commonly done on the surface of hot drums or by spraying the paste into heated air. This results in dry, precooked products which form smooth gruel when they are mixed with water. Roller drying method reduces the formation of resistant starch as compared to other methods like pressure cooking and boiling (Parchure and Kulkarni, 1997).

Supplementation with non-fat dried milk or soy protein and vitamin-mineral mixtures improves the nutritional quality. Appropriate packaging is very important so that it can ensure to protect the product from moisture, insects and microbes. Some pros and cons of the following processes used for preparing weaning foods are as follows:

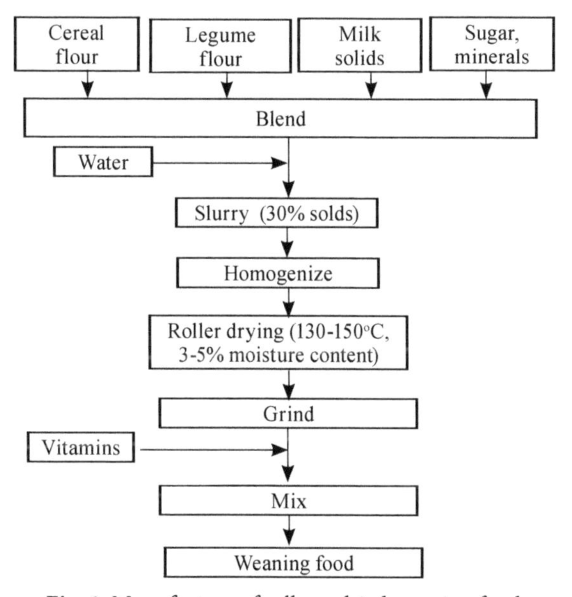

Fig. 1: Manufacture of roller – dried weaning foods

Vermicelli process

- The process allows full cooking and gelatinization of the constituents.
- High moisture content is required for extrusion with a hand press.
- A long drying time is required.
- The process causes the least damage to nutritive value.
- The product has high viscosity.

Puffing by high-temperature, short-time roasting

- This is basically a dry-condition process.
- It can be practiced even in areas where climate is cloudy, rainy and humid in most of the year.
- By roasting cereals puffing is achieved using hot medium (salt or sand) for less than one minute.
- It does not permit removal of bran, particularly from the cereal component.
- This low moisture formulation can be preserved for longer duration
- This is high viscosity product

- This particular process is suitable for the formulation based on millets like sorghum, bajra, ragi and pulses like bengal and green gram.
- Food products prepared by this method is more suitable for older children
- A fine aroma is developed during puffing.

High-temperature, short-time roasting with flaking

- This process is best suited to rice-based weaning food, as rice flaking is practiced both commercially and in households.
- It allows the mother to buy flakes and roasted legumes from the market and formulate own mixture for baby feeding by simple mixing of ingredients.
- The roasting develops the proper aroma and can bring moisture to a low level. No drying facilities are required
- There is a possibility of damage to Iysine during the roasting.
- The product has high viscosity.

As can be seen, each method has some specific merits that may make it suitable for particular localities or situations. When the formulation is made at home where skim milk powder is not available, fresh milk in small quantity can be added in grain based weaning foods. Almost 75 g of these foods would supply about 300 calories to an infant. The cost of these weaning foods will be one quarter of the conventionally prepared one.

(iii) Roasting

Roasting is a process that dry cooks a cereals, oil seeds and legumes. These dry roasted products can be grounded and mixed with water or milk and also sugar and oil to prepare weaning food gruel. Roasted products have limited ability to absorb water so they are nutrient dense . High temperature roasting process produce a pleasant roasted flavor that improves taste , palatability and also reduce anti nutritional factors and inactivates enzymes which denatures heat-labile vitamins. The process of roasting involves shaking the grains in a heated thick bottom pan or by agitating grains in hot salt or sand, that requires only a moderate level of skill to operate . Roasting loosens the outer cover of seeds which helps to easily remove the skin before grinding it. Roasting dries grains and destroys surface micro flora, hence increases the shelf life of weaning foods prepared by using this method (Table 3).

Table 3: Microbiological Specifications for dried weaning formula(UNICEF)

Class/group	Organism	Limit aimed at	Maximum Level tolerated
I (pathogenic)			
1	Salmonella arizona	Absent in 15 g	Absent in 1g
	Escherichia coli	Absent in 10g	Absent in 1g
2	Staphylococcus aureus	Absent in 1 g	Absent in 1 g
3	Bacillus , per 1g	$<10^2$	$<10^3$
4	Mold spore per 1 g	$<10^2$	$<10^2$
5	Entero bacteriaceae	Absent in 1 g	Absent in 0.1g
II (indicator organisms)			
6	Total aerobic count per 1g	$<10^4$	$<10^5$
7	Lancefeild streptococci	$<10^2$	$<10^3$
8	Clostridium species per 1g	<10	$<10^2$

Source: Harper, J. M., & Jansen, G. R. (1985).

(iv) Milling

Milling, is a primary processing technique which cleans and separates the components of grains (germ, bran, endosperm) and reduces their size. Cleaning steps in the milling can remove insect and microbiological contamination of raw materials. Milling method is beneficial in reducing bulk and lowering fibre content but in this course it also reduces amount of minerals and vitamins in the flours. However, the milled products are more susceptible to the damage caused by insects if it is not packaged appropriately. Growth of bacteria is not a concern in these products if 15% moisture content is maintained. There are several advantages of milling process as it reduces the phytate amount in the bran of most cereals, pulses and legumes. Phytates inhibits absorption of minerals (calcium) and trace elements. Tannins, another anti nutritional factor, which is present in many cereals and seed coats, can also inhibit the digestion of protein and starch, if these are not removed during the milling process.

Milling by itself does not produce a supplementary food. Instead, milled ingredients are more suitable and convenient for formulating supplemental foods. Milled flours cooked with water form thick gruels with low nutrient density and protein content. Beans, legumes, lentils, oil and sugar can be used to improve taste and nutrient quantity in the weaning foods.

(v) Malting

Malting processing method is the only possible method, which can reduce dietary bulk in weaning food without reducing their nutritional value. Malting is a traditional processing method, which reduces viscosity due to amylolytic breakdown of starch and thus reducing the total bulk. Only by adding 5%

drum dried malted barley flour to the gruel mix can reduce its viscosity. However, when amylase is added prior to drum drying, it is observed that the biological value protein and lysine availability reduces. Usually at commercial level 0.2% of fungal amylase is used instead of malted flour (Malleshi, 1988).

Roller drying method reduces the formation of resistant starch as compared to other methods like pressure cooking and boiling (Parchure, 1997). Although fat can add more calories, the fat content cannot not be increased beyond certain limit as it may affect taste, texture , difficult to digest and shelf life of the product. Studies on evaluation of food intake in young children (Rutishuser, 1973) have shown that the weaning food must be energy dense to about 1.0 kcal/g to meet the energy requirements (Joint, 1985).The energy density of human milk is about 0.7 kcal/g (Macy, I.G.,1961) and hence, it is reasonable to assume that energy density of weaning food should be around 0.7kcal/g but preferable the desired amount in the gruel should be about 1.0 kcal/g (Mosha, 1990). Thus, 100 ml of gruel would provide about 100 kcal (Rathod, P. 1991). Finger millet is important millet in southern part of India. It has the uniqueness of being used for malting next only to barley in the tropical world. The millet is excellent source of calcium and good source of carbohydrate and protein. Therefore, the malt produced by this millet is very good source of protein, carbohydrate, calcium and many other micronutrients besides hydrolytic enzymes. Malted ragi flour is very nutritious and hence can be used in preparation of weaning foods,infant foods, geriatric food and medical food along with milk. Also, it can be used to reduce the dietary bulk as Amylase Rich Food (ARF). By products like seed coat & rootlets can be utilized in Cattle and Poultry feed formulations.

Process of ragi malting

Finger millet → Cleaning → Soaking → Germination → Drying → De-vegetation → Green malt → Kilning → Moist conditioning → Grinding → Sieving→Malted→Ragiflour

Some of the merits and demerits of the malting process used for preparing weaning foods are as follows:

- This process allows pre digestion of starch and protein.
- Germination process can reduce the viscosity of weaning food to desired level which is very desirable for young babies.
- A pleasant aroma develops during the kilning process
- Phytase hydrolyses the phytin to available phosphate.
- Vitamin C and Lysine increases in cereals.
- A long processing duration and adequate sundrying or other dehydrating facilities are required.

- Debranning of cereals and legumes can be easily carried out after germination process.
- Chapati making process can be applied to any composite mix based on cereals, legumes, millets, etc.
- Highly acceptable chapati products are obtained with sorghum- based and wheat-based formulations.

The normal process of chapati making practiced at home can be extended. Moisture reduction can be effected by sun drying, by toasting at low heat, or by air currents. The process is easily applicable in households, particularly in non-rice areas where chapati- making is common. The product has high viscosity.

(vi) Germination

Whole soaked pulses or cereals can be sprouted prior to cooking to increase vitamin levels, to increase bioavailability, reduce the molecular weight of the carbohydrates that are present, and increase the nutritive value and bioavailability of the essential amino acids (Wang, & Fields, 1971).

The amylases, released during germination hydrolyze starch to shorter-chain carbohydrates and sugars, resulting in reducing viscosity and facilitating digestibility. Drying them can increase shelf life of sprouted grains. Roasting and heating of the sprouted dried grains increases flavor and acceptability. The benefits of the germination process include partial pre digestion and lowering the viscosity of weaning food. The proper aroma is developed during the kilning process. Phytase enzyme hydrolyses the phytin to available phosphate. Vitamin C and Iysine is enhanced in many cereals. A long processing time and adequate sun drying or other mechanical drying facilities are required. Debranning of the cereal or legume can be assured after the germination process (CFTRI, 1982). Malting process of the finger millet (ragi) is explained by CFTRI. This millet is very rich in calcium and can be used in the formulations and preparation of geriatric food also.

(vii) Baking

Baked products are usually made out of butter, cereal and sugar so they tend to be nutritionally dense. Biscuits are baked with fat, sugar, vegetable and animal proteins. Biscuits can be made into small pieces and mixed either with milk or water to prepare a gruel forming a required consistency can be used as an ideal weaning food. This provides control and promises food safety and hygiene needed to prepare weaning foods. Older children eat biscuits directly as a ready-to-eat finger food providing supplementation to older infants (specially when it is fortified with micro nutrients).

Individual biscuits gives a degree of portion control and acceptable to all the children. Baked products are digestible and also micro nutrient fortification increases the amount of minerals, protein and vitamins. Biscuits baked at high temperatures are dry and can be stored for long time period. To enhance the shelf life, packaging is required to reduce insect infestation and moisture uptake. Baked products normally require expensive refined ingredients such as shortening, flour, leavening and sugar; this increases the cost and may limit their production.

(vii) Extrusion

Extruded products are prepared from cereals, legumes and oil seeds or combination of all. These are popular products and are completely precooked for easy reconstitution. It can be fortified with micro nutrients (Harper,1985). This process has been used for successful production of nutritious products that have been distributed in dry packaged form through both commercial and governmental programs. Extrusion heat processes dries food ingredients by using friction between high-speed screw and food. Low-cost extruders, which process foods at moistures of less than 20 percent, have the lowest capital and operating costs and can produce fortified, packaged, stable food products for an additional cost of 30-40% cost of raw ingredients. Mechanical disruption of the cell walls and starch content of plant occur while preparing a product using extrusion method.

This method breaks down the starch and reduces the viscosity of gruel and increases the digestibility and absorption of nutrients. Extruded products also enhances calorific value of the food hence, suitable for the formulation and preparation of weaning foods.The high-temperature heat treatment effectively pasteurizes the product. Six-month storage requires packaging to provide resistance to moisture and insects.

(viii) Fermentation

A variety of fermentation processes have been used with cereals to increase digestibility, palatability, and shelf life . Many of these products have been used as weaning foods(Abdel Gadir et al., 1983). These processes normally begin by soaking the whole grain for 24–72 hours. Wet grinding followed by removal of hull and germ part. Fermentation at 30–50 percent moisture requires another 24–72 hours at approximately 30°C, using a mixed culture of acid-forming bacteria. Before consumption, water is added to give a 7–10 percent solids concentration, and the mixture is brought to a boil to produce a gruel with suitable consistency for weaning food. These type of foods are more common in Africa, but similar processes are used in most countries. Control of time, temperature, strain of micro organism alters the pH and flavor of the finished product. The final pH ranges between 3.4 and 3.8 because of the fermentation of sugars to lactic, acetic and other short-chain acids. Since most bacteria grow best at about pH 7.0 and few grow at lower than pH 4, the lowered pH inhibits bacterial growth and extends the shelf life to approximately one day. Fermentation produces a strong acidic flavor, increases protein digestibility and relative nutritional value. Fermentation can also reduce cyanide toxicity in cassava and sorghum, trypsin inhibitors in soybeans, and the anti nutritional character of phytate and tannins.A significant disadvantage of the fermentation process is the lengthy preparation time . Also product have low caloric density and low protein quantity and quality. These deficiencies can be overcome by hydrolysis of the starch by enzymes from sprouted grains and addition of plant proteins such as lentils, legumes, or oil seeds before fermentation. Shelf-stable fermented products require drying, which significantly increases complexity and cost, but it adds to the convenience and the ability to expand distribution. Yogurt and souring of milk are other examples of beneficial acidic fermentations, which extend the shelf life, and the utility of milk products as supplemental foods.

Such fermentation also reduce the lactose in the food product , which is beneficial for lactose intolerance cases. Acidification of milk with lactic acid and lactobacilli was studied in Gambia and it was observed that to inhibit bacterial growth to a slight degree, counts of *E. coli* in acidified or non acidified milk samples reached similar high levels after 4–8 hours (Gabriel, & Akharaiyi, 2007). Fermenting soy with fungi produces variety of fermented products such as tofu, tempeh, and miso. Acidification of soybeans, either by addition of lactic acid or fermentation prevented growth of *B. cereus* in the production of fermented food products such as tempeh (Nout, R. 1993). These products have higher concentrations of protein but often use salt to control the micro flora in the fermentation. This characteristic, however, diminishes the product's value as a supplemental food for rehabilitation of children with diarrhea.

Conclusion

Right kind of processing method is needed for preparation of weaning foods to make it digestible, absorbed and thus can help in maintaining normal growth when breast milk quantity is not enough to meet the increasing nutrient demands. Nutritious and calorie dense foods can be used to treat under nourished children after the episodes of infectious diseases like diarrhea and also it improves their nutritional status and immunity. Many food processes exist which can make weaning food suitable for infant's consumption by improvising digestibility, viscosity and texture. They may vary in their characteristics and ability to support growth from the perspective of caloric density, nutrient composition, and storage stability. Cooking is the first ever-processing method, which improves digestibility, palatability and texture. Fermentation, germination and milling process can enhance sensory attributes of weaning foods. Processing methods such as roasting and extrusion that require moisture are the low-cost small-scale technologies, which are not readily available to home application. Home preparation is time-consuming but it reduces cost and increases safety and familiarity.

Geriatric foods

Introduction

In this modern society, people desire both good health and longevity and hence demand nutritious and functional food that promotes their wellbeing, enjoyment and active life style (Deliza, et al., 2003). Convenient and value added foods which provides health benefits to elderly are high demand food product development in the food industry. Dietary supplement production is also increasing as elderly started filling the gap of nutrients by adding supplements to their diet. It has been given immense increase in the new food formulation and manufacturing of health category food for the geriatric population. Recent food technologies has initiated to develop healthy foods which is conveniently obtained and suitable to the geriatric needs. It is challenging because geriatric consumers tend to reduce rather than increase their food consumption with increasing age. More over, loss in appetite causes risk of deficiency which leads to malnutrition .Promoting intake of nutrient in elderly is challenge now than ever, because life expectancy is increasing and the world population is greying rapidly. Improving the diets of elderly consumers will likely provide value addition to their quality of life.

Foods for elderly

The elderly consumers have been shown to be motivated by convenience, sensory appeal and price in making food choices, they differ strongly in the perceived importance of health, natural content, familiarity and weight control. This heterogeneity regarding their requirement of the specific food , calls for market segmentation to study the consumer behavior which is driven by the following traits common in geriatric population. Due to special needs their preferred food type is different (Fig .2).

- **Poor appetite:** Due to illness, pain or nausea, depression or anxiety, social isolation bereavement or other significant life event, food aversion, resistance to change, lack of understanding linking diet and health, beliefs regarding dietary restrictions, alcoholism, reduced sense of taste or smell.
- **Inability to eat:** Due to confusion, diminished conscious- ness, dementia, weakness or arthritis in the arms or hands, dysphagia, vomiting, COPD, painful mouth conditions, poor oral hygiene or dentition, restrictions imposed by surgery or investigations, lack of help while eating for those in hospitals and rest homes.
- **Lack of food:** Major reason being poverty, poor quality diet (home, hospital or rest home), problems with shopping and cooking, ethnic preferences not catered for, particularly in hospitals and rest homes.

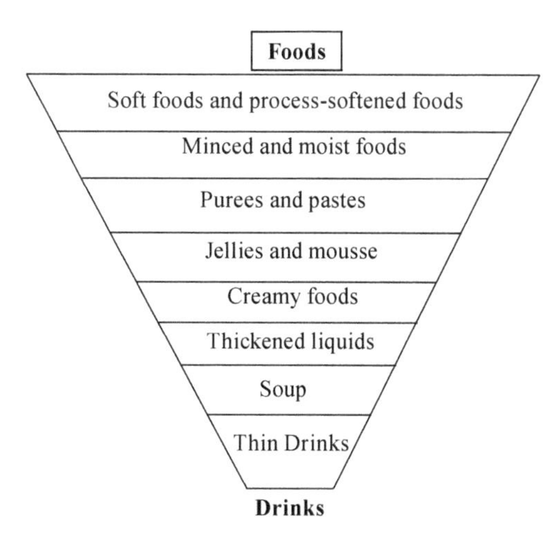

Fig. 2: Types of food preferred by geriatric population

- **Medicines:** Medicines consumed in the old age related issues can alter nutritional status in numerous ways, e.g. anorexia, decreased or altered taste, dry mouth, confusion, gastrointestinal disturbance including nausea, vomiting, diarrhoea, constipation, dyspepsia. Incorrect use of medicines may also cause problems, e.g. hyper metabolism with thyroxine and theoph Impaired digestion and/or absorption

Medical and surgical problems affecting stomach, intestine, pancreas and liver, cancer, infection, alcoholism

- **Altered requirements:** Increased or changed metabolic demands related to illness, surgery, organ dysfunction or treatment.
- **Excess nutrient losses:** Vomiting, diarrhoea, fistulae, losses from nasogastric tube and other drains.
- **Illness related Malnutrition:** Some disease conditions also maximize risk of under nutrition for example chronic all gastrointestinal, liver and kidney diseases, cancer, HIV, AIDS, stroke and surgery (Hsieh & Ofori, 2007).
- **Surgery:** The metabolic changes caused by surgery, the increased demands required for successful healing, sepsis and the stress of the surgical procedure itself, all increase energy needs (Hsieh, P. Y. H., & Ofori, J. A. 2007). To supply this energy, protein stored, as muscle is broken down to amino acid.

(i) Increasing nutrient intake

Trying to motivate the elderly to consume more quantity of food is unlikely to be successful, as their tendency to eat less is the core problem. To overcome this problem one of the strategies, which are going to be helpful, is to develop nutrient dense and nutrient enriched food products for elderly, a type of functional food. Enriched foods look just like normal food products, but their nutritional value is more than the traditional food either by addition of extra nutrients or by increasing existing nutrient levels in the product. Enriched foods are relatively nutrient-dense, they can increase nutrient intake in the elderly if consumed in small quantity, and as usual observation is loss of appetite at this age. Geriatric consumers are conscious and ready to spend in healthy and innovative products, like enriched foods, which promises to satisfy their nutritional requirements. Most of the time enriched foods are more acceptable than the traditional food supplements. Across the geriatric population, only about 15% of new product ideas and 60% of new products that are introduced in the market are commercially successful (United Nations Population Ageing and Development, 2012). Although technological advances make it possible to enrich virtually any food product, but not all functional ingredients are well- received by consumers .For example, yoghurt enriched with fish oil may fulfill the needs of consumers who need to reduce their cholesterol level, but the expected impact of fish oil on the taste of the product is likely to cause product rejection. In developing any new enriched foods, it is important to know the preferences of consumers (Hsieh, & Ofori, 2007).

(ii) Technology development for geriatric food

Mostly elderly are wise and kind but at the same time they are also seen to ne slow and helpless. Illness and physical limitations makes them more susceptible to under nutrition.

As age increases their cognitive capacity tends to decline, making it more difficult for elderly people to process complex information. For example, the elderly point out their interest in knowing what is good or bad for their health and struggle to keep up with the latest nutritional guidelines. This also shows the way functionality of enriched food is displayed which is easy to understand and can relate to their common health problems.

Over the period of time many technologies in food processing have been researched and considered following aspects in designing healthy foods for elderly that can address the common health issues related to the old age (Fig-3)

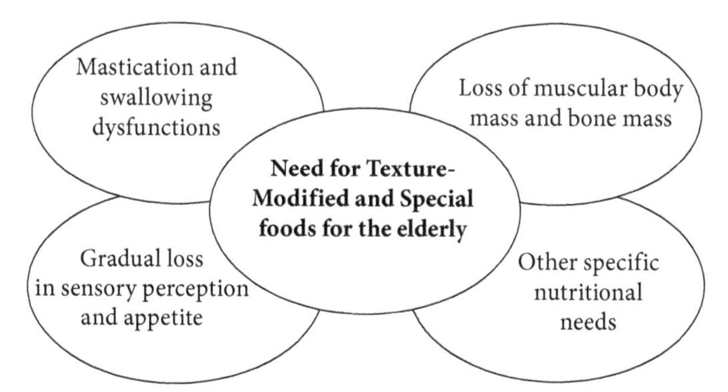

Fig. 3: Main aspects to be considered in the design of healthy foods for elderly

Technologies involved in ingredient modification

Following are the technologies, which has been developed keeping specifically the requirement and the health conditions of elderly.

(i) Decaffeination

Caffeine is an alkaloid which is naturally present in coffee, cocoa beans, cola nuts and tea leaves. However, excess intake of caffeine is associated with causing health problems, for example heartburn, acidity and indigestion (Babu, et al., 2005). More than normal consumption of caffeine is reported to cause cardiac arrhythmias and increased heart output in geriatric population. It is also known to accelerate osteoporosis and bone demineralization. For the mentioned detrimental effects, an effort was made by researchers to develop caffeine-free beverages. Decaffeination prevents the related health risks and also offers some health benefits. There are three conventional methods of caffeine removal: 1) water decaffeination 2) solvent extraction and 3) super critical carbon dioxide extraction. There are two types of water decaffeination -Swiss Water and French Water decaffeination techniques and both techniques depends on temperature and solubility of caffeine in water to remove caffeine from coffee without using chemical solvents. This decaffeination processes involves soaking the coffee beans in hot water, which dissolves caffeine and flavor substances in the water. Then extracted beans are then sprayed with the same flavor-laden water, allowing beans to reabsorb the flavor, so that decaffeination can be attained without compromising flavor loss. Solvent extraction method is based on the solubility of caffeine in various organic solvents such as methylene chloride, ethyl acetate, ethyl alcohol, acetone and ethyl ether and proceeds either by direct solvent extraction of the beans or indirectly through water extraction of the beans followed by solvent extraction of the caffeine from the water extract. More recently, super critical carbon dioxide extraction offers a means of removing only the caffeine, leaving the other flavor components in place. The process involves first compressing carbon dioxide to above 50 times atmospheric pressure to transform it from the gaseous state into a dense liquid. Pre-moistened beans are then treated with the liquefied carbon dioxide to extract the caffeine. The benefit of this method is that does not involve any hazardous chemicals, the product remains of superior quality and the amount of other coffee soluble components extracted along with the caffeine is minimal. There are other decaffeination methods which also have been developed to overcome short falls in the traditional method like water decaffeination is not efficient because of the low solubility of caffeine in water, while solvent extraction procedures leaves the ill effects of chemicals. For example, methylene chloride though not demonstrated to be carcinogenic in human being but was found carcinogenic in mice at some levels, and also causes depletion of the ozone layer.

The super critical carbon dioxide procedure, despite its other advantages, is very capital intensive (Gokulakrishnan, et al., 2005). Microbial decaffeination as an alternative procedure have been produced. There are certain bacteria and fungi degrading caffeine are employed for decaffeination by spraying suspensions of these microorganisms onto caffeine bearing plants. Caffeine degradation by bacteria proceeds more rapidly than the equivalent process by fungi, and unlike fungal is not inhibited by the presence of an external nitrogen source. The bacterial *Pseudomonas* species and the fungi *Aspergillus* and *Penicillium* are efficient degraders of caffeine.

These microorganisms owe their caffeine de- grading potential to enzymes, namely de methylases and oxidases, and an effort has been made to isolate and purify these enzymes for use specifically for caffeine degradation purposes.In view of health risk associated with consuming excess amounts of caffeine and considering its addictive nature, developing improved techniques for the economical removal of caffeine from beverages is therefore of continuing interest to the beverage industry (Gokulakrishnan, et al., 2005).

(ii) Fat replacement

Chronic diseases such as cardiovascular disease, hypertension, diabetes, and cancer in geriatric population have been associated with the high dietary fat content and obesity. Hence, it is important to decrease consumption of dietary fat which is recommended to reduce weight also. Studies also shows that reducing intake of fat is difficult food behavior change (Hahn, 1997).Fat provides desirable sensory and functional qualities to foods, so low fat foods or fat-free foods compromises flavor, texture and mouth feel. Therefore, the fat replacers play important role in replacing fats in the traditional foods that provide the similar characteristic flavor, other organoleptic characteristics and mouth feel of fat but reduces the calories and risk factors related to fat (Kuller, 1997).

The idea is to offer consumers low or fat-free foods that retain the characteristic sensory qualities traditionally attributed to fat. There are several ingredients currently available for replacing small amount of fat in prepared foods. There are three categories of fat replacers : 1) carbohydrate-based 2) protein-based and 3) fat-based. Carbohydrate-based replacers are obtained from cereals, grains and plants and are modified to provide fat-like textures and mouth feel in food products. They provide a reduced caloric product as a outcome of low energy density as compared to fats. Protein-based replacers are derived from milk, egg, whey, or vegetable proteins (soy) and it also provides reduced energy content. Fat-based replacers which gives characteristic flavor , texture and mouth feel offers same functional and sensory quality of those native fat they replace (Hahn, et al., 1997).

Fat-based replacers are of two types : modified fats and synthetic fats. One example of a modified fat is salatrim, an acronym for short and long chain acid triglyceride molecules, which is produced by reconfiguring a triglyceride to include certain mixtures of stearic acid (a long chain fatty acid) and acetic, butyric, or propionic acids (short chain fatty acids) on the glycerol backbone. Because short chain fatty acids are energetically less dense than longer chain acids and stearic acid is only partly absorbed, salatrim provides fewer calories than a typical type of fat. This can be used to replace fat in chocolate, confectionary products, dairy products frozen desserts, and cookies. The FDA has granted a GRAS (Generally Recognized As Safe) status for salatrim.

Example of synthetic artificial fat made from sucrose and edible vegetable oils is olestra and its chemical configuration does not occur in nature. Unlike normal fats, which are made up of one molecule of glycerol attached to three molecules of fatty acids, olestra are produced replacing the glycerol molecule with sucrose and attaching six, seven or eight fatty acids.

With these many fatty acid digestive enzymes are not able to hydrolyze so olestra remains undigested and contributes no calories or fat to the diet. In 1996, Olestra was approved by the Food and Drug Administration (FDA). However, along with many beneficial effects of synthetic fat replacers, several health concerns have been observed that is interfering with absorption of fat-soluble vitamins, gastrointestinal or other specific side effects, decreasing motivation to undertake lifestyle behaviors like physical exercise. Thus, there is need for more research to address these concerns in order to make fat replacement technology more applicable and acceptable for elderly.

(iii) Enzyme technology

Recently use of enzyme in the food products have been shown nutritional and health benefits , that will be beneficial for the elderly population. For example, oxidoreductases have been used as catalysts in food systems to convert cholesterol to the non-toxic coprostanol. This usage makes foods free from the detrimental effects associated with cholesterol. Enzymes are also used to improvise the nutritional content of foods. One of the example is of phytic acid that is an anti –nutritional content in many cereal grains, legumes and oil seeds which interferes with the absorption of micro nutrients like calcium , iron and zinc. Exogenous supplementation of such foods with phytases enhances their micronutrient availability. Some oligosaccharides, such as raffinose, stachyose and verbascose, are anti-nutritional factors present in legumes that are not metabolized by humans, causing flatulence, diarrhea and indigestion in geriatric and normal population. These oligosaccharides are linked by α-D-galactosidic bonds which are resistant to cooking and other processing steps, but are hydrolysable by α-D- galactosidases. Thus α-D- galactosidases have been exploited as food additive in production of processed legume- based food to hydrolyze the heat-resistant oligosaccharides. Because enzymes exist in natural sources and are perceived as non-toxic, the catalytic activity of enzymes has been exploited in large scale processes in food production as a preferred method than the chemical processing method as compared to enzymes derived from plant and animal , microbial enzymes are often more useful source as they offer wide range of catalytic activities, high yields, easy genetic manipulation, rapid growth in cheap media and also regular supply(Table-4). Industrial food enzymes fall into four categories, namely hydrolases, oxidoreductases, isomerases and lyases and each performs specific functions in food processing. For example, in the dairy industry, sulfhydryl oxidase is used to correct flavor defects due to the thiols formed in UHT-preserved milk and in the starch industry, hydrolases and isomerases are used to produce sweet high-fructose syrups from starch (Adler-Nissen, et al., 1987). Chitinases are produced by plants as a defense mechanism against invading fungal pathogens. These enzymes are also active against human pathogens such as *Listeria monocytogens, Clostridium botulinum, Bacilluscereus, Staphylococcus aureus* and *Escherichia coli.* High levels of chitinase activity are present in germinating soybean seeds and in the other legumes and research shows that enzymes (Table 4) can be used as anti microbial and in addition to reduce levels of anti-nutritive food components can be used for convenient food preparation for elderly (Scott, 1988).

Table 4: Conventional technologies used to soften traditional foods and meals.

Technology	Foods	Principle / Claims
Enzymatic treatments	Beef, chicken (eventually other foods)	Impregnation of foods with enzymes which breakdown cell wall components and/or structural tissue leading to bland textures
Freeze - thawing infusion	Bamboo shoots, roots, fish, mushrooms, etc.	Impregnation of substances (e.g:- enzymes) into foods combined with slow freezing nad vacuum. Softens the food while keepingflavors

Technology	Foods	Principle / Claims
High - pressure processing	Several foods (meat, fruits, salads, ready - meals, etc.)	High pressures soften cellular foods but retain flavours and nutrients and may improve bio- availability of bioactive compounds
Pulsed electric fields and sonication	Several cellular foods, meats (possibly other foods)	Tissue softening in fruits and vegetables is induced by cell membrane electroportation. Loosening of collagen fiber arrangement is caused by cavitation and shear

(iv) Fermentation

Process of fermentation involves production of food using microorganisms that produces enzymes such as amylases, proteases and lipases which hydrolyze the carbohydrates, fats and proteins present in food gives enhanced flavor, aroma and texture. Fermentation is an inexpensive procedure and it can be easily done at local household level with low cost involved. This method is a suitable technique for geriatric population as it increases the nutritional and health benefits of traditionally fermented foods. Fermentation technology was used to make alcoholic beverages through yeast fermentation, vinegars through fermentation by *Acetobacter*, yogurt and pickles through fermentation by *Lactobacilli* . Majorly Lactic acid bacteria are used in traditionally fermented foods as a preservative, as the lactic acid is released it makes the food more digestible and also prevents the growth of pathogenic micro organisms . Many studies have reported the apparent health benefits of fermented milks, including the lowering of serum cholesterol and anticancer activity (Abdel Gadir et al., 1983).

The recent use of fermentation in food processing has emphasized food production with health benefits and better nutritional content. Fermentation process decreases the levels of anti-nutritive compounds like tannin and phytate, to increase the bioavailability of essential nutrients as iron (Towo, et al.,2006).

This also reduces anti nutritional factors and the occurrence of natural toxins such as cyanide in root cassava, to minimize the level of non-digestible carbohydrates and hence reduce negative side effects like abdominal distress and flatulence associated with non-digestible carbohydrate which increases the bioavailability of vitamins like thiamine, riboflavin, niacin or folic acid. Beneficial aspects of fermentation processes for human and animal health are clear but certain risk factors are also associated with these foods. For example, in the process of fermentation there is production of some pathogenic organism and toxic compounds such as mycotoxins and biogenic amines. Further research on the development of harmless cultures studying their safety concerns would reap the maximum benefit from fermented food products.

(v) Food biotechnology

Biotechnology uses biological systems, living organisms, or components of organisms to make or modify products or processes for specific uses as per Codex Alimentarius Commission. Biotechnology is defined as the application of 1) in vitro nucleic acid techniques, including recombinant deoxyri- bonucleic acid (DNA) and the direct injection of nucleic acid into cells or organells, or 2) the fusion of cells beyond the taxonomic family that overcome natural physiological reproductive or recombination barrier.

(vi) Encapsulation

Encapsulation has been used majorly in past years for pharmaceutical use and chemical industries, and has now gone on to find new applications in the food industry. The type of micro encapsulation used in the food industry involves the incorporation of food ingredients, enzymes, cells, nutrients and/or other bio-ingredients in small capsules (microcapsules), enabling to introduce active ingredients in the food products and allow them to be released gradually at a controlled time and rate.

The encapsulated core material is protected from moisture, heat or other extreme conditions to enhance its stability and maintain viability. Encapsulation is also employed to mask odors or tastes, to control interactions of the active ingredient with the food matrix, and to control the release of the active agent (Ubbink, and Krüger, 2006). Probiotics, which are living microorganisms, exert beneficial effects in the gut by controlling undesirable microorganisms in the intestine. The encapsulation of probiotics therefore offers a way to introduce them into food systems that produces the desired effect associated with live microorganisms .

Commonly used encapsulating agents are carbohydrates (due to their ability to absorb and retain flavors), cellulose (based on its permeability), gums (which offer good gelling properties and heat resistance), lipids (based on its hydrophobicity), proteins (usually gelatin, which is non- toxic, inexpensive and commercially available), emulsifiers, and fibers. Some combination of these encapsulating agents is commonly used. Controlled release is also achieved by way of diffusion through thin wall of encapsulating agent, which serves as a semi-permeable membrane. Active component in either liquid or solid forms are encapsulated and added to the food product to enhance sensory qualities of the food (for example, improve flavor or mask odor and taste) to improve the nutritional content of food , or both. Encapsulation of fish oils provides a means of protecting them against oxidation and enabling them to be incorporated into a larger variety of foods. Probiotics living microorganisms exert beneficial effects in the gut by controlling undesirable microorganisms in the intestine.

The encapsulation of probiotics therefore offers a way to introduce them into food systems that produces the desired effect associated with live microorganisms. Efforts to encapsulate active ingredients require careful planning. Protecting the ingredients during processing and then delivering and releasing them in a highly complex food matrix depends on various factors such as the composition and structure of the encapsulating material, the production conditions (temperature, pH, pressure, humidity), and the effectiveness of the encapsulated particles.

(vii) Nanotechnology

Nanotechnologies involve the study and use of materials (nano materials) at nano scale (sizes of 100nm or less) sizes and dimensions exploiting the fact that some materials at these ultra small scales have different physiochemical properties from the same materials at a larger scale(Fig-4).

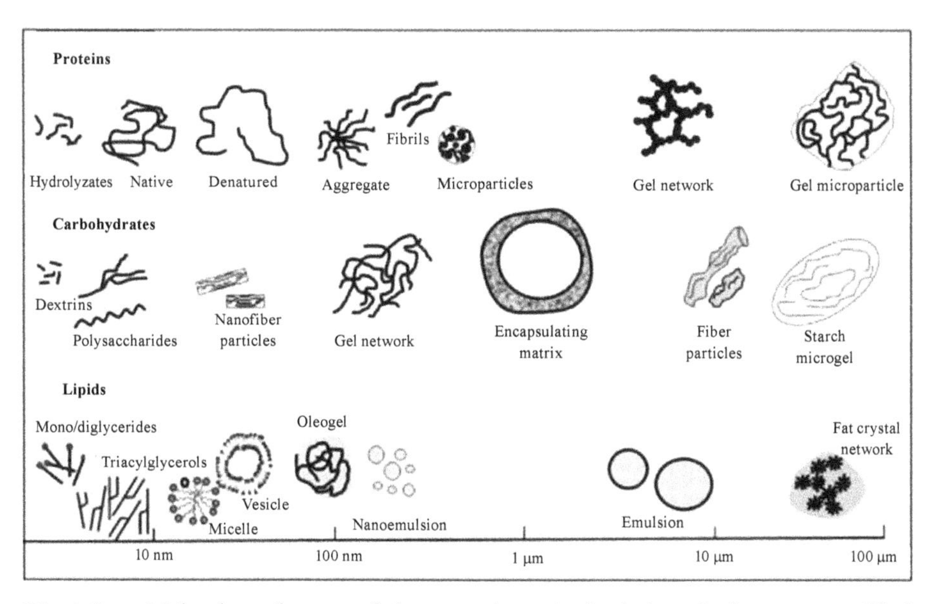

Fig. 4: Some Molecules and structural elements relevant in the design of soft texture-modified foods for the elderly (approximate scales)

Nano- materials are produced using two building strategies, either a “top down” or a “bottom up” approach. With the former approach, nano materials are created by breaking up bulk materials using such means as milling, whereas with the latter approach the nano materials are built from individual atoms or the molecules that have the potential to self assemble. Most of the recent applications of nanotechnology are in the areas of electronics, medicine, pharmacy and materials science. Nanotechnology also offers exciting possibilities for detecting chemical, biological radiological and explosive (CBRE) agents, and for protecting lives from and neutralizing CBRE agents. However, the field is under the extensive study and has potential applications in the area of agriculture and food (Morrissey, 2006). Probiotics, which are living microorganisms, exert beneficial effects in the gut by controlling undesirable microorganisms in the intestine. The encapsulation of probiotics therefore offers a way to introduce them into food systems that produces the desired effect associated with live microorganisms.

Efforts to encapsulate active ingredients require careful planning. Protecting the ingredients during processing and then delivering and releasing them in a highly complex food matrix rely on various factors as structure and composition of the encapsulating material, the production conditions (temperature, pH, pressure, humidity), and the effectiveness of the encapsulated particles.

Conclusion

Enriched foods can support elderly consumers in managing their nutrient intake, however, some precautions are useful in successfully approaching the elderly population with these foods. Elderly consumers are very diverse in what they want. Market segmentation can help to better understand these diverse wants and can give guidance for developing products that meet the needs and wants of a larger proportion of the elderly population.

Scientific evidence has prompted consumers to increasingly opt for low calorie and low fat food as well as other foods that promise health benefits. Food processors are eagerly adding value to their products based on nutrient information to meet the current consumer demand for healthier food products. These added values include removing or reducing anti-nutritive components that are present naturally in food matrix; reducing food components such as fat, caffeine or calories; adding bioactive ingredients that offer health benefits; and increasing the amount of essential nutrients present in food through the various technologies like encapsulation, nanotechnology etc. Various food technologies must work together to achieve the goal of manufacturing healthy food products while at the same time maintaining their sensory qualities. With continuing advances in food technology incorporating functional ingredients and latest technologies, there is lot of scope for the development of novel food products for the health benefits of geriatric population. Nutrition information is being translated into consumer products at an accelerated pace with the aid of food technology. Research into the creation of new health foods promises to be exciting futures that can immensely benefitting the geriatric population.

References

Abdel Gadir, A. M., Mohamed, M., Abd-el-Malek, Y., Ahmad, I. H., & Arroyo, P. T. (1983). Indigenous fermented foods involving an acid fermentation: preserving and enhancing organoleptic and nutritional qualities of fresh foods. *Microbiology Series* (USA).

Adler-Nissen, J. (1987). Newer uses of microbial enzymes in food processing. *Trends in Biotechnology,* 5(6), 170-174.

Aykroyd, W. R., Doughty, J., & Walker, A. F. (1982). Legumes in human nutrition (FAO Nutritional Studies No. 19). preparation, Rome, FAO.

Babu, V. S., Patra, S., Thakur, M. S., Karanth, N. G., & Varadaraj, M. C. (2005). Degradation of caffeine by Pseudomonas alcaligenes CFR 1708. *Enzyme and Microbial Technology,* 37(6), 617-624.

Bahlol, H. E. M., Sharoba, A. M., & El-Desouky, A. I. (2007). Production and evaluation of some food formulas as complementary food for babies using some fruits and vegetables. *Ann. of Agric. Sc., Moshtohor,* 45(1), 147-168.

Balasubramanian, S., Kaur, J., & Singh, D. (2014). Optimization of weaning mix based on malted and extruded pearl millet and barley. *Journal of Food Science and Technology,* 51(4), 682-690.

Barrell, R. A. E., & Rowland, M. G. M. (1980). Commercial milk products and indigenous weaning foods in a rural West African environment: a bacteriological perspective. *Epidemiology & Infection,* 84(2), 191-202.

Bender, A. E. (1973). Nutrition and dietetic foods. Leonard Hill Books (Intertext Publishing Ltd.)..

Bressani, R., Harper, J. M., & Whickstrom, B. (1984). Processed and packaged weaning foods: development, manufacture and marketing. Improving the Nutritional Status of Children During Weaning Period. MIT. Cambridge, MA, EEUU, 51-64.

Central Food Technological Research Institute, Mysore, India, (1982). Food and nutrition bulletin, The united nations university press Food and Nutrition Bulletin Volume 04, Number 4, UNU, 85 pages.

Chandrasekhara, H. N., & Ramanatham, G. (1983). Gelatinization of weaning food ingredients by different processing conditions. *Journal of Food Science and Technology,* 20(3), 126-128.

Dahiya, S., & Kapoor, A. C. (1993). Nutritional evaluation of home-processed weaning foods based on low cost locally available foods. *Food chemistry,* 48(2), 179-182.

Deliza, R., Rosenthal, A., & Silva, A. L. S. (2003). Consumer attitude towards information on non conventional technology. *Trends in Food Science & Technology,* 14(1-2), 43-49.

Desikachar, H. S. R. (1982). Technology options for formulating weaning foods for the economically weaker segments of populations in developing countries. *Food and Nutrition Bulletin,* 4(4), 1-4.

Gabriel, R. A. O., & Akharaiyi, F. C. (2007). Effect of spontaneous fermentation on the chemical composition of thermally treated jack beans (*Canavalia ensiformis* L.). Int. *J. Biol. Chem,* 1(2), 91-97.

Gokulakrishnan, S., Chandraraj, K., & Gummadi, S. N. (2005). Microbial and enzymatic methods for the removal of caffeine. *Enzyme and Microbial Technology,* 37(2), 225-232.

Guthrie, H. A. (1975). Infant nutrition. Introductory nutrition. 3rd ed. St. Louis, Mo, USA: CV Mosby, 394-9.

Hahn, N. I. (1997). Replacing fat with food technology. *Journal of the Academy of Nutrition and Dietetics,* 97(1), 15.

Harper, J. M., & Jansen, G. R. (1985). Production of nutritious precooked foods in developing countries by low-cost extrusion technology. *Food Reviews International,* 1(1), 27-97

Hotz, C., & Gibson, R. S. (2007). Traditional food-processing and preparation practices to enhance the bioavailability of micronutrients in plant-based diets. *The Journal of Nutrition,* 137(4), 1097-1100.

Hsieh, P. Y. H., & Ofori, J. A. (2007). Innovations in food technology for health. Asia Pacific *Journal of Clinical Nutrition,* 16(S1), 65-73.

Imtiaz, H., BurhanUddin, M., & Gulzar, M. A. (2011). Evaluation of weaning foods formulated from germinated wheat and mungbean from Bangladesh. *African Journal of Food Science,* 5(17), 897-903.

James, J., Simpson, B. K., & Marshall, M. R. (1996). Application of enzymes in food processing. *Critical Reviews in Food Science & Nutrition,* 36(5), 437-463.

Jenkins, D. J., Jenkins, A. L., Wolever, T. M., Thompson, L. H., & Rao, A. V. (1986). Simple and complex carbohydrates. *Nutrition Reviews* (USA).

Joint, F. A. O. (1985). Energy and protein requirements: Report of a joint FAO/WHO/UNU Expert Consultation. In Technical Report Series (WHO) (No. 724). World Health Organization.

King, J., Nnanyelugo, D. O., Ene-Obong, H., & Ngoddy, P. O. (1985). Household consumption profile of cowpea (*vigna unguiculata*) among low-income families in Nigeria. *Ecology of Food and Nutrition,* 16(3), 209-221.

Kuller, L. H. (1997). Dietary fat and chronic diseases: epidemiologic overview. *Journal of the American Dietetic Association,* 97(7), S9-S15.

Macy,I.G and Kelly,H.J. (1961). Human milk and cow's milk in infant nutrition. In Milk: the mammary gland and its secretion (pp. 265-304). Academic Press.

Malleshi, N. G. (1988). Weaning foods. Mysore, India: Regional Extension Service Centre (Rice Milling) Ministry of Food Processing Industries, Government of India, and Discipline of Grain Science and Technology, Central Food Technological Research Institute, 1-40.

Malleshi, N. G., & Desikachar, H. S. R. (1982). Formulation of a weaning food with low hot paste viscosity based on malted ragi (*Eleusine coracana*) and green gram (*Phaseolus radiatus*). *Journal of Food Science and Technology,* 19(5), 193-197.

Malleshi, N. G., Daodu, M. A., & Chandrasekhar, A. (1989). Development of weaning food formulations based on malting and roller drying of sorghum and cowpea. *International Journal of Food Science & Technology,* 24(5), 511-519.

Martinez, B., Rincón, F., Ibáñez, M. V., & Bellán, P. A. (2004). Improving the nutritive value of homogenized infant foods using response surface methodology. *Journal of Food Science,* 69(1), SNQ38-SNQ43.

Morales, E., Lembcke, J., & Graham, G. G. (1988). Nutritional value for young children of grain amaranth and maize-amaranth mixtures: effect of processing. *The Journal of Nutrition,* 118(1), 78-85.

Morrissey, S. (2006). Nanotechnology in food and agriculture. *Chem Eng News,* 84, 31-31.

Mosha, A. C., & Svanberg, U. (1990). The acceptance and intake of bulk-reduced weaning foods: the Luganga village study. *Food and Nutrition Bulletin,* 12(1), 1-6.

Nelson, A. I., & Steinberg, M. P. (1978). Whole soybean foods for home and village use (No. Folleto 6203).

Nout, M. J. R. (1993). Processed weaning foods for tropical climates. *International Journal of Food Sciences and Nutrition,* 43(4), 213-221.

Obizoba, I. C., & Egbuna, H. I. (1992). Effect of germination and fermentation on the nutritional quality of bambara nut (*Voandzeia subterranea* L. *Thouars*) and its product (milk). *Plant Foods for Human Nutrition,* 42(1), 13-23.

Olkku, J., & Rha, C. (1978). Gelatinisation of starch and wheat flour starch—a review. Food Chemistry, 3(4), 293-317.

Parchure, A. A., & Kulkarni, P. R. (1997). Effect of food processing treatments on generation of resistant starch. *International Journal of Food sciences and Nutrition,* 48(4), 257-260.

Pszczola, D. E. (1997). Ingredients for Fat Replacement. *Food Technology,* 51(1), 82-88.

Ramalakshmi, K., & Raghavan, B. (1999). Caffeine in coffee: its removal. Why and how?. *Critical Reviews in Food Science and Nutrition,* 39(5), 441-456.

Rasane, P., Jha, A., Kumar, A., & Sharma, N. (2015). Reduction in phytic acid content and enhancement of antioxidant properties of nutricereals by processing for developing a fermented baby food. *Journal of Food Science and Technology*, 52(6), 3219-3234.

Rathod, P., & Udipi, S. A. (1991). The nutritional quality and acceptability of weaning foods incorporating amaranth. *Food and Nutrition Bulletin*, 13(1), 1-8.

Rutishauser, I. H., & Frood, J. D. L. (1973). The effect of a traditional low-fat diet on energy and protein intake, serum albumin concentration and body-weight in Ugandan preschool children. *British Journal of Nutrition*, 29(2), 261-268.

Saeeda, R., Safdar, M. N., Amer, M., Nouman, S., Khalid, N., & Muhammad, A. (2009) Preparation and quality evaluation of nutritious instant baby food from indigenous sources. *Pakistan Journal of Agricultural Research*, 22(1/2), 50-55.

Sangronis, E., & Machado, C. J. (2007). Influence of germination on the nutritional quality of *Phaseolus vulgaris* and *Cajanus cajan*. *LWT-Food Science and Technology*, 40(1), 116-120.

Scott, D. (1988). Antimicrobial enzymes. *Food Biotechnology*, 2(2), 119-132

Sefa-Dedeh, S. (1984). Hunger, Technology and Society: An old Processing Method, a New Protein Food. *Food and Nutrition Bulletin*, 6(1), 1-5.

Steve, I. O. (2012). Influence of germination and fermentation on chemical composition, protein quality and physical properties of wheat flour (*Triticum aestivum*). *Journal of Cereals and Oil Seeds* Vol. 3 (3), 35-47.

Svanberg, U. S. O., Fredrikzon, B., Gebre-Hiwot, B., & Taddesse, W. W. (1987). Sorghum in a mixed diet for preschool children. I. Good acceptability with and without simple reduction of dietary bulk. *Journal of Tropical Pediatrics*, 33(4), 181-185.

Towo, E., Matuschek, E., & Svanberg, U. (2006). Fermentation and enzyme treatment of tannin sorghum gruels: effects on phenolic compounds, phytate and in vitro accessible iron. *Food Chemistry*, 94(3), 369-376.

Ubbink, J., & Krüger, J. (2006). Physical approaches for the delivery of active ingredients in foods. *Trends in Food Science & Technology*, 17(5), 244-254.

UNICEF. (2012). Infant and young child feeding, nutrition section program. New York.

United Nations Population Ageing and Development. (2012). Available from: http://www.un.org/en/development/desa/population/publications/pdf/trends/WPP2012_Wallchart.pdf

Wang, Y. Y. D., & Fields, M. L. (1978). Germination of corn and sorghum in the home to improve nutritive value. *Journal of Food Science*, 43(4), 1113-1115.

World Health Organization (2005). Modern food biotechnology, human health and development: an evidence-based study.

10

Progression of Biotechnology and its Application in Food Processing

Janifer Raj Xavier[1*] and *Gopal Kumar Sharma*[2]

[1]*Fruits and Vegetables Technology Division, Defence Food Research Laboratory Defence Research and Development Organisation, Siddharthanagar Mysuru-570 011, Karnataka, India*

[2]*Head, Grain Science Technology Division, Defence Food Research Laboratory Defence Research and Development Organisation, Siddharthanagar Mysuru-570 011, India*

Introduction

Biotechnology is defined as the 'application of scientific and engineering principles to the processing of material by biological agents to provide goods and services'. The term 'biotechnology' is also used when a technique involving live organisms are employed to make or modify a product, improve plants or animals or develop microorganisms for specific uses. According to the definition of the Codex Alimentarius Commission (CAC 2001), modern biotechnology is defined as the application of (i) *in vitro* nucleic acid techniques, including recombinant deoxyribonucleic acid (DNA) and direct injection of nucleic acid into cells or organelles, or (ii) fusion of cells beyond the taxonomic family, that overcome natural physiological reproductive or recombination barriers, and that are not techniques used in traditional breeding and selection. Scientific developments of present world owe to processes and products of the field of biotechnology. Modern biotechnological advancements have played a major role in important branches of science such as agriculture, food processing, medicine and veterinary applications. Biotechnological interventions of the downstream processing applications include primary production areas namely breeding of crops and livestock, production of additives for food and feed and pharmaceuticals along with diagnostic tools. Inputs for food processing applications from biotechnology include biocatalysts such as enzymes, tracing food ingredients such as Genetically modified (GM) ingredients in end products of food and feed and food quality testing applications for food borne pathogens. Applications of modern biotechnology in form of genetically modified organisms have gained much importance in comparison to the downstream applications during food processing and food safety (Zhu, 2017). Food processing operations include transformation of harvested crop or animal produce into produce which are marketable with long shelf life and minimises waste and losses of raw produce in food chain and ensure their continuous availability in super market shelves. Food safety is an important area which assures of the food being safe for consumption. Food technological applications of modern biotechnology includes food enzymes that are important in various food processing applications in reactions involving lipids, carbohydrates and proteins, food fermentations to improve properties such as taste, flavour, texture, nutritional and shelf stability, food additives in form of bio preservatives, flavours, fragrances, genetically modified starter cultures, genetically modified crops, modern rapid diagnostic kits for food safety are discussed in detail.

Enzymes and their applications in food processing

Enzymes are natural substances capable of increasing the rate of specific biochemical reactions in living systems without being used up during the process. Though enzymes catalyze various reactions in living organisms, they also act *in vitro* and find applications in various industrial applications. The first successful commercial use of enzyme in food processing was cheese-making and the importance of enzymes in food processing is well documented. Their specificity, efficacy, safety, suitability and ease of usage are major reasons which are attributed to the usage of enzymes in various food processing applications. The food enzymes market size was estimated at US $1,944.8 million in 2018 and is expected to reach US $3,056.9 million by 2026, registering a CAGR of 5.6 percent from 2019 to 2026 (Food Enzymes Market Outlook – 2018). Enzymes are natural products, without which the food and beverages processing would not be possible. The exogenous enzymes facilitate the process, enhance quality, make processing shorter, and are more constant. The widespread use of enzymes for food and beverages processing is a well-established approach but the latest techniques for the designing of new biocatalysts are built to make the useful applications of biological agents more precious. Food enzymology dates back to ancient times when alcoholic fermentation of malted grains derived fermented products and the word enzyme is derived from term 'in yeast'.

Enzymes catalyse reactions by reducing the energy barrier for transformation of reactants into desirable products. Specific catalytic mechanisms of enzymes include approximation, covalent catalysis, general acid-base and molecular strain. Enzymes of broad categories such as oxido reductases, transferases, hydrolases, lyases, isomerases and ligases find importance in exogenous food applications. Food uses of enzymes include transformation of carbohydrates, proteins, lipids and other miscellaneous applications. Glycosyl hydrolases or glycosidases act on glycosidic bonds and used in production of sweeteners such as corn syrups, maltose glucose syrups, and dextrins. Amylases are used to produce thin starch suspensions by hydrolysis of starch into simpler sugars as well in variety of applications in baking and production of digestive aids. Recombinant endo amylase hydrolysed starch to produce maltooligosaccharide with amylopectin as preferred substrate (Wang et al 2019). Malted cereals and fungi are important sources of recombinant amaylases and are used extensively in baking. Polydatin-b-D-glucosidase, an enzyme used in resveratrol preparation has been recently reported (Zhou et al 2016).

Enzymes play a major role in various fruit and vegetable processing such as pre peeling, pulp washing, juice extraction and clarification. Olsen (1995) listed the following reasons such as increase in juice yield, maximum utilization of raw material by liquefaction, to bring out desirable aroma and color, breakdown of insoluble sugars into simpler sugars and juice clarification to justify the importance of enzymes during fruit and vegetable processing. Various enzymes are required for the above mentioned processes due to the presence of varied biomolecules in plant tissues. Hydrolysis of polysaccharides and extraction of juice from pulp known as liquefaction technology uses enzymes such as polygalacturonases, pectin lyases, pectinesterases and cellulases releases D-glacturonic acid and neutral sugars (D-arabinose, D-galactose, L-rhamnose and D-xylose) from complex pectic substances. Thus the pressing processes become easier and maximum amount of juice extracted from the fruit pulp is maximized. Pectinases act up on cloudy juices and yield clarified juices with improved consumer acceptance in terms of taste, appearance, stability and texture while xylanases are used along with cellulases and pectinases for clarification and liquefaction of fruit juices (Grassin and Fauquembergue, 1996). Citrus juices are prone to bitter taste due to presence of limonin and naringin in citrus peel, which can be successfully debittered using enzymes such as limoninase and naringinase. Formation of limonin is prevented by limonoate dehydrogenase by catalyzing the oxidation of lactone A-ring to 17-dehydrolimonoate a precursor of limonin (Kola et al 2010). Naringinase consists of α-rhamnosidase (EC 3. 2.1.40) and flavonoid-β-glycosidase (EC 3. 2.1.21) produced by solid state fermentation of *Aspergillus niger, Aspergillus usamii, Aspergillus oryzae, Cochiobolus miyabeanus, Penicilium decumbens, Phanopsis citri, Rhizopus nigricans and Rhizotonia solani. A. niger* produces naringinase which is commercially used and it is reperted to have the pH optimum of 3.0 to 7.0 and 4.0 to 6.0 for rhamnosidase and β-glycosidase activity, respectively (Puri and Banerjee,2000). Fruit and vegetable processing industries use enzymes to increase the yield and soluble solid content. Enzyme addition in the auxiliary tank and clarification stage are important to remove the insoluble pectin and to reduce the viscosity, degrade residual starch, and facilitate the flocculation of insoluble molecules. 3 to 20 g of commercial pectinase per 100 kg of fruit and from 0.2 to 2.0 g of commercial cellulase per 100 kg of fruit with temperatures ranges from 30°C to 50°C is used approximately in auxiliary tank while 1.5 to 3.0 g of commercial pectinase per 100 kg of fruit and 0.5 to 2.0 g of commercial amylase per 100 kg of fruit with temperatures from 25°C to 45°C are used in clarification stage (Luciana et al 2016).

Proteinases or proteases are group of enzymes hydrolyzing proteins and belong to four major categories namely serine proteases, aspartic acid proteases, cysteine proteases and metalloproteases. Major food applications of proteases include meat tenderization, milk clotting, beverage processing, dough conditioning, debittering, texturization by crosslinking and flavor development. A novel milk-clotting enzyme produced using *Bacillus licheniformis* BL312 is useful in *Monascus*-fermented cheese production as a substitute for traditionally used rennet (Zhang et al 2019). Protein hydrolysates produced using proteases find applications in infant formula and development of foods with anti allergenic properties due to reactive proteins present in egg whites, legumes and cereals (wheat and soy), peanuts, milk (casein, globulin), fish and seafood. Hydrolysis of αs-casein and whey protein concentrate using serine protease obtained from yeast *Yarrowia lipolytica* reduced immunoreactive epitopes of milk protein (Dąbrowska et al 2020). Meat tenderization by proteases is achieved using sulf hydryl endopeptidases such as papain, ficin and bromealin which are capable of hydrolyzing collagen and elastin. Calpain 1, a calcium-dependent cysteine protease, is prominently used to tenderize meat at refrigerated conditions (Morton et al, 2019).

Lipases are enzymes capable of acting on lipids and produce acids and alcohols. Endo acting lipases in the presence of water aid in quick degradation of lipids by hydrolysis of acyl glycerol groups. Lipases are widely used in food industry to enhance flavor and generate specific aromas in products based on meat, cereals, fruit, vegetable and alcoholic beverages. Lipases acting on cereal grains improve bread volume and uniform crumb formation in addition to delay in staling of bread and production of quality noodles in terms of colour and soft texture (Raveendran et al 2018). *Lactobacillus plantarum* produced an alkali tolerant lipase used for production of sour dough, vegetable sausages and cheese (Esteban-Torres et al 2015). Phospholipases acting on phospholipids hydrolyse phospholipids of egg yolk aid in preparation of mayonnaise and eco friendly removal of phospholipids from vegetable oils (Horchani et al 2012). Pérez et al (2018) reported omega-6 and -9 fatty acid production from vegetable oils by using immobilized lipases.Novel enzymes with associated unique properties such as temperature and pH adaptance, tolerance to various metal ions, pressure are essential in food bioprocessing. Designer enzymes are being produced using protein engineering techniques such as site-specific mutagenesis to design enzymes catalyzing specific reactions.

Food fermentation

Production of fermented foods using various kinds of microorganisms is carried out from time immemorial. These microorganisms can be used to ferment thermally treated foods and raw materials of plant and animal origin to bring out newer products. Food fermentations are an essential part of the food processing industry bring out characteristic flavor, aroma, texture, aesthetics, extension of shelf life, through enzymes or microorganisms capable of producing bacteriocins and other compounds such as hydrogen peroxide, mitigate natural toxins such as hydrogen cyanide mycotoxins and also anti-nutritional substances such as enzyme inhibitors, flatulence factors, glucosinolates, lectins, tannins, polyphenols, phytic acid and saponins [68].Moreover, production of colour and flavor by biotechnology for improvement of organoleptic properties such as flavour, taste, aroma is gaining importance. Commercial production of more than 100 molecules for aroma and flavor enhancement is products of biotechnology. Acceptance of food mainly depends on these organoleptic properties and tools of biotechnology are being used to impart organoleptic qualities along with chemical and nutritional attributes.

Biotechnology has played an important role in development of improved strains of bacteria, yeast, and moulds, with desirable properties which could be used as 'starter cultures' consisting of single or mixed cultures for the production of fermented foods. Genetic engineering techniques are being widely used to modify starter cultures and design strains to improve their antibiotic resistance against enteropathogens, impart properties such as anti-cholesterolemic, anti-inflammatory, anti-carcinogenic. Diverse range of fermented food products are produced by a group of genetically varying and functionally similar microorganisms called lactic acid bacteria. Fermented dairy products such as buttermilk, cheddar cheese, cottage cheese, camembert cheese and sour cream are produced by controlled fermentation using lactic acid producing bacteria such as *Lactococcus lactis* ssp. *lactis* and *L. lactis* ssp. *cremoris*, *Leuconostoc mesenteroides* and *L. dextranicum*. *Lactobacillus delbruekii* sp. *bulgaricus*, *L. helveticus* and *Streptococcus thermophilus* are organisms capable of producing flavor compounds such as diacetyl and used for commercial production of high valued emmental, gruyere, Italian cheese, flavoured yoghurts and milk. Individuals with lactose intolerance and vegetarian population could chose alternative solid and liquid products based on cereals and pulses.

Preparation of fermented products based on cereals, pulses, fruits, vegetables, meat, and dairy using modified strains of microorganisms such as bacteria, yeast, and moulds for specific properties is currently a growing trend to mitigate degenerative and life style diseases. Bioavailability of nutrients and improved palatability of fermented foods based on substrates such as barley, maize, millet, oats, rice, rye, sorghum and wheat are reported. Red yeast rice is used in China to prepare rice wine and as food colourant and flavour. In the West, due to its medicinal properties it has led to the development of a product known as cholestin for use as a dietary supplement to reduce cholesterol levels in the body. Rice fermented using *Monascus purpureus* and *Saccharomyces bouldardii* resulted in the production of Monacolin K or lovastatin, reduced cholesterol biosynthesis in human body. The red rice along with Monacolin also contains substances that inhibit HMG-CoA reductase such as β-sitosterol, camesterol, stigmasterol and sapogenin, in addition to isoflavones and isoflavone glycosides and monounsaturated fatty acids. Weber et al (2013) have showed that 204 g/day of red rice caused substantial lowering in LDL-cholesterol without affecting the HDL-cholesterol content in human trials. Amazake (Japan), Pozol (Mexico), Kvass (Russia) and beer are popular fermented beverages based on staple crops. Gluten free products based on sour dough fermentations are preferred for their characteristic texture, flavor and nutritional quality. Fruit and vegetable fermented products are rich in substances such as antioxidants, minerals, vitamins and fiber. Vegetables such as cauliflower, cabbage, carrot, green peas and bell peppers are used to produce fermented products namely kimchi, sauerkraut etc. Whole fruits such as apples, lemons, mangoes, palms, papaya and pears, pulps and juices of banana and grapes are suitable for fermentation by lactic acid bacteria. Cider, vinegar and wine are famous fermented fruit based products. Apple cider vinegar has been reported to lower cholesterol and triglyceride levels by arresting oxidation of LDL cholesterol particles.

Recently fermented teas produced using bacteria, yeasts or fungi are gaining importance as they are rich in poly phenols, vitamins, organic acids and caffeine and are useful in treatment of gastrointestinal infection and neurodegenerative diseases such as diabetes, cardiovascular diseases and cancer [58]. Fermented olives and ginseng are reported to impart specific functional properties for treatment of Alzheimer's disease and performance improvement during *in vivo* studies. Fermented meat based products such as fermented sausages have volatile compounds such as short chain fatty acids, acetaldehyde, diacetyl, acetoin, ethanol, acetone, 3- butanediol which are responsible for the characteristic aroma. Prolonged ripening and drying of fermented sausages impart concentrated flavor, denser nutrients, firm texture and low moisture content. Fermented fish as whole and derived products are produced using lactic acid producing gram positive cocci such as *Streptococcus, Pediocccus* etc. Fermented sushi of Japan namely *Nare-zushi* made using mackerel, salmon, crucian carp or sandfish in raw form and fish sauces made using anchovy, small shrimp, icefish, sand lance, tuna, sardine and squid are popular for their characteristic flavor and taste [53].

Biotechnology for improving nutrition

A (GM) or transgenic plant is a plant that has a novel combination of genetic material obtained through the use of modern biotechnology. The genetically transformed plant contains a gene or more genes that were artificially introduced instead of plant acquiring them through pollination. Genetic engineering provides powerful tool to enhance the modification of plants with a view to express a novel trait that is normally not found in the given species. Genetic manipulation to improve crops is used by farmers for thousands of years through conventional selective breeding techniques. Domestication of corn by Native Americans from its wild relative named teosinte with short, thin ear and tiny kernels through selective breeding is a simple form of genetic manipulation to increase productivity. The extended tools of genetic manipulation allowing multiple genes to be transferred or modified include chemical mutagenesis, radiation mutagenesis, embryo rescue and somaclonal variation. Advances of genetic engineering techniques have paved the way for development of crops with precise and specific single gene transfers, which requires meticulous evaluation for food or environmental safety prior to introduction into environment and market. Transgenic plants developed till date fall under major categories with traits such as insect and disease resistance, herbicide tolerance, pollination control, drought tolerance and modified product quality. Major GM crops commercialized globally are soybean (90.7 million ha), maize (55.2 million ha), cotton (25.1 million ha), canola (9.0 million ha) and other minor crops such as sugar beet, alfalfa, papaya, squash and poplar (1.4 million ha). Herbicide tolerant crops are deployed in 102.6 million hectares of biotech crops, insect resistance at 27.4 million hectares and stacked traits at 51.4 million ha.

Many of our common food crops could be improved to better meet the nutritional requirements of humans or animals. Protein, starch and oil composition as well as micronutrient content can be improved to make foods and feeds more nutritious. The transgenic technology promises to produce plants with output traits such as increased nutritional value, medicinal properties, industrial utility, novel taste and aesthetic appeal. Transgenic potatoes developed with 30–60 percent high starch content absorb less fat during frying. Specialty vegetable oils containing favorable fatty acid profiles such as high oleic acid have been developed by enzyme biotechnology (Stark et al 1992; Muller-Rober et al 1992; Geigenberger et al 1998). Antisense technology was used to develop the first transgenic crop 'Flavrsavr' tomato in 1992 with slow ripening trait, by silencing a gene producing polygalacturonase. Vitamins and mineral elicit biologic responses and have positive effects on health. Genetically engineered rice 'Golden Rice' was developed with β-carotene, using three genes for three enzymes in the phytoene synthase pathway, to decrease malnutrition and blindness associated with vitamin A deficiency. Phytochemicals such as carotenoids and related pigments are important to reduce the risk of neurodegenerative diseases like cancer and mascular diseases. GM papaya, tomato, kale and spinach has been developed with improved quantity of carotenoids, lycopene and leutin which on consumption may elicit positive biologic responses and improve health and well being of human beings (DellaPenna, 1999).

Processed foods with additional health benefits along with their normal nutritional value are called as designer foods the term first used in Japan in 1980's. Designer foods are gaining advantage over normal foods as they are fortified with health promoting ingredients (biofortification) using fermentation by genetically modified microorganisms or recombinant DNA technologies (Kinney and Knowlton, 1998). Dairy products such as eggs with modified fatty acid composition in terms of omega-3 fatty acids, conjugated linoleic acid (CLA) and high availability of nutrients such as vitamin E, beta-carotene and minerals such as selenium, improved antioxidant potential of people consuming the same. Milk containing healthier fatty acids such as CLA and omega fatty acids, improved protein levels through modified amino acid profiles, lactose and β-lactoglobulin free has been developed using modifications by biotechnological tools. Mineral enriched chicken, beef and pork has also been obtained by modified diet containing selenium to poultry and farm animals.

Other phytonutrients with supposed health benefits such as glucosinolates, phytoestrogens and phytosterols could selectively be over expressed to therapeutic levels for preventative health care applications. Methods are being developed to produce vaccines in plants by introducing genes that express a protein antigen in crops such as corn, potatoes and bananas. These antigens elicit an immune response by producing antibody specific to the antigen on eating the fruit or stem tubers obtained from the transgenic crops. The feasibility of this approach was demonstrated in animal studies against a subsequent challenge from pathogens. In spite of the powerful tools provided by genetic engineering to enhance the modification of plants, for potential benefit of society, careful considerations of effects of utilizing these tools is essential to ensure that the results will be a net benefit to society. Till date, over 50 biotechnology crops such as corn, potatoes, tomatoes and squash have passed the regulatory review process and have been commercialized. Although, the need for experimental evidence and scientific finding to assess the risks versus benefits of transgenic technology is inevitable, in order to overcome the biological, cultural and dietary barriers pertaining to the genetically modified crops (Falk et al 2002).

Biotechnological approaches for food testing

Food safety is a major public concern worldwide and food consumption has been identified as the major pathway for human exposure to certain environmental contaminants in comparison to inhalation and dermal routes. Harvesting and transportation operations render raw produce bruised resulting in release of plant nutrients, and thereby, providing substrates for microorganisms present on the surface of the vegetables to grow. In addition, the processing of fresh produce into products may alter or increase the number and type of pathogens present on the surface of the product. During the last decades, the increasing demand of food safety has encouraged research regarding the risk, associated with consumption of foods contaminated by food borne pathogens, pesticides, heavy metals and toxic residues which causes potential health risks to consumers (Rahman and Noor, 2012). Biotechnological interventions are being recently used as a tool in diagnostics in order to monitor food safety, prevent and diagnose food-borne illnesses and verify the origins of foods. The most essential hazardous factors such as biological, chemical or physical are being detected and monitored, specifically and sensitively using latest inventions of proteomics and genomics to ensure the safety of foods. As classical methods for detection of contamination during production processes and final product, are considered labour intensive and time consuming, modern and rapid techniques developed using tools of biotechnology are a need of the hour. Specific and sensitive methods pertaining to detection of microbial contaminants and toxic compounds technologies based on proteomics and genomics are important to meet modern trends in food technology. Expansion and utilization of powerful technologies such as gene cloning, transformation, polymerase chain reaction, molecular markers, MALDI-TOF for rapid detection of high risk food borne pathogens and toxins for food safety applications is highly relevant in the present context. Serology based test kits are being used for diagnostics and lab on a chip technologies such as microarrays based on nucleotides and proteins on chips opens up wide opportunities in the field of detection and diagnostics as a whole. On-site detection of food contaminants and toxins during food processing operations and food fermentation processes is made easy by lab-on-chip based technologies to monitor the contaminants.

Biosensors based on bacteriophages are gaining importance for rapid detection of food spoilage microorganisms that are resistant to antibiotics (Ahmed et al 2014). Principle behind the specific impedimetric detection of bacteria using bacteriophages work under the principle of retardation of changes in impedance, due to changes caused by addition of bacteriophage to specific target organism. Conductivity of the growth medium undergoes a major shift due to the breakdown of large molecules without charge such as carbohydrates and proteins to smaller molecules carrying charge such as acids and amino acids (Schmelcher and Loessner, 2014). Bacteriophage based detection systems are highly likeable due to reasons such as their specificity towards target, ability to differentiate between viable and non-viable cells, signal amplifiers, easy to propagate and versatility. They also possess various affinity molecules which could be used for variety of applications. Techniques involved, methods employed and assays using bacteriophages in the field of biosensors has been reviewed in detail (Zourob and Ripp, 2010; Singh et al 2013).

Bacteriophages could be applied in two ways: ones based on the proteins isolated from bacteriophages and the lytic phages as whole (Henry and Debarbieux, 2012). Dip stick based biosensor for rapid detection of various food borne pathogens with a detection limit of 10-50 CFU/ml in combination with quantitative real time PCRs was reported. Endolysins with enzymatic activity (EAD) and cell-wall binding domains (CBD) are found useful in detection of *Listeria monocytogenes*. Direct visualisation of vegetative cells of *Clostridium tyrobutyricum* in cheese samples was carried out using fluorescent labeled CBDs and the CBDs were capable of binding with clostridial spores and aided in rapid detection of spores of clostridium in milk and in prevention of late blowing defect of cheese.

Food biopreservation preservation using food additives derived from microorganisms

Biopreservatives are antimicrobial metabolites derived from microorganisms which prevent or control the growth and survival of food borne pathogens and contaminants without affecting the quality of food products. They are one among the twenty five categories of food additives, which are added purposefully in smaller quantities during food processing procedures to impart specific properties such as improving the shelf life, altering properties in terms of colour, structure, texture, flavour without modifying the nutritional value of foods (Lidon and Silvestre, 2007). Organic acids, aldehydes and ketones, low molecular weight proteins and bacteriocins are among important metabolites produced by microorganisms with antimicrobial properties. Use of antibiotics to curtail food spoilage causing micro organisms may lead to development of antibiotic resistant microorganisms which lead to the application of bacteriocins, the antimicrobial peptides for food preservation.

Bacteriocins are hydrophobic peptides composed of 30 to 50 amino acids which are capable of providing anti microbial activity against a broad spectrum of microorganisms and are digested by proteases present in gastro intestinal tract during digestive process in human beings. They are produced by lactic acid bacteria and are reported to be non-immunogenic, non-toxic and heat resistant to pasteurisation temperatures. They are very suitable for modifications of their intensity and spectrum of action (Nath et al 2014). As the bacteriocins are produced by lactic acid producing bacteria, they are generally considered as safe (GRAS). The first bacteriocins named colicine was discovered by Andre Gratia in 1925 as reported by Jacob et al (1953). Lactic acid producing bacteria capable of producing bacteriocins could be used as starter cultures for food fermentation to impart the desired antimicrobial action in foods and prevent the spoilage of food products along with prevention of butyric acid fermentation and white fungus. Nisin producing bacterial strains prevented growth of *C. botulinum* in cheese and *L. monocytogenes* in Camembert cheese when used as started cultures for cheese production. *Pediococcus pentosaceus* DT016 cultures suppressed growth of *L. monocytogenes* significantly during storage of vegetables at 4°C such as fresh lettuce, rocket salad, parsley and spinach when used as active cultures and as pediocin solution in comparison to chlorine wash, a commonly used surface treatment to reduce contaminating microorganisms (Ramos et al 2020). Meat and meat products subjected to non-thermal methods of food preservation require biopreservatives such as nisin, enterocin AS-48, enterocins A and B, sakacin, leucocin A and pediocin PA-l/ach in combination to impart protection from food spoilage causing microorganisms. Refrigerated vacuum packed sea foods are protected using nisin in combination with micogard against Gram-negative microorganisms such as *C. botulinum* in cheese and *L. monocytogenes*. Pediocin PA-1/AcH effectively controls Gram positive bacterial strains belonging to genera of *Leuconostoc, Lactobacillus, Carnobacterium, Brochothrix* and *Clostridium* infecting vacuum packed products of fruits and vegetables, dairy and meat. Cured and sliced meat products stored at refrigerated temperatures are protected from Listerial infections using active cultures of *Lactobacillus sakei*. Natamycin or pimaricin (E235) is also used as a food bio preservative against yeasts and fungi. Surface treatment of cheese and dry sausage against the formation of fungi.

Endolysins are a group of enzymes produced by bacteriophages at the end of the virulence cycle capable of hydrolysing peptidoglycan present in bacterial cell wall, resulting in cell wall damage and rupture. Efficacy of bacteriophages for biopreservation has been recently confirmed by large number of reports by removal of food spoilage causing bacteria (Harada et al 2018). Virulent bacteriophages which undergo lytic cycle are found to be effective for biocontrol of food spoilage bacteria. Recent reports on endolysins, their scope and biotechnology for improvement of food safety have opened venues for safe application of bacteriophages and reduce chemical preservatives in the food chain (Jhamb, 2014). Bacteriophage therapy for treatment of livestock and crops may reduce the risk of pathogen contamination before processing of plant and animal based products. Lytic phages are capable of destroying bacteria effectively and may help in reduction of dangerous pathogens in comparison to regular chemical preservation methods of processed food. Sillankorva et al (2012) reported elimination of fatal bacteria *Escherichia coli* O157:H7 in processed foods by pre and post harvest treatment of raw material. Myoviridae bacteriophage (pSs-1) a virulent strain of phage was reported as a biocontrol agent for water contaminated by *Shigella flexneri* and *Shigella sonnei* (Jun et al 2016). Commercially available phage based biocontrol systems for food use is ListShield® (Intralytix) or LISTEX® (Micreos Food Safety), EcoShield® (Intralytix) and SALMONELEX® for *L. monocytogenes*,*E. coli* O157:H7 and *Salmonella* sp. respectively (Rodríguez- Rubio et al 2016).

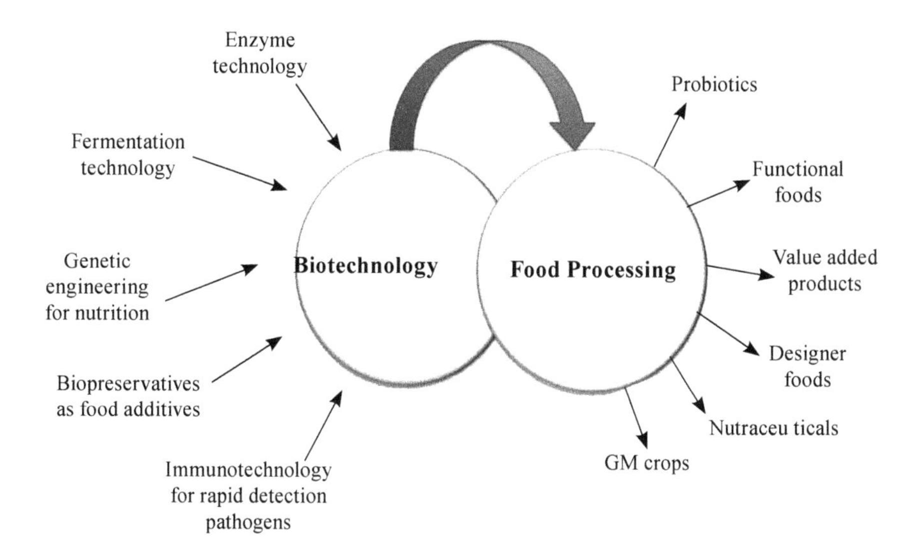

Fig. 1: Relationship between biotechnology and food processing

Future prospects

Globally, food processing research and its adoption in terms of industrial scale-up amounts to only less than one percent of the turnover. Apart from food processing giants such as DuPont and Monsanto, smaller industries are growing in a faster pace in the current scenario due to the growing entrepreneurial interests in modern biotechnological tools in food industry. As the awareness of the applications of biotechnological tools in food processing and food safety increases, research skills and integrated growth of biotechnology into food research disciplines would emerge. Nutrigenomics, a study of the response of organisms to food and food components using genomics, proteomics and metabolomics tools and nutrigenetics which refers to genetically determined variations in how individuals react to specific foods, would pave way in matching nutritional requirements to human genotypes to reduce the risk of diseases and provide optimum health through personalized nutrition. Services in form of tailored nutritional advice are not far in future where companies involved in genetic testing and nutritional expertise would play major role for well being of individuals.

References

Ahmed, J.V. Rushworth, N.A. Hirst, P.A. Millner (2014) Biosensors for whole-cell bacterial detection *Clin. Microbiol. Rev.*, 27 (3), 631-646

Dąbrowska Anna, Joanna Bajzert, Konrad Babij, Marek Szołtysik, Tadeusz Stefaniak, Ewa Willak-Janc, Józefa Chrzanowska, (2020) Reduced IgE and IgG antigenic response to milk proteins hydrolysates obtained with the use of non-commercial serine protease from *Yarrowia lipolytica, Food Chemistry*, 302, 125350

DellaPenna, D. (1999) Nutritional genomics: manipulating plant micronutrients to improve human health. *Science* 285: 375–379.

Falk Michael C., Bruce M. Chassy, Susan K. Harlander, Thomas J. Hoban, IV, Martina N. McGloughlin and Amin R. Akhlaghi. (2002). Food Biotechnology: Benefits and Concerns. *American Society for Nutritional Sciences*. 1384-1390.

Food Enzymes Market by Type (Carbohydrase, Lipase, Protease, and Others), Application (Bakery Products, Beverages, Dairy Products, and Others), and Source (Microorganisms, Animals, and Plants): Global Opportunity Analysis and Industry Forecast, 2019–2026 https://www.alliedmarketresearch.com/food-enzyme- market

Food Enzymes Market: Global Trends and Forecasts to (2018). Available at http://www.researchandmarkets .com/research/3xgvqh/food_enzymes.

Geigenberger, P., Hajirezaei, M., Geiger, M., Deiting, U., Sonnewald, U. & Stitt, M. (1998) Overexpression of pyrophosphatase leads to increased sucrose degradation and starch synthesis, increased activities of enzymes for sucrose-starch interconversions, and increased levels of nucleotides in growing potato tubers. *Planta* 205: 428–437.

Grassin, C. and Fauquembergue, P. *Fruit Processing, Oberhonnerfeld*, 1996, 12, 490-495.

Harada Liliam K, Erica C.Silva, Welida F. Campos, Fernando S. Del Fiol, Marta Vila, Krystyna Dąbrowska,Victor N.Krylov, Victor M.Balcão, (2018) Biotechnological applications of bacteriophages: State of the art Microbiological Research, 212–213.

Henry, L. Debarbieux Tools from viruses: bacteriophage successes and beyond (2012). *Virology*, 434 (2) 151-161

Jhamb, S., 2014. Biopreservation of food using bacteriocins, bacteriophages and endolysins. *Bombay Technol.* 64 (1), 9–21.

Kinney, A. J. & Knowlton, S. (1998) Designer oils: the high oleic acid soybean. In: Genetic Modification in the Food Industry (Roller, S. & Harlander, S., eds.), pp. 193–213. Blackie, London.

Morton James David, Zuhaib Fayaz Bhat, Alaa El-Din Ahmed Bekhit, (2019). Proteases and Meat Tenderization, Editor(s): Laurence Melton, Fereidoon Shahidi, Peter Varelis, Encyclopedia of Food Chemistry, Academic Press, Pages 309-313.

Olsen, H. S. 1995. Use of enzymes in food processing. In: *Biotechnology, 2nd Ed., Vol. 9, Enzymes, Biomass, Food and Feed*, H.-J. Rehm, G. Reed, A. Puhler and P. Stadler (eds.), pp. 663–736, Weinheim: VCH Verlagsgesellschaft mbH.

Pérez Malena Martínez, Enrico Cerioni Spiropulos Gonçalves , Jose Carlos Santos Salgado Mariana de Souza Rocha, Paula Zaghetto de Almeida, Ana Claudia Vici, Juliana da Conceição Infante, Jose Manuel Guisán , Javier Rocha-Martin, Benevides Costa Pessela and Maria de Lourdes Teixeira de Moraes Polizeli (2018). Production of Omegas-6 and 9 from the hydrolysis of açaí and buriti oils by lipase immobilized on a hydrophobic support molecules 23, 1-18 ; doi:10.3390/molecules23113015

Puri, M. and Banerjee, U.C. *Biotechnology Advances*, 2000, 18, 207–217.

Ramos Bárbara, Teresa R.S. Brandão, Paula Teixeira, Cristina L.M. Silva (2020). Biopreservation approaches to reduce Listeria monocytogenes in fresh Vegetables Food Microbiology 85 103282

Schmelcher, M.J. LoessnerApplication of bacteriophages for detection of foodborne pathogens *Bacteriophage*, 4 (2014), 28137

Singh,A, S. Poshtiban, S. EvoyRecent advances in bacteriophage based biosensors for food-borne pathogen detection (2013) *Sensors*, 13 (2) 1763-1786

Stark, D. M., Timmerman, K. P., Barry, G. F., Preiss, J. & Kishore, G. M. (1992) Role of the amount of starch in plant tissue by ADP glucose pyrophosphorylase. *Science* 258: 287–292.

Wang J, Guleria , Koffas MAG, Yan Y (2016). Microbial production of value- added nutraceuticals. *Curr Opin Biotech* 37: 97-104.

Weber JM, Reeves AR, Seshadri R, Cernota WH, Gonzalez MC and Wesley RK (2013). Biotransformation and recovery of the isoflavones genistein and daidzein from industrial antibiotic fermentations. *Appl. Microbiol Biotechnol.* 97: 6427-6437.

Zhanga Yao, Yongjun Xiaa, Zhongyang Dingb, Phoency F. H. Laia, Guangqiang Wanga,Zhiqiang Xionga, Xiaofeng Liua, Lianzhong Ai (2019). Purification and characteristics of a new milk-clotting enzyme from *Bacillus licheniformis* BL312 LWT113: 108276

Zhou L, Li S, Zhang T, Mu W and Jiang B (2015). Properties of a novel polydatin-β-D-glucosidase from *Aspergillus niger* SK34.002 and its application in enzymatic preparation of resveratrol. *Journal of the Science of Food and Agriculture* 96(7):2588-95. doi: 10.1002/jsfa.7465.

Zhu YY (2017) Application of modern biotechnology in food inspection. *Modern Food* 23: 67-69.

11

Smart Packaging in Food Sector

Rajeshwar S. Matche* **and** ***Monica Oswal***

Department of Food Packaging Technology, CSIR- Central Food Technological Research Institute, Mysuru 570 020, Karnata, India

Introduction

Recent advances in food technology have made our lives virtually easier and efficient. Changing lifestyle has led to the emergence of packaged food that provides convenience to the consumers. The authentic packaging of food implements consumers with the role of communication, containment, convenience, and protection. Any typical food package provides us with the information about the product and its nutritional fact, thus an efficient way of marketing/ branding. It protects the food product from the external environment and adulteration, thus keeps it safe. It allows the consumers to relish the food as per their convenience (reheating and direct consumption) and holds the product of any size, shape and restrains during handling and transportation (Robertson, 2006).

Food packages can be altered or modified according to one's requirement with features like single-serving dishes and portability. However, traditional packaging is lacking adequacy to meet unending consumer demands and product intricacies. Therefore, a contemporary concept of packaging with higher proficiency is needed to meet various consumer requirements. For instance, a packaging that can - enhance product's shelf-life with lesser or no preservatives, that maintains freshness, that meets permissible regulatory requisites and enables tracking throughout the product life-cycle. Emerging technologies like smart packaging have the expertise to invigilate a product and its milieu and pursue any modifications into these. Moreover, smart packaging helps in gaining global market recognition, and meets international food safety standards. In recent years, there has been rapid development in food packaging that includes active packaging, intelligent packaging, or smart packaging. These are the technologies that deal with the packaging of food, beverages, cosmetics, and pharmaceutical products. Moreover, these terminologies are apparently identical but, each has a different function to deal with and is as follows:

Active packaging is "packaging in which subsidiary constituents have been deliberately included in or on either the packaging material or the package headspace to enhance the performance of the package system" (Robertson, 2006).

Intelligent Packaging is "a packaging system that is capable of carrying out intelligent functions (like detecting, sensing, recording, tracking, communicating, and applying scientific logic) to facilitate decision making, to extend shelf-life, enhance safety, improve quality, provide information, and warn about possible problems" (LaCoste et al., 2005)

Smart packaging is "one that possesses the capabilities of both intelligent and active packaging. Smart packaging provides a total packaging solution that on the one hand monitors changes in the product or the environment (intelligent) and on the other hand acts upon these changes (active)" (Vander roost et al., 2014).

Smart packaging is captivating industrialists and researchers due to its high impact on food packaging and supply chain. This chapter mainly deals with recent advances in smart packaging, its trends, and applications in industries.

Active packaging

The packaging is said to be active only when it provides additional function apart from barrier properties to the food from detrimental factors (Rooney, 2005). This technique lodges active ingredients into the packaging that either releases or absorbs constituents into or from the food material or from the surroundings. Thus, monitors the quality, sensorial properties and extends the shelf-life of the products (Arvanitoyannis et al., 2012). The active ingredients can be embedded in the form of sachets, pads, or in label forms, sometimes also coated or laminated or extruded with the packaging material. The preservation of the food along with prolonged shelf-life is achieved using various strategies that include control of temperature, moisture, lowering or removing O_2 levels, use of antimicrobials, acids, ethylene or CO_2 scavengers, flavor absorbers/releaser or combination of these (Restuccia et al., 2010). This has impacted positively on the chemical interactions within the food, for instance, delayed oxidation into meat muscles, controlled respiration rate into fruits and vegetables, balanced pH into fruit juices (Biji et al., 2015).

Active packaging also enables one to exploit the selectivity to customize the amount of atmospheric gases into a package using coating, blending, extrusion, lamination and micro perforation (Brody et al., 2008). The application of smart packaging lowers the amount of migration from film to products, reduces localized activity, and processing conditions that would trigger the growth of microbes into perishable goods (Bolumar et al., 2011).

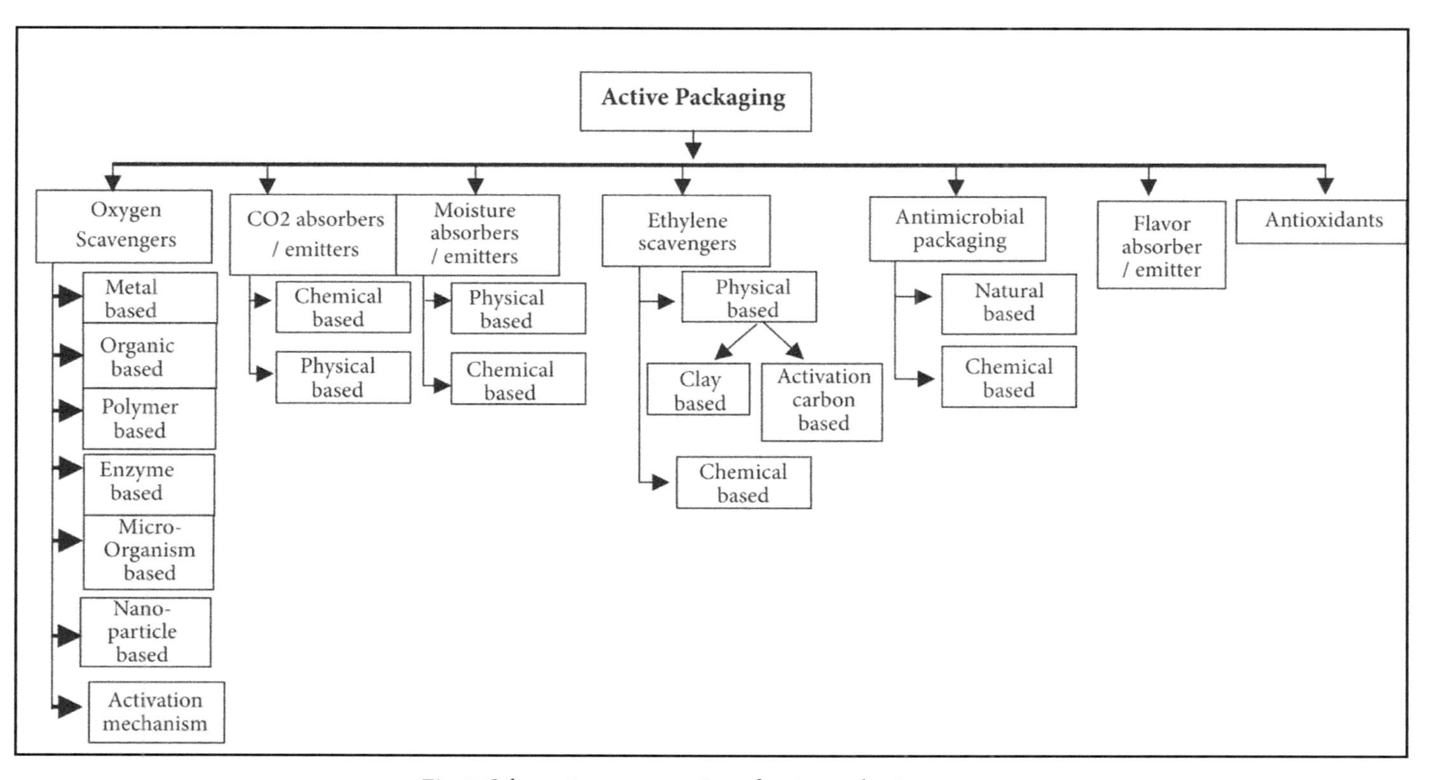

Fig. 1: Schematic representation of active packaging system

Table 1: Selected examples of active packaging systems

Active Packaging System	Examples	Food applications
Oxygen scavengers	Metal based (Iron, Zinc, cobalt(II), Iron nanoparticles, Palladium, Platinum, Magnesium), Organic (Ascorbic acid, Tocopherol, pyrogallic acid), Inorganic (Sulfites, bisulfites,), Polymer based (Oxidation- reduction resins), Enzyme-based (Glucose oxidase, oxalate oxidase, catalase, laccase), Microorganisms based (Yeast, aerobic bacteria like Bacillus amyloliquefaciens, Kocuria varians)	Bread, cakes, cooked rice, biscuits, pizza, pasta, cheese and dairy products, spices and seasonings, cured meats and fish, coffee, snack foods, egg products, dried foods and beverages
Carbon dioxide scavengers	Iron oxide, calcium hydroxide and ferrous carbonate/metal halide	Strawberry, pear, brinjal, kimchi, soy paste, mushrooms
Carbon dioxide emitters	Sodium bicarbonate, sodium glycinate, calcium oxide, activated charcoal and ascorbate	Coffee, fresh meats and fish, nuts and other snack foods and sponge cakes
Ethylene scavengers	Potassium permanganate, activated carbon and activated clays/zeolites	Fruits and vegetables
Antimicrobial packaging	Organic acids, silver zeolite, spice and herb extracts, antibiotics, metal oxides, ethyl alcohol, allyl isothiocyanate and chlorine dioxide	Cereals, meats, fish, bread, cheese, snack foods, fruits and vegetables
Antioxidants	BHA/BHT antioxidants, carotenoids, selenium, vitamin E antioxidant, and sulfur dioxide	Seafood, meat, milk, nuts, wine, fruits and vegetables
Moisture absorbers	Poly (vinyl acetate) blanket, activated clays and minerals and silica gel	Fish, meats, poultry, snack foods, cereals, dried foods, sandwiches, fruits and vegetables
Ethanol emitters	Encapsulated ethanol, alcohol spray	Pizza crusts, cakes, bread, biscuits, fish and bakery products
Flavor/odor absorbers	Cellulose triacetate, acetylated paper, citric acid, ferrous salt/ascorbate and activated carbon/clays/zeolites	Fruit juices, fried snack foods, fish, cereals, poultry, dairy products and fruits

Oxygen scavengers in food packaging systems

Oxygen plays a major role in the deterioration of the food products in various ways which includes oxidizing the lipid contents causing rancidity, undesirable flavour and odour, growth of mold, aerobic bacteria, insects and development of off colors. The main causes of spoilage in food are the chemical reactions within the constituents of the food, microbial growth, and the action of the enzymes present in the food (Choi et al., 2016). Therefore, removal or minimizing the oxygen content in the packaging of the food product will help in enhancing the shelf life of the food and also conserve the nutritional value of the food (Gaikwad et al., 2018).

The action of oxygen on food

Molecular oxygen can be reduced to a variety of reactive species such as superoxide, hydroxyl radical, hydrogen peroxide, and water, when they come in contact with certain compounds and undergo a chemical reaction. Carbon-carbon double bonds are particularly susceptible to the reactions with these intermediate reactive species. The free radical nature of the reactive oxygen species makes the oxidative reactions in which they participate autocatalytic. Products that contain oxygen reactive compounds or the product which contain complex organic compounds have carbon-carbon double bonds and thus undergo oxidative reactions affecting the quality and shelf life of the product (Zenner et al., 2002).

The main causes of spoilage in food are the chemical reactions within the constituents of the food, microbial growth, and the action of the enzymes present in the food (Choi et al., 2016). Some of the reasons for these may result from the presence of the oxygen around the food. Oxygen plays a major role in the deterioration of the food products in various ways which includes oxidizing the lipid contents causing rancidity, undesirable flavour and odour, growth of mold and aerobic bacteria and results in the deterioration of the organoleptic, nutritional and keeping quality of the food (Gaikwad et al., 2018).

Advantages of the oxygen scavengers: (Pereira et al., 2012; Ozdemir et al., 2004).

- Prevention of the oxidation phenomena, thus prevention of the rancidification of fats and oils and the consequent emergence of off-odors and off-flavors, loss or change of colors characteristic of food, loss of oxygen-sensitive nutrients (vitamins A, C, E, unsaturated fatty acids,etc.).
- Prevent the growth of aerobicmicroorganisms.
- Reduce or eliminate the need for preservatives and antioxidants in food by incorporating the added value of "fresh" or"natural."
- An economical and efficient alternative to the use of a modified atmosphere and vacuumpackaging.
- Slow down the metabolism of food and extend the shelf life of thefood.

Requirements of the compound or material to besat is factorily used as oxygen scavenger in the food industry must be (Rooney,1995):

- Harmless to the humanbody
- Absorb oxygen at an appropriaterate
- Should not produce toxic substances, unfavorable gases orodours
- Must possess constant quality and performance

- Compact insize
- Maximum oxygen absorptioncapacity
- Costeffective.

Types of oxygen scavengers

Metal based oxygen scavengers

The use of iron, zinc or magnesium powders as oxygen scavengers to remove oxygen from the headspace of packaged cans are most widely used agents. The metalpowders' used as oxygenscavengers paved the way for the subsequent development of the commercially available metal based oxygen scavengers. Most activated metal powders scavenges the oxygen by entering the oxide state and binds the oxygen. These are based on the principle of oxidation in the presence of moisture or LewisacidslikeFeCl_3 orAlCl_3 (Rollick, 2010). Iron powder-basedoxygen scavengers are the most commonly used oxygen scavengers for the commercial purposes. The reactions which are involved in the oxidation of the iron are:

$$Fe \rightarrow Fe_2+ + 2e- \quad (1)$$

$$1/2_2+H_2O+2e- \rightarrow 2OH^- \quad (2)$$

$$Fe^{2+} + 2\ OH^- \rightarrow Fe\ (OH)_2 \quad (3)$$

$$Fe\ (OH)_2+14O_2+12H_2O \rightarrow Fe\ (OH)_3 \quad (4)$$

Iron based scavengers may pose potential problems like accidental metal contamination, inhibiting heating in the oven, and inadvertently setting off online metal detectors (Gaikwad et al., 2018). Platinum and palladium efficiently catalyze the conversion of hydrogen and oxygen into water (Yu et al., 2004). Thus, they can be used asoxygen scavengers in food packaging in and the modified atmosphere containing hydrogen (Nyberg et al., 1984). Metal based oxygen scavengers can be included in the food packages in the form of sachets, films, labels or strips. Other metals like zinc and cobalt can also be used as oxygens cavengers by suitable activation processes (Gaikwad et al., 2018). Food samples like sausages were packaged with iron powder containing polymer films which showed a scavenging capacity of 33cm^3 O_2/m^2film in 4 days. It was observed that with an increase in temperature, the absorption of oxygen increased (Gibis et al., 2011).

Organic compounds based oxygen scavengers

Organic oxygen scavengers are such a low molecular weight organic compounds whose side chains reacts with oxygen or the backbone breaks apart when polymer in which they are incorporated reacts with oxygen (Ching et al., 2000). Ascorbic salts are known for the iroxygen scavenging abilities. Transition metals like Cu, Fe, Zn, or Coare used as catalysts to activate the ascorbic salts in the presence of moisture

(Brody et al., 2011). The mechanism is based on the oxidation of the ascorbate to dehydroascorbic acid. Cu^{2+} is reduced to Cu^{+} by ascorbic acid to form dehydroascorbic acid, then O_2 forms acomplex with the cuprousion to cupricion Cu^{2+}, superoxide anionic radical. This radical leads to the formation of $H_2O_2$and O_2in the presence of copper ions. Further without the formation of OH, which is a highly reactive oxidant, copper ascorbate complex converts $H_2O_2$to H_2O (Cruz et al., 2012). The reaction can be depicted as follows:

$$AA+ \frac{1}{2} O_2 \rightarrow DHAA + H_2O \quad (5)$$

In a study combination of ascorbic acid+ Iron + Zinc particles were incorporated into LLDPE for packaging of bread & bun and obtained the following (Matche et al., 2011):

The oxygen scavenging capacity was:

ascorbic acid + iron- was 47.6 ml of O_2 in 750 h

ascorbic acid + zinc was 37.4 ml of O_2 in 750 h.

The CFU/g of microbes for bun was

No packaging –70 ;ascorbic acid + iron-10; ascorbic acid + zinc-0

The CFU/g of microbes for bread was

No packaging –400; ascorbic acid + iron-41; ascorbic acid + zinc-40

α-tocopherol can be used as a natural free radical scavenger in food packaging applications (Hamilton *et al.*, 1997). α-tocopherol acts as biodegradable oxygen scavenging system when loaded into the PLA microparticles by an oil-in-water emulsion solvent evaporation method (Scarfato et al., 2015). Eggyolkproteinhyd roly satesprepared by enzymatichydrolys is of the fat-free egg-yolk protein showed the oxygen scavenging ability in a dose dependent manner. They suppressed discoloration of β-caroten eandinhibited the thio-barbutyric acid reactive substances (TBARS) formation from tuna homogenates and ground beef. Thus suggesting they are the good source of natural antioxidants and have the potential to be used as oxygen scavengers (Sakanaka et al., 2006).

An active film containing α-tocopherol was used to extend the life of salmon fish and it was observed that the lipid oxidation in films reduced by 70% due to storage in the films (Pereira et al. 2013). In a study α-tocopherol incorporated in PLA microparticles showed the following observations (Scarfato et al., 2015):

Rate of oxygen scavenging was found to be:

0.12 ml O_2/g.day - pure α-tocopherol

0.11 ml O_2/g.day - α-tocopherol loaded PLA microparticles

Polymer-based oxygen scavengers

Polymer-based oxygen scavengers such as oxidation-reduction resins and polymer-metallic complexare gaining importance. These include a base polymer that is modified with an unsaturated side chain attached to its backbone and which is suitable for packaging applications (Gauthier, 2015). Oxidation of the unsaturated hydrocarbons can actasan oxygenscavengermechanisminapolymermatrix (Ferrari*etal,* 2009) Oxygen barrier can be provided by incorporating unsaturated hydrocarbons to the polymerssuch as LDPE and PET (Galdi et al., 2008). 1, 4- polybutadiene can be used for the oxygen scavengingpurposealongwithcobaltneo-decanoateasthecatalysttoenh anceitsability (Li et al., 2012)

$$O_2 + RH \rightarrow ROOH \quad (6)$$

$$Co2+ + ROOH \rightarrow Co3+ + OH- + RO \quad (7)$$

$$Co3+ + ROOH \rightarrow Co2+ + OH- + ROO- \quad (8)$$

where RH refers to allylic carbon-hydrogen bonds of the polymer which shows the susceptibility to undergo oxidative degradation. The scavenging rate increased with the increase in catalystcon centration. A sample containing 400 ppm of cobaltneo-decanoate scavenged 2 mg of O_2/100 mg of polymer whereas the sample containing 1000 ppm of cobalt neo-decanoate scavenged 9 mg of O_2/100 mg of polymer (Li et al., 2012). Scavenging capacity was also observed to increase with an increase in temperature (Cahill et al., 2000). Active oxygenscavengers such as polyamideor polyolef in are also used to develop oxygen scavenging films that provide oxygen barrier properties to the walls of the package (Fava et al.,2013).

These types of polymer-based systems may pose a problem, as the reaction between the polyunsaturated molecules and oxygen may release the undesirable byproducts such as aldehydes, ketones or organic acids which may affect the quality ofthe food by the development of off flavours, off odour and color (Li et al., 2012). To prevent this, photo initiators are commonly added to prevent the premature oxidation of the scavengers during processing and storage; function albarriers are used toimpede the migration of undesirable oxidation products (Gaikwad et al.,2018).

Cut apples were packed with an oxygen scavenging layer consisting of polyamides and polyethylene terephthalate as the barrier layer. It was found that the oxygen transmission rate decreased with the addition of the oxygen scavenging layer. There was no change in colour of the cut apples, suggesting the ability of the system as an effective oxygen scavenger (Di et al., 2015).

Enzyme based oxygen scavengers

The oxidation of certain enzymes is another approach in food packaging to regulate the oxygen concentration. Enzymes react with the specific substance to scavenge incoming oxygen (Gaikwad et al., 2018). Enzymes are tested as high-barrier coatings for active food packaging. The results showed that the developed active coatings delay the oxidation reactions and the rancidity of packaged food (Jarnstrom et al., 2013). Glucose oxidase is the most commonly used enzyme in the oxygen scavenging systems. The mixture of two enzymes, glucoseoxidase and catalase are widely used asoxygens cavengers (Vermeiren et al., 1999). Glucose oxidase, oxidizes glucose into gluconic acid and H_2O_2 in th epresence of the moisture. H_2O_2 form edduring the reaction is broken down by the introduction of the catalaseenzyme. The following mechanism depicts the oxygen scavenging action of these enzymes (Cruz et al., 2012):

$$2\ \text{glucose} + 2O_2 + H_2O \rightarrow \text{D-glucono-2gluconic acid} + 2H_2O_2 \quad (9)$$

$$2H_2O_2 \rightarrow 2H_2O + O_2 \quad (10)$$

The enzymes, oxalate oxidase and catalase can also form effective oxygen scavenging systembyco-immobilization.Theconversionofoxalicacidandoxygentocarbondioxide and hydrogen peroxide is catalyzed by oxalate oxidase and the co-immobilized catalase degrades the generated H_2O_2which results in the increased shelf life of the food items (Winestrand et al., 2013). Enzyme laccases are the copper containing oxido-reductases that can scavenge the oxygen. Enzyme laccase contains phenolic hydroxyl groups which catalyze the one-electron oxidation to phenolic radicals and in the presence of water oxygen is reduced. The phenolic radicals will form quinines or polymerization products subsequently (Johansson et al., 2014).

Enzyme based scavenging systems provide advantages in microwave applications and of detection for metal detectors in the food production line. The enzymes can be coated on polymers such as polystyrene, polyethylene, and polypropylene by the process of immobilization. To activate the redox reaction, these immobilized enzymes should be in direct contact with the food which restricts the use of these enzymes-based films (Brody *et al.*, 2011).In addition, enzymes aresensitivetothechangesinthepH, temperature, wateractivity, salt concentration and other factors, also they require the addition of the water which restricts their use in low moisture foods (Graf, 1994).

Micro-organisms based oxygen scavengers

Micro-organisms entrapped in solid matrix can be used for oxygen scavenging due to their inexpensive and environment friendly nature. Yeast immobilized on a solid surface such as wax, paraffin or cloth can be used as food-grade oxygen scavenger (Floros et al., 1997). Various media like water or ascorbic acid is used to activate the yeast, after which yeast consumes oxygen for respiration (Nezat, 1985). It was found that the oxygen concentration decreased from 1.8 ppm to 1.2 ppm in 19 days for 40 mg yeast-wax, when used for the packaging of the beer and the wax to yeast was used in the ratio of 20:3 (Edens et al., 1992). The entrapped aerobic micro-organisms such as *Bacillus amylolique faciens, Kocuria varians and Pichia subpelliculosa* which are capable of consuming oxygen and can be used as scavengers (Anthierens et al., 2011). This type of scavenging system is useful for high moisture containing food as the scavenging activity is initiated when water originating from the packaged food product isabsorbed.

Activation mechanism of oxygen scavengers

Most of the oxygen scavenging systems needs activation by using various mechanisms before they can start to remove the oxygen from the food packaging. The oxygen scavengers can be classified according to different characteristics or properties related to their activity or their form (Schroeder, 2001). These can be classified based of the activation mechanism (Bharadwaj et al., 2019):

Auto activated systems

Some oxygen scavenging compounds start absorbing oxygen instantly, when exposed to oxygen or air at the normal ambient temperature and humidity. Severalunsaturatedpolymericscavengersorthenaturalbiologicalscavengers such as enzymes, vitamins, and fatty acids do not need any triggering step for the initiation of the reaction. They are normally characterized by high scavenging capacity and fast reaction. However, the auto activated systems need special care duringhandling, processing, and storage for avoiding the considerable reduction in their scavenger effectiveness by premature oxidation (Bharadwaj et al., 2019).

Water activated systems

The essential component of many oxygen scavenging reactions mechanisms is water, which can act as a solvent or as a swelling agent. When the chemical reagents are coated withalow-permeability material or they are incrystalline form, the water has to dissolve there agents in order to advance the oxidation. Differences in scavengers may arise with the source of the water itself. The addition of water as vapor is required for many scavengers to begin the absorption of oxygen. These scavengers are frequently designed such that they should be incorporated into the retortable plastics packaging which will undergo steam retorting at around 120°C. Water vapor permeation from the out side is adequate to let the process to start in the secases. For othe rsystems, to activate the scavenger, it is essential that the liquid substance is filled in the package. These oxygen scavenging systems are mainly used for the beverage applications such as juice, beer, tea, etc. as in this case, the product itself will be the potential source of water (Bharadwaj et al., 2019).

UV activated systems

Several oxidation reactions need continuous or brief exposure to light or ultra violet radiation to begin. The UV activated systems are based on the photosensitized oxygen scavenging reactions and their technology consists of a photosensitive dye coated or impregnated on a polymeric film. In UV activated systems, irradiation is done by UV light which results in the activation of O_2in its singlet state. But in this system, the limiting factor is UV radiation. UV processing has to be done after completion of food packing, considerings ensitivity factor of the food products into account as this radiation may alter the properties of the food product. Even though the photo activators are low molecular compounds that can easily migrate through the sachet or the lidor the polymer structure, these materials may not be suitable for the food contact after activation (Bharadwaj et al., 2019).

Other mechanisms

There are several researches going on in the filed of exploring new active packaging systems. For example, Toppan Printing Co., Japan, for example, has patented a system that uses magnetic fields to activate scavenging compounds (Fukuyoshi,2001).

Biodegradable thermoplastic starch films were extruded with a binary oxygen scavenger consisting of ascorbic acid and iron powder (Fe) or copper chloride ($CuCl_2$) as catalysts which showed the oxygen scavenging property (absorb 13.5mL of O_2 per gram of dry film after 15 days at 80% RH) and can be activated by increasing water content (Mahieu et al., 2015).

Active, oxygen-scavenging, low density polyethylene (LDPE) films were prepared from a non-metallic-based oxygen scavenging system (OSS) containing different concentrations of gallic acid and potassium chloride.The film impregnated with 20% organic oxygen scavenging material showed an effective oxygen scavenging capacity of 0.709 mL/cm^2 at 23 °C. Relative humidity triggered the oxygen scavenging reaction (Ahn et al., 2016).

Pyrogallol (PG) has a high oxygen scavenging ability in alkaline environments and is coated on LDPE film which is modified using sodium carbonate and activated by moisture. The LDPE/PG 20% film showed an effective oxygen scavenging capacity of 0.443 mL/cm^2 at 23 °C (Gaikwad et al., 2018).

From past to the modern days,many companies are commercially producing the oxygen scavenging systems in various forms such as sachets, films, labels, plastic bottles, bottle caps and crowns and few examples are mentioned in Table 2.

Nanoparticle-based scavenging system

Recently, varieties of compounds such as the use of natural compounds are being checked for their oxygen scavenging ability and their incorporation into different types of films. Because of their suitability and active functions, nanomaterials are already being implemented in the food packaging industry. Currently, the food processing industry is concentrating on creating new nanomaterials to enhance the mechanical and barrier properties of packaging (Peter et al. 2012). In a study nanoiron-containing, kaolinite were incorporated into HDPE films as an active oxygen-scavenging film, along with polyolefin nanocomposites (Busolo et al., 2012). The active kaolinite clay has a high oxygen scavenging performance. Nanoparticles of metallic iron with an average particle diameter of about 115 nm show high oxygen uptake kinetics at 100% relative humidity (RH). An oxygen scavenging system (OSS), composed of oxygen scavenging nanoparticles and iron chloride (II) was incorporated into warm-water fish gelatin film. The oxygen scavenging fish gelatin film had a good oxygen scavenging capacity, 1969.08 cc $O_2/m^2/mil$, and moisture was used as the activator to trigger the oxygen scavenging reaction (Byun et al., 2011).

In another study nanoparticles of a polymerizing monomer were dispersed in the presence of a platinum group metal catalys to obtain a polymer oxygen scavenging system (Akkapeddi et al., 2014). The incorporation of metal-based micro- and nanostructured materials, including Fe, were investigated into food-contact polymers to enhance the mechanical and barrier properties and prevent the photodegradation of the plastics (Llorens et al., 2012). Additionally, heavy metals are effective compounds for preparation as elemental nanoparticles. They can be incorporated for food preservation purposes but can also scavenge oxygen and extend food shelf life.

Various nanoparticles such as iron particles, zinc particles, palladium nanoparticles, and titanium nanoparticles are also gaining importance as oxygen scavengers. Oxygen scavenging films made of poly(3-hydroxybutyrate) (PHB) containing palladium nanoparticles (PdNPs) prepared by electrospinning followed by annealing treatment at 160 °C were modified with surfactant hexadecyl trimethyl ammonium bromide (CTAB) to optimize their dispersion and distribution in PHB showed good oxygen scavenging performance (Cherpinski et al.,2018).

Table 2: Commercially available oxygen scavengers

Manufacturer	Commercial name	Scavenger	Product form	Picture
Albis	Shelfplus O_2	Iron based	Film	
Multisorb Technologies, Inc	Fresh Max, FreshPax	Iron based	Sachet/label	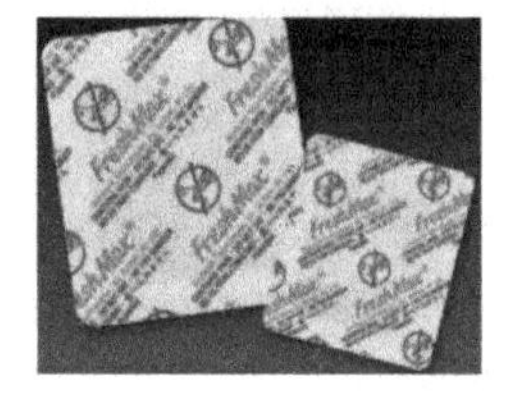

Aptar CSP Technologies	Activ-Film	UV activated scavenger	Film	
W.R. Grace & Co.	Darex	Ascorbate/ Sulphite	Bottle Crown/ Bottle	
BP Amoco Chemicals	Amosorb 3000	Photosensitive dye	Film/Bottle	
Clariant Ltd.	Oxy-Guard	Iron based	Sachet	

Moisture absorber/ controller/ scavenger

The presence of moisture and water activity plays an important role in the microbial spoilage of moisture-sensitive foods such as fruits, vegetables, fish, meats, poultry, snacks, cereals, dried foods, etc (Bhardwaj et al., 2019). The loss of water in foods take place due to normal respiration or evaporation from the product followed by permeation through the packaging material that results in quality losses and reduction in their shelf-life hence, water-sensitive food products should be packaged with high water barrier materials (Brody et al., 2011). During the metabolism of fats and carbohydrates, water is produced in respiring foods that wet the food giving high water vapour pressure and leads to condensation, which is common in many packed foods (Svensson, 2014). In condensation or sweating, one part of the package becomes cooler than another and water vapor condenses as liquid droplets in the cooler areas which in turn results in worsen package appearance, while a moisten food surface results in increased surface mold growth and therefore diminished shelf life of the food product (Brody *et al.*, 2011). The excess moisture content in the pack leads to softening of dry crispy products, caking of hygroscopic powdered products like milk powder, instant coffee powder, sweets, and so on. Controlling moisture accumulation is essential to increase the shelf life of the foods, which can be done effectively by the use of moisture absorbers or scavengers (Biji et al., 2015).

Moisture absorber or controller helps in binding water either in the form of vapour or liquid by preventing the food product from retaining water and preventing spoilage are usually prepared in the form of sachets, pads, sheets or blankets and used for packaging of dried foods. The use of various absorbers or desiccants will lead to the reduction of excess moisture and in turn, helps in maintaining food quality and extends the shelf life (Bhardwaj et al., 2019). Some of the moisture control strategies used in packaging are - moisture prevention by use of barrier packaging, moisture elimination by applying desiccant or absorber and moisture reduction by using MAP technology or by vacuum packing through the removal of humid air in the headspace. Relative humidity (RH) controllers that scavenge humidity in the headspace, such as desiccants, and moisture removers that absorb liquids are grouped as active moisture scavengers (Yildirim et al., 2017)

The moisture removers like desiccants are used to control humidity in the packaging headspace to protect the sensitive products against water and humidity which are commonly placed into packages in the form of sachets, microporous bags, or are integrated in pads. Some of the desiccants like silica gel (maintain dry conditions within packages of dry foods, down to below 0.2 water activity) clays, molecular sieves (such as zeolite, sodium, potassium, calcium alumina silicate), humectant salts (such as sodium chloride, magnesium chloride, calcium sulfate), and other humectant compounds (such as sorbitol), as well as calcium oxide are used to absorb moisture that enters or remains in a package (Yildirim et al., 2017).

Zeolites absorbs up to 24% of their weight in water, which has a high affinity for moisture. Whereas cellulose fiber pads are used to remove water, which can surrender water when they are saturated (soaker pads in the bottom of the meat, poultry, and fresh produce trays) (Brody et al., 2011). Moisture-drip absorbent pads, sheets, and blankets are also usedfor absorbing liquid water in foods having higher moisture content such as fruits, vegetables, meats, fish and poultry. Humidipak (Humidipak Inc.) that works by absorbing and releasing moisture in response to the surrounding environment, is one of the commercially sold humidity control sachet which uses a saturated solution of substances to maintain the optimum humidity in the package (Bhardwaj et al.,2019).

Some of the known moisture absorbers that have been used commercially to extend the shelf life of food products are Toppan sheet™ (Toppan Printing Co. Ltd Japan), Thermarite™ (Thermarite Pty Ltd, Australia), Luquasorb™ (BASF, Germany) (Bhardwaj et al., 2019). Japan's Chefkin (a pocket created by overwrapping of a duplex of two sheets: the external sheet which is a water-vapor barrier and the inner sheet is a water-vapor-permeable), Pitchit film (Japan's Showa Denko Company) use to decrease water activity at the food surface, a packaged sandwich composed of two sheets of polyvinyl alcohol (PVA) film sealed along the edge and a layer of humidifying agent propylene glycol between the two sheets. The PVA film is very permeable to water vapor but is a barrier to the propylene glycol. Others like Japan's "Dai Nippon", "Kyushi" a drip tray (Japan's Dai-ichi Plastic Industry and Dai Nippon) "Crisper F" (Japan's Kagaku Kogyo), etc. (Brody et al., 2011).

Carbon dioxide absorbers or emitters

Addition of CO_2 gas into food products has impacted with promising results. Modified atmospheric packaging is one of the prime examples in which the packages are flushed with known concentrations of carbon dioxide and thus prolongs the freshness and shelf-life of perishable products (Jung et al., 2012). The accumulated carbon dioxide through the respiration of the product must be equated with the amount that egress from the package. Inclusion of CO_2 along with optimal O_2 levels lowers the metabolism rate and reduces ethylene emission of horticultural produce (Rodriguez et al., 2009). The released metabolites from the produce triggers the release of CO_2 from the sachets and prevents the collapse of the food package (Loez et al., 2004). Carbon dioxide has a fascinating property of dissolving at a much lower temperature in both fatty and non-fatty foods and creates a partial vacuum. Therefore, the application of CO_2 has higher antimicrobial potency at chilled conditions for the non- respiring foods susceptible to microbial spoilage and oxidation (Chaix et al., 2014). The inclusion of this gas into the food package environment reduces package volume due to its dissolving properties into food product and thus balances the pressure between headspace and external environment. This property makes it beneficial for the modified atmosphere packaged product maintained at a much lower temperature.

However, the amount of CO_2 to be flushed must be customized owing to the properties of food products and packaging material. Since the high dissolution of CO_2 into the food matrices can sometimes cause package collapse and can affect the textural and sensorial properties of the food product (Lopez-Rubio et al., 2004). Increased CO_2 levels (5-9 %) often causes' tissue injury, staleness and blemishes into fruits and vegetables (Lee *et al.*, 2016). Fermented products (dairy & roasted coffee) releases and accumulates higher CO_2 levels during storage and increases package volume (Lee et al., 2016). Therefore, CO_2 absorbers can be an effective measure in achieving desired levels of gases into food package with extensive shelf-life. The absorbers can be in form of sachets/pads, labels or laminates, blended with the packaging film (Watkins, 2000). Nevertheless, a proper selection of packaging material for MAP of horticultural produce is strenuous to maintain the balance of gases. However, different designs and devices are evolved to obtain a suitable modified atmosphere with preferable amounts of oxygen and carbon dioxide has been progressed (Mangaraj et al., 2009). To elude the excess accumulation of CO_2 levelsmicro-perforations are included intothe packages with high respiring produce. In order to achieve a desired gas environment, gas scavengers and emitters are employed for prolonged shelf-life.

Carbon dioxide absorbers

The CO_2 absorbers are mainly divided into two categories –

i. Chemical absorbers
ii. Physical absorbers

Chemical CO_2 absorbers

There are several forms of chemical salts and alkaline available to scavenge the CO_2 and the most commonly employed is calcium hydroxide- $Ca(OH)_2$ into the food packaging system (Rodriguez et al., 2009). The following is the CO_2 scavenging reaction of $Ca(OH)_2$:

$$Na_2\,CO_3 + CO_2 + H_2O \rightarrow 2NaHCO_3 \quad (12)$$

This salt is safe to use in direct contact with food products and the above-mentioned reaction is thermodynamically spontaneous and occurs in normal environmental conditions inside the food package. The mass-based CO_2 absorption capacity of $Ca(OH)_2$ is 1.35×10^{-3} mol $g^{-1.}$ It does not require any catalyst and produces calcium carbonate and releases water which is absolutely safe water.

The other alkaline salt preferably used in the moist condition is sodium carbonate that reacts with CO_2 and produces sodium bicarbonate:

This reaction occurs in preferably the presence of water i.e. either moist foods or humidified conditions (Shin *et al.*, 2002). The scavenging activity of Na_2CO_3 relies on particular environmental conditions and the reaction can be hastened with a regular supply of moisture. Beside the sachets containing the salt needs to have variable moisture permeability for better CO_2 scavenging capacity. In comparison with $Ca(OH)_2$, Na_2CO_3 has a lower mass-based CO_2 absorption capacity of about 9.43×10^{-3} mol g^{-1} and is due to a higher molecular mass of Na_2CO_3.

Another, chemical-based CO_2 scavenging system includes the use of amino acid in solid form as sodium glycinate. In a study, sodium glycinate was used as flavor absorber ingrained in an agar film to scavenge a CO_2 as in the below-mentioned reaction:

$$NaCOOCH_2NH_2 + CO_2 + H_2O \rightarrow NaCooCH_2NH_3 + HCO^-_3 \quad (13)$$

This reaction appears in a moisture absorbing food products or in an aqueous phase. Here, the mass-based CO_2 absorption capacity is 1.03×10^{-2} mol g^{-1}. However, the absorption capacity and rate of reaction depend on the sodium glycinate concentration and CO_2 pressure levels (Lee et al., 2011).

Chemical carbon dioxide emitters

A commonly utilized CO_2 releasing technology involves two active substances, namely sodium bicarbonate ($NaHCO_3$) and an organic acid. Citric acid is in many cases the acid of choice in such CO_2 releasing systems (Yildirim et al., 2015). The reaction begins when liquid from the food product comes into contact with the active ingredients and dissolves them. The acid lowers the pH of the system to a value in which the sodium bicarbonate buffering system is shifted towards the formation of un-dissociated carbonic acid and carbon dioxide. According to Le Chatelier's principle, when liquid is introduced into the system, the pH will drop, and the production of carbon dioxide starts.

The other chemical-based scavenging systems are: Calcium Oxide, Cao, for CO2 absorption, and is only applicable in terms of a large quantity of fresh produce during transportation.Magnesium oxide and magnesium hydroxide have a CO_2 scavenging system but are less efficient and are generally not used due to their unpopularity (Charles et al., 2006).

Conversely, the O2-absorption capability of iron-based scavengers is hampered by a CO_2-rich atmosphere (Rooney, 2005). The following equation represents a recently proposed mechanism of reaction by which a hydroxylated iron oxide surface with adsorbed moisture absorbs CO_2 gas (Baltrusaitis et al., 2005):

$$Fe - OH + H_2O\ (a) + CO_2\ (g) \rightarrow Fe - CO^-_3 + H_3O^+ \quad (14)$$

Physical CO_2 absorbers

Zeolite and activated carbon are the most commonly used physical carbon dioxide absorbers. These absorb CO_2 through physical action and also involve some extent of chemisorptions. Physical absorbers include zeolite, activated carbon, or silica gel (Wang et al., 2015). However, the amount of CO_2 absorbed relies on the dynamic equilibrium and varies with a forward and backward shift along with the atmospheric conditions. The amount of CO_2 absorbed from the package depends on pore size, hydrophobicity, the surface to volume ratio, compatibility of adsorbent with the gas (Marx et al., 2013). These absorbers are either in the form of granules or powder or sheets or beads packed in porous sachet or packet.

The activated carbon has a higher surface area about 500 to 2500 $m^2\ g^{-1}$ with its bulk density up to 600 Kg m^{-3}.It has an amorphous porous structure, hydrophobic, and is marginally affected by the presence of moisture (Xu et al., 2013). However, the amount of carbon dioxide absorbed increases with increased CO_2 pressure and in the presence of small amounts of moisture (Marx et al., 2013).

Zeolites are 3D crystallites of aluminum silicates composed of Al and Si tetrahedral units. The assembly of these units consequents into an open structure with varied configuration and these openings absorbs gases. These are often in various forms including powder, beats, and pellets and the powder form has lower bulk density (Lee, 2016).

Application of CO_2 scavengers into food products

A product like Kimchi generally produces a large amount of CO_2 gas and the volume of the package was decreased using salts of $Ca(OH)_2$ or Na_2CO_3 or zeolites (Lee et al., 2003). $Ca(OH)_2$ sachet placed in soybean paste and red pepper paste reduced the internal pressure and CO_2 levels (Jang et al., 2000).

In a study, the strawberries packed using CO_2 scavenger stored at refrigerated conditions remained fresh throughout 4 weeks (Aday et al., 2011). The use of CO_2 scavengers in dairy and meat products with exclusion of the O_2 environment has proved to be sufficient to impede the growth of *E.coli*, and other pathogens. In another study, the incorporation of CO_2 up to8 % reduced internal the browning of pears (Nugraha *et al.*, 2015). Similarly, a combination of zeolite, sodium carbonate, or calcium hydroxide has resulted in controlled CO_2 levels resulting in constant package volume in fermented products like kimchi and coffee (Crump et al., 2015).

The MAP cod fillets (70% CO_2, 30% N_2) with CO_2 emitter had reduced the CO_2 levels to 40% and had a shelf life of 11 days with ammonia-like odor compared to 7 days for the control. Use of citric acid in combination with sodium bicarbonate has been utilized in MA packaging and storage study of cod, reindeer meat and chicken that had an extensive shelf-life compared to control (Holck et al., 2014).

Concerning commercialization, there are different emitters already in the market today. Emitters based on sodium bicarbonate and citric acid include CO_2 Freshpads (CO_2 Technologies, Urbandale, Iowa, U.S.A.) (Kerry et al., 2014), Super Fresh (Vartdal Plastindustri AS, Vartdal, Norway), and the Active CO_2 pad (CellComb AB, Saffle, Sweden). There are also examples of combined O_2 scavengers and CO_2 emitters in the market, such as Ageless RGE (Mitsubishi Gas Chemical Company, Inc., Tokyo, Japan) and FreshPaxR M (Multisorb Technologies Inc., Buffalo, NY, U.S.A.). These systems are based on either ferrous carbonate or a mixture of ascorbic acid and sodium bicarbonate (Coma, 2008).

Antimicrobial packaging

An antimicrobial packaging system is a form of active packaging technology in which the antimicrobial agents are either directly incorporated into the food or into the polymer films to suppress or inhibit the microbial growth present in the food or package itself (Appendini et al.,2002). Antimicrobial packaging systems extend the lag phase and reduce the growth phase of microorganisms to delay the shelf life of the product. Antimicrobial agents are used in reducing spoilage in food by inhibiting or killing spoilage and pathogenic microorganisms thus maintaining safety and quality of the product inside the package by extending the shelf life (Cooksey et al., 2005). Antimicrobial packaging systems are mainly used in meat, fruits and vegetables. Antimicrobial agents or antimicrobial polymers have three types of mode: (i) release system in which the antimicrobial agents migrate into the food or head space of the package and retard the growth of microbes. (ii) absortion system microbial agents absorsb the essential agents from food thus inhibiting the growth of microorganisms. (iii) Immobilisation system suppress the growth of microbes at the contact surface (Han J H et al., 2003). Natural antimicrobial agents like extracts from spices such as cinnamon, clove, oregano, thyme, and plant extracts like onion, garlic, and mustard. Silver substituted zeolite has been incorporated into plastics as antimicrobial agents. in silver zeolite the surface atoms are replaced by silver and it is laminated as a thin layer on the inner surface of the polymer in which the food comes into contact; when the foods aqueous solution comes in contact with laminate silver ions are released into the food. Commercially available zeolite AgIONTM Silver Ion Technology has been approved by the FDA for all food pollymers (Quintavalla et al., 2002). The use of antimicrobial agents is regulated by several regulatory agencies like FDA, EPA, and USDA in the USA.

Flavor absorber or emitter

Packaged food products respire and emit various levels of gases, the release of O_2 into the food package is the prime agent for food deterioration. The fat-rich food reacts with the released O_2 and causes rancidity and releases free radicals causing obnoxious odour & flavor. The foods with higher protein contents release amines, aldehydes, and pile up at the package headspace. These, along with microbial metabolites liberate hydrogen sulfide gas causing putrefacation. Besides this, the odour barrier and migration properties of the food package also influence the quality of the product. Therefore, it becomes necessary to eliminate the unpleasant odour-causing agents using odour/flavour absorbers or emitters. The flavor absorber/emitter in form of sachets/pads are placed inside the package and monitors the gradual release of flavors to maintain the product freshness, improves its aroma, compensates the natural flavor loss. These also scavenge the unpleasant odour from various chemical reactions occurring inside the food package and thus prolong shelf-life (Almenar et al., 2007). The putrefied odor from fish amines and in poultry spoilage can be eliminated by incorporating citric acid/ascorbic acid along with ferrous salt that oxidizes the amines or use of sulfide scavengers to eliminate H_2S gas from the package (Mexis et al., 2014).

Usually, packaging material with polyethylene terephthalate/ polyethylene is used as odour proof materials in the transportation of certain fruits to stop the odor migration. demonstrated In a study, the incorporation of naringinase enzyme into the food package had reduced the bitterness in citrus juices caused by naringenin (Mexis et al., 2014). Incorporation of triacetate cellulose into the packaging films have proved to be limonene absorbers and reduced bitterness into bitter oranges. Absorbent pad consists of activated carbon and antimicrobials into vacuum packaged food. This pad hinders the growth of microbes that reacts with food and release amines and simple sugars. The activated carbon traps the odor inside the package and thus maintains freshness and product longevity (Versteylen et al., 2015). With increasing consumer demands of better tasting foods with utmost convenience, it becomes challenging and necessary to develop food packages for absorption and migration of flavours.

Antioxidants

High-fat content foods like oils, nuts, butter, meats, fish, bakery goods are highly prone to oxidative rancidity and affect nutritional quality, textural property in food products. This can be inhibited with the application of antioxidants into the food package either by coating or blending or melting technology (Gómez-Estaca et al., 2014). Antioxidants can retard or stop the oxidative activity that causes food spoilage (Pereira et al., 2010). Consequently, Gómez-Estaca et al. in 2014 delineated two step reactions : i) release of antioxidants and ii) their action by absorbing the O2 or metal ions from the food package system. A chemical antioxidant like butylated hydroxyl toluene and t-butylated hydroxyquinone were blended or coated with packaging films to reduce oxidative rancidity and extend shelf-life. The coating of PE film incorporated with TBHQ prolonged shelf-life of fried extruded products (Koontz et al., 2016). Longer shelf-life was perceived with the application of BHT into the HDPE packaging of Oat Cereal. Perhaps, some studies have reported the accumulation of BHT into human adipose tissue (Wessling et al,. 1998).Hence, natural antioxidants are incorporated into food packages and those include Vitamin E, ascorbic acid, catechin, quercitin, essential oil extracts from green tea and spices (Phoopuritham et al., 2012).

EVA film blended with quercetin and catechins eliminated production of hexanal in fried peanuts and extended shelf-life of sunflower oil (Carol Lopez di-castillo et al., 2012). An active packaging film developed by blending rosemary extract with PP delayed the oxidation of meat and stabilized myoglobin than PP film alone (Cristina merin et al., 2006). Incorporation of green tea extracts into packaging film for preserving mushrooms that maintained white color and freshness throught their shelf-life (Wrona et al., 2015). The oxidation of turkey meat products placed in PET trays sprayed with citrus extracts were prone less to oxidation than those placed in unsprayed citrus extracts PET trays. Chitosan and green tea extract blended film showed lower moisture migration, antioxidant activity and improved mechanical properties into food package systems (Siripatrawan et al. 2010).

Edible antimicrobial agents are also into pictures due to their biodegradability, product compatibility, and barrier properties and includes those prepared from proteins, carbohydrates and fats. Whey proteins assimilate an ample quantity of antimicrobials like lysozyme, nisin, polysorbates (Lopez rubio et al., 2004). Nisin was applied to cellulose-based packaging of ham, cheese and turkey for prolonged shelf-life and freshness (Ming et al., 1997). The O_2 scavengers prevent oxidation of food by just absorbing O_2 from the packaging environment, but antioxidants retard the activity of free radicals to block the oxidation process. However, certain antioxidants exhibit toxicity and transfer to food from the package. Hence, more studies are required in this aspect with respect to concerning safety and prior to commercialization.

Ethylene scavenging

Ethylene (C_2H_4) is a volatile, pure unsaturated hydrocarbon found in nature. It is a plant growth hormone with significant functions such as sprouting of plant seedlings, the growth of plants, and the growth of fruit, which helps in accelerating the ripening of fruit and senescence of plants (Gaikwad et al.,2019). Ethylene production is a biochemical process, independent of respiration that occurs in each living cell to produce energy. Based on the product, the respiration rate of fresh produce varies, which is characterized by the quantity of carbon dioxide emitted. The respiration rate increases with an increase in the concentration of ethylene in the surrounding environment of the product, since the effect of respiration rate, depends primarily on relative concentration compared to the ethylene emission of the plant, rather than on the absolute ethylene concentration (Brody et al., 2011).

The three main approaches for ethylene regulation are

- Ethylene reduction by MAP
- Through perforated packaging material to permeate gases
- Use of ethylene scavenger where ethylene is removed or the use of ethylene absorbers which physically absorb and hold ethylene molecule (Gaikwad *et al.*, 2019).

Potassium permanganate based Systems

Potassium permanganate ($KMnO_4$) is one of the most commonly used ethylene scavenging systems. Itis a reliable agent that chemically scavenges C_2H_4 by an oxidation process, which is usually contained in sachets placed inside the package (Lo′pez-Rubio et al., 2004). It is reinforced on inert matrices such as silica gel or alumina to expand the surface area of C_2H_4 absorption, where it oxidizesgaseous ethylene with a colour change from purple to brown.The aluminafunctions mainly as the absorptive surface to hold themolecules of ethylene gas and is a carrier for the permanganate. $KMnO_4$ is a wide-spectrum oxidizing agent that reactswith ethylene along with othercontaminant gases. When it reacts by oxidizing ethylene to ethylene glycol, a visible-colour change takes place (Brody et al., 2011; Gaikwad et al., 2019). First, the color change occurs on the pellet's surface and eventually penetrates the core, indicating the reactive capacity is nearing exhaustion, where ethylene isinitially oxidized to acetaldehyde, which in turn is oxidizedto acetic acid, and acetic acid and further oxidized to carbondioxide and water (Gaikwad et al., 2019).

$CH2CH2 + 12KMnO4 \rightarrow 12MnO2 + 12KOH + 6CO2$

Blankets, tubes and sachets are the different forms of packaged ethylene scrubbing media. Blankets and tubes are commonly employed in transport vehicles, whereas sachets are utilized in individual boxes of fruits or vegetables (Brody *et al.*, 2011). Some of the commercially available $KMnO_4$ based scavengers are Chemisorbant (Purafil, Inc., Doraville GA, USA), MM-1000 MULTI MIX ® MEDIA (Circul-Aire Inc., Montreal, Canada), Bi-On® SORB (Bioconservacion S.A., Barcelona, Spain) and SofnofilTM (Molecular Products Limited, Essex, UK) (Gaikwad *et al.*, 2019).

Clay-based systems

Ethylene adsorbing capacity has been seen in several clays, which are hydrous covered alumino-silicates composed of two layers, including tetrahedral (consists of areas of Si4+ and most common Al3+) and octahedral (consist of Mg2+or Al3+) layers. Ethylene is possibly removed by physical adsorption on materials with active surfaces, which can be contained into an ethylene-permeable sachet, or fine particles of such clays can be incorporated into the packaging film via an extrusion process. Zeolite as an adsorbent has received significant attention in both industry and agriculture applications for ethylene elimination. Incorporation of zeolites allows high gas permeability of packaging materials through their crystalline porous three-dimensional structure (Gaikwad *et al.*, 2019).Other natural clays having a porous structure have increased much consideration as ethylene scavengers for fresh produce active packaging applications such as halloysite nanotubes, montmorillonite, Cloisite, and Japanese Oya. Films containing Oya-Stone have been extruded with polyethylene into pouches used to contain ethylene-generating fruit or vegetables (Brody et al., 2011).

Activated carbon-based systems

Activated charcoal impregnated with a palladium catalyst, placed in paper sachets effectively removes ethylene by oxidation from packages of minimally processed fruits. Activated carbons are non-crystalline porous structure, and it is a form of carbon the acquired by pyrolysis of carbonaceous substance. These may be in the form of granular, powdered or fibre, where studies represent granular form is most ideal because of its more natural regeneration and adaptability. The granular form is the best ethylene adsorption capacity of activated carbon compared with powder form and fiber forms (Gaikwad et al., 2019).

Table 3: Commercially available active packaging material

Trade Name	Manufacturer	Principle	Type
Ageless	Mitsubishi Gas Chemical Co. Ltd.,	Japan based	Oxygen scavenger
Freshilizer	Toppan Printing Co. Ltd., Japan	Iron based	Oxygen scavenger
Freshmax, Freshpax, Fresh Pack	Multisorb Technologies, USA	Iron based	Oxygen scavenger
Oxyguard	Toyo Seikan Kaisha Ltd., Japan	Iron based	Oxygen scavenger
Zero 2	Food Science Australia, Australia	Photosensitive dye	Oxygen scavenger
Bioka	Bioka Ltd., Finland	Enzyme based	Oxygen scavenger
Dri-Loc®	Sealed Air Corporation, USA	Absorbent pad	Moisture absorber
Tenderpac®	SEALPAC, Germany	Dual compartment system	Moisture absorber
Biomaster®	Addmaster Limited, USA	Silver based	Antimicrobial packinga
Agion®	Life Materials Technology Limited, USA	Silver based	Antimicrobial packing
SANICO®	Laboratories STANDA,	Antifungal coating	Interleavers
Neupalon	Sekisui Jushi Ltd., Japan	Activated carbon	Ethylene scavenger
Peakfresh	Peakfresh Products Ltd., Australia	Activated clay	Ethylene scavenger
Evert-Fresh	Evert-Fresh Corporation, USA	Ativated zeolites	Ethylene scavenger

Intelligent packaging

"Materials and articles that monitor the condition of packaged food or the environment surrounding the food" can be called intelligent packaging (IP) (European Commission, 2009). Intelligent packaging IP are the systems which are capable of sensing, detecting, recording information, tracing and communicating the recorded information to the manufacturer, retailer and consumer the about state of the packed product (Yam et al., 2005). It provides assurance of quality, tamper evidence and safety of the pack to the consumer before the consumption of the product (Day et al., 2008).The purpose of active packaging is to extend the shelf-life and improve the quality of the food whereas intelligent packaging is to indicate and monitor the freshness of the food (Han et al., 2000). Intelligent packaging system do not have the intention to release their components into the product, these also help to improvise the hazard analysis critical control points(HACCP) and also quality analysis critical control points(QACCP) systems (Heising et al., 2014). IP is basically of two types: i) one measures the condition of the package on the outside, the other directly measures the quality of the product from inside these are either directly in contact with the food or placed in the headspace (Yam et al., 2005). Intelligent packaging conveys the information regarding the quality of the pack whether the product inside the package is still fresh or in consumable condition or expired directly to the consumer thus ensuring the safety (Robertson et al., 2013) Intelligent packaging can be combined with active packaging to yield "smart packaging"(vanderoost et al., 2014).

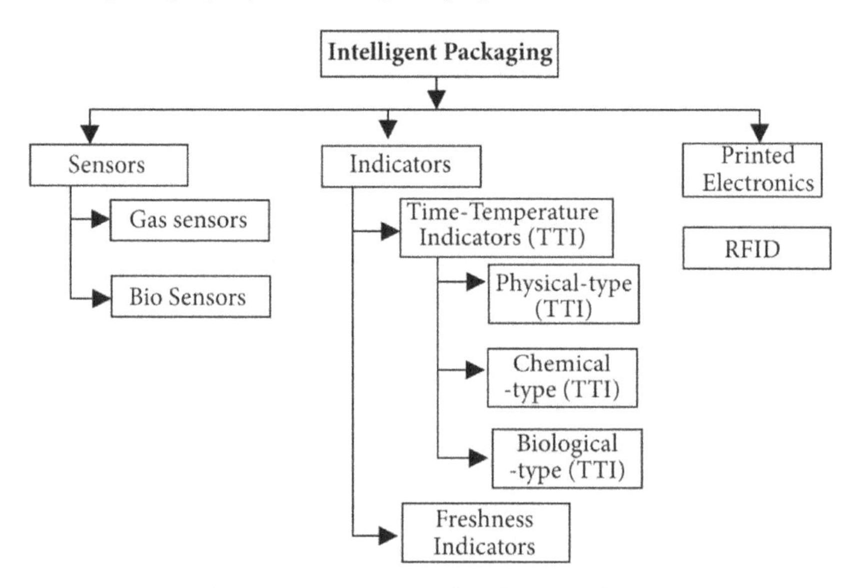

Fig, 2: Schematic representation of intelligent packaging system

Time-Temperate Indicator (TTI)

Time Temperature Indicator (TTI) is an economical device that can attach externally to the package surface of the product which reflects fully or partially the time-temperature history of the foodstuff. It is mainly functional in reflecting the time-temperature history of chilled or frozen food which is sensitive to temperature, like seafood, milk, frozen fish. TTI was also used to review the sterilization process in thermally processed milk (Claeys et al., 2003).

Based on the working principle, TTI's subdivided into three types: i) chemical-based system, ii) biological-based system and iii) physical-based system. Therefore, TTI response must match the quality degradation, must fulfill for the successful application of TTI's of the food products. The activation energy (Ea) of temperature-sensitive TTI response correlated with activation energy (Ea) of quality losses of the food product should match (Wanihsuksombat et al., 2010).

The applications of TTI time-temperature indicator technology have brought about a significant advancement in food quality and safety control, (Taoukis *et al.,* 2003). Throughout the distribution and storage of food products, temperature management is also an essential factor in quality loss. With the appropriate temperature management during storage and transportation it can be possible deliver high-quality food products. There are many types of TTI's classified which are working on different principle had shown below:

Classification of time-temperature indicator

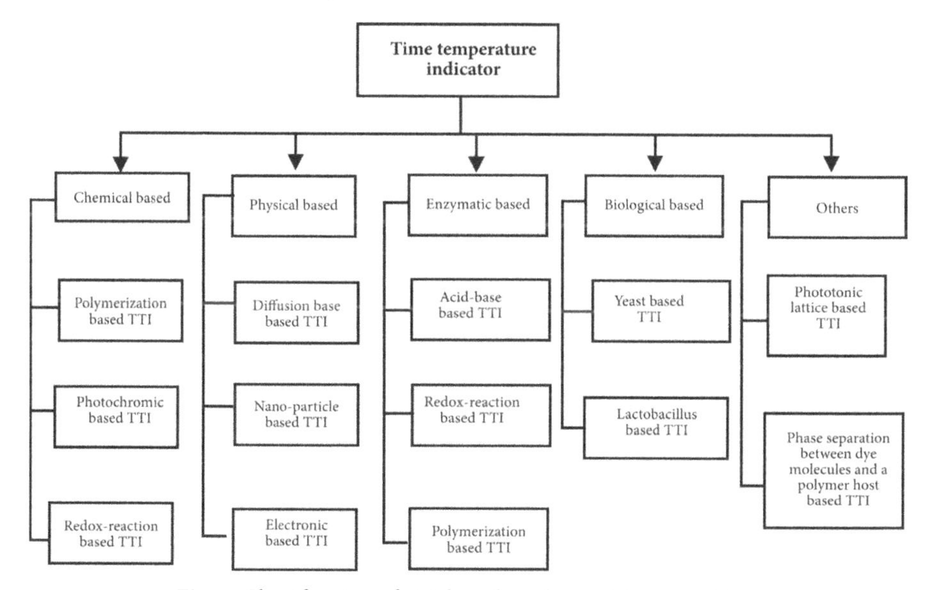

Fig. 3: Classification of TTI based on the working principle

TTI's allow indicating the remaining shelf-life of perishable products and the recording thermal history through the colour change by the accumulative effect of time and temperature (Wang et al., 2015). TTIs are cost-effective and user-friendly devices to monitor, record, and translate the overall effect of temperature history (Taoukis 2003; Pavelkove, 2013). TTI's generally categorized as chemical, physical, and biological-based on the principles they make use of to achieve colour change. TTI's subdivided into three types:

Physical-type diffusion-based TTIs

Lactic acid-based TTI prototype based on the vapour diffusion of lactic acid could be applied to show the time-temperature history of some foods (Wanihsuksombat et al., 2010). A prototype isopropyl palmitate (IPP) diffusion-based TTI system showed potential for monitoring the microbial quality of non-pasteurized angelica (NPA) juice based on temperatureabuse (Kim et al., 2016).

3M Monitor Mark® *(3M Co., USA)* is a diffusion-based indicator label and indicated by temperature-dependent permeation and a blue-dyed fatty acid ester diffusing along with a wick. The response range of this TTI is 48 hours for -15°C, 48 hours for 5°C, 48 hours for 10°C, and one week for 31°C (Pavelková 2013; Wang et al., 2015; 3M United States 2017). Time strips® (Timestrip UK Limited, UK) uses a special porous membrane. The squeezing of a start button leads to move the liquid to directly contact the membrane, and then the liquid diffuses through the membrane (Kuswandi et al., 2011). The time strips response range is limited to 2 hours to 7 days with an available temperature range of -20 to 30°C. Physical-type diffusion-based TTIs would be problematic because of the exudation of the colour material and the fact that ageing porous substances would cause an adverse impact on safety and accuracy (Wang et al., 2015).

Chemical-type TTIs

A colour indicator based on the bacterial strain (sk22) isolated from commercial cod developed for confirmation of temperature abuse in the cold chain (Ohta *et al.*,2008). The colour in this indicator changed for 32-72 h at 12°C.) A laccase based TTI prototype could be applied to predict losses of food quality ascribed to enzymatic changes, hydrolysis, and lipid oxidation (Kim et al., 2012). A simple indicator developed using red cabbage dye with sodium hydrogen carbonate, sodium carbonate, and lactose was used with a change in color for 36, 48, and 96 hours at 12, 10, 4°C,respectively.

A novel on-package colour indicator based on bromo-phenol blue was used for real-time visual monitoring of freshness state of packaged guavas through the colour indicator gradually changed colour from blue to green after five days in room temperature (Kuswandi et al., 2013). Maillard reaction-based TTI was useful for chilled temperature distribution and had validated the accuracy of colour change for alerting the growth of *Listeria monocytogenes* (Rokugawa, 2015). The laccase-based TTI, including NaN3, was composed of two parts of an enzyme solution and a substrate solution and could predict the Pseudomonas fragile growth (Park et al.,2013).

Fresh-Check® TTI (Temptime Co., USA) is using a highly coloured polymer by a solid-state polymerization reaction. The polymer gradually darkens on the colour that tends to reflect the cumulative exposure to temperature. If the inner colour is darker than the outer colour, it means that the product has reached the end of the shelf-life. The indicator is required to maintain -24°C before application as indicators could be activated above the storage temperature (Pavelková 2013; Wang et al., 2015). However, the range of use gets limited due to the storage temperature. Anther disadvantage is the potential toxicity of polydiacetylene compounds. The accuracy is affected by the selected compound and the presence of sunlight or bright direct light, which can accelerate the polymerization reaction (Wang et al., 2015).

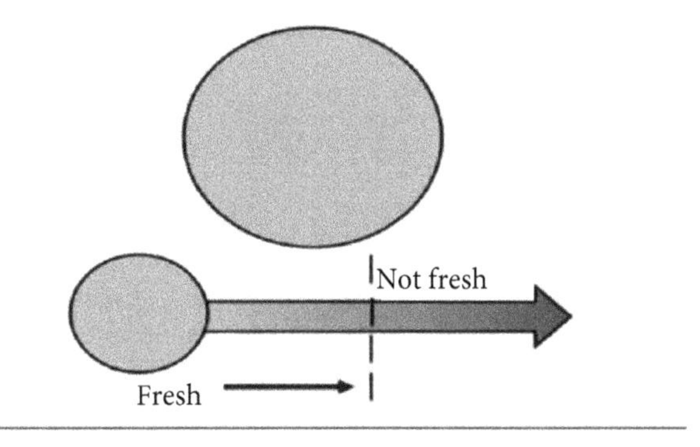

Fig. 4: Freshness indicator Source: O'grady et al. (2008)

OnVuTM TTI (Ciba Specialty Chemicals & Fresh point Inc., Switzerland) a solid-state reaction TTI, is based on the colour change of photosensitive compounds and organic pigments by temperature. Additionally, this TTI gets activated by the UV irradiation, changes from colourless to blue. The rate of change of colour is proportional to temperature (Pavelková 2013; Wang et al., 2015). The flexible use for various foods limited because of the fixed rate of change in at an identical temperature.

Biological-type TTIs

The colour change in a microbial TTI prototype is based on the growth and metabolic activity of *Lactobacillus sakei* strain was similar with the lactic acid bacteria (LAB) growth in fresh ground meat stored under modified atmosphere packed (MAP) conditions (Vaikousi et al.,2009). A novel colourimetric indicator label for monitoring the freshness of intermediate-moisture dessert spoilage used pH-sensitive dyes, bromothymol blue, methyl red, and carbon dioxide (CO_2), and the indicator response correlates with microbial growthpatterns (Nopwinyuwong et al.,2010).

A microbial TTI using the Weissella cibaria CIFP 009 (psychrotrophic lactic acid bacterium) could be predicted accurately about aerobic mesophilic bacteria (AMB) counts, lactic acid bacteria (LAB) counts, and freshness (Kim et al., 2012)

The commercially available TTIs, Check Point® TTI (Vitsab A. B., Sweden) based on a colour change by the enzymatic system. The pH gets decreased by controlled enzymatic hydrolysis of a lipid substrate, and the pH decrease has occurred the colour change of a pH indicator from deep green to bright yellow to orange-red (Kuswandi *et al.*, 2011; Pavelkove, 2013). eO® TTI (CRYOLOG, France) is based on pH change by controlled microbial growth selected strains of lactic acid bacteria and is stored in a frozen state of -18°C to prevent the bacterial growth. The colour of this TTI is changed to red by temperature abuse, or when the product reaches its use-by date (Pavelkove, 2013). It could be used successfully as a quality management tool for meat products. However, it considered that flexible usu for various foods gets limited due to the difficulty of rate control for microbial growth (Ellouze et al., 2010).

Table 4: Types of TTIs with different working and experimental method

Types Chemical	Principle	Working	Example	Experimental	References
Chemical based	Polymerisation based TTI	Monomer and polymer of diacetylene can change the colour during polymerization and rate of polymerization and the rate of polymerization increase with temperature and time.	Fresh-Check®	A mixture of PVC and polyvinyl acetate used as an ink for the thermochromic label (PDA material) which can record thermochromic transition photographically in the range of 20-65°C.	Phollookin et al. (2010).
			HEAT-marker®	A crystallised diacetylene compounds like alkylurea (2,4-hexa-diyn-1, 6- bis) with solvent (to dissolve) used in TTI which irreversibly show changes in colour as the environment changes. It directly depends on the solvent system in which the colour change is related to the solubility of the diacetylenic monomer in the solvent system when temperature and time change.	Baughman et al. (2009).
	Photochromic based TTI	It is a thermally induce fading reaction process in which photochromic compound shows different colours after activation at different wavelengths. The fading process change with the accumulation of time and temperature.	OnVu™	The acetylene compound (2,4-hexadiyn-1,6-diethyl urea) get synthesised by propargylamine and isocyanate, which also shown colour changes from white to blue, and even to black as the polymerisation rate increases.	Song et al. (2013).
				The formation of TTI based on PDA vesicle with having an amphiphilic polymer solution which monitors the temperature range 0–50°C.	Zheng et al., (2013)
				New photochromic liquid- carbosilane dendrimer developed in which the thermal back-isomerisation reaction rate kH is much smaller than photochemical backisomerization positive/reverse reaction rate kt/kc.	Zhang et al., (2005)
				To form H-aggregation, a spiropyran derivative attached to the side hains of polyacrylates or polysiloxane which shows the absorption spectrum had a blue shift. Other factors such as polarity of the solvent, compound property, the pH, metal ions of the solution affect the rate of thermally induced fading reaction. The thermally induced close- loop reaction of the modified open-loop spiro aromatic slows down obviously, and the reaction duration may be extended.	Krongauz et al., (1988)

Types Chemical	Principle	Working	Example	Experimental	References
	Redox-reaction based TTI	It is light-induced redox-reac tion Tblased TTI, which shows a colour change when the compound comes in contact with air, gets react with the oxygen is directly related to the accumulation of time and temperature.		Another spiro-pyran thio alkyl or aryl thio is a kind of derivative that is challenging to be bleached by ambient light but has high stability and printability when stored at room temperature before its activation.	Salman et al., (2010)
				The colourless or slightly coloured spiro-aromatic compound in the thermodynamically stable state and the coloured in the meta-stable state. When the colour changes to a pre-determined meta-stable state, the light induction can be removed to prevent photo bleaching.	Tenetov et al., (2012)
				The different spiro-pyran compound identified and its photochromic reaction under UV radiation at diff erent temperatures confirms that either it can be us indication for cold chain products at low temperature.	Kreyenschmidt et al., (2010)
				Diarylethene is a compound based on valence isomerism converts the open loop or valence isomer in the isomerism. The open loop more stable in thermodyn form while closed loop is in coloured or meta stable form.	Levy et al., (2007)
				The derivative of anthraquinone used in printing the gets decomposed into crimson particles. When the p react with the oxygen, the colour starts changing, influenced by both time and temperature.	Galagan et al., (2008)
Physical-based	Diffusion-based TTI	it based on diffusion where the diffusion of the coloured chemicals, i.e., fatty acid esters, phthalates, specific polymers through the porous wick which made of specially designed material, from one reservoir to another.	Monitor Mark™	TTI which is working based on dyes diffusing through a dye compatible polymer composition when the polymer composition is above the determined temperature	Ezrielev et al., (1995)
			Tempix TTI	The viscoelastic material migrates into a diffusely light-reflective porous matrix at a rate that varies with temperature to change the light transmissivity of the porous matrix progressively and thereby provide a visually observable indication.	Arens et al., (1997)
				The amorphous material in used contact with a porous matrix.	Spevack et al., (2003)

Types Chemical	Principle	Working	Example	Experimental	References
		I		The dyeing chemical material with low-temperature melting point moves along the capillary/ porous diffusion tube.	Ye et al., (2004)
				Printable TTI which could be printed directly on the substrate materials or at the interlayer between substrate materials. This indicator includes a protective layer, a diffusion layer, two reagents on the substrate or the protective layer or the diffusion layer, an optional outer protective layer and an optional external base layer.	Koivukunnas et al., (2008)
				The volatile property of materials. In the process (heat–evaporation– adsorption) of real transformation, the colour gets changes. During the storage of food, the volatile dye is absorbed by the absorption layer after heating, and the number of volatile dyes positively correlated with the cumulative thermal effect. It targeted at specific heat-sensitive products, the kind and amount of dyes can be adjusted in combination with other methods to control the rate of dye evaporation, so that the colour response of TTI can be consistent with the food quality.	Deng et al., (2013)
	Nano-particle based TTI	Nanomaterial used, which has a perfect thermochromic property		The silver-based nanoplates used as a thermochromic material. After absorbing heat, the surface morphology of Ag nanoparticle changes, which results in the shift of wave numbers to a visible region.	Zeng et al., (2010)
	lectronic-based TTI	The thermal sensor that converts temperature signals to electrical signals and then converts electrical signals to final visual output.		The gelatin-templated AuNPs (gold particle) kept at 30°C at different times, and they stored at -20°C. Distinct colour development can be observed in as early as six hours at 30°C, and the colour intensity is proportional to the duration of exposure.	Lim et al., (2012)
				Kinetically programmable Ag overgrowth on Au nanorods. In the reaction of epitaxial overgrowth of Ag shell on Au nanorods, which shows a sharp colour change, which can be dynamically adjusted.	Zhang et al., (2013)

Types Chemical	Principle	Working	Example	Experimental	References
		.		The electronic TTI includes a high-frequency temperature logger to collect information that will pass to a computing device that is a microprocessor, and then the food's thermal history will recorded. It is used to monitor the quality of insulin.	Zweig et al., (2005)
				The electronic label which includes a temperature sensor and a oscillator. When the external temperature of food changes, the temperature sensor transmits the information to the oscillator, and then the frequency of oscillator changes. There is a counter in the TTI that records the complete cycles of the oscillator.	Jensen et al., (2013)
				A conductor layer, an insulator layer, a transparent conductor layer and a doped polymer layer in which the dopant can be an acid, base, light latent acid or light latent base. It gets activated when the temperature reached a particular value, the acid or base component will start corroding the conductor layer, which changes the conductor layers electrical. The change can visualized by using ink containing metal particles as the conductive layer, which means that the TTI can utilize without the support of professional equipment.	Haarer et al., (2012)
Enzymatic based	Acid-base reaction based TTI	The lipid substrate is hydrolyzed by lipase in controlled conditions, which results in a decrease of pH. Consequently, the colour of the pH indicator changes. The rate hydrolysis of the lipid substrate directly proportional to the temperature. The continuous colour change can also be measured, and thus, the product quality can be estimated by observing the colour change.	Check-Point™	An enzymatic TTI prototype (xylanases) that used in the temperature range of 100–130°C.	Gogou et al., (2010)
				The urea substrate, phenol red indicator, urease solution, disodium hydrogen phosphate–potassium dihydrogen phosphate used as a buffer solution. The buffer solution gets mixed in order to ensure catalytic reaction occurs in a specific range of pH. Urease catalyzes the decomposition of urea to produce ammonia, which causes pH to change and consequently brings about the colour change of the phenol red indicator.	Wu, (2005)
				The components of the system are alkaline lipase solution, mannose solution, Gly-NaOH solution also, bromine thymol blue solution (pH indicator).	Lu et al. (2012)

Types	Principle	Working	Example	Experimental	References
Chemical					
	Redox reaction based TTI			New enzymatic TTI based on laccase, in which the components are laccase solution, guaiacol (substrate for laccase), bovine serum albumin (enzyme stabiliser) and sodium acetate buffer. Laccase catalyses guaiacol oxidation discolouration to indicate the cumulative time-temperature exposure.	Kim et al., (2012)
				Developed a glucoamylase-type TTI, in which glucoamylase catalyses both the hydrolysis of dextrin and iodine solution works as the indicator. The extent of a colour change indicates the cumulative time-temperature effect.	Qian et al., (2012)
				Based on starch and amylase based reaction in which blue substance gets formed when iodine solution reacts with starch.	Cai et al., (2006)
Biological-based	Yeast-based TTI	Yeasts and lactic acid bacteria have been widely used in biological TTI systems recently. Other bacteria such as Streptococcus can also be available in the TTI systems.	TRA-CEO® eO®	The TTI system based on the anaerobic respiration of yeast to generate acid in certain circumstances, especially at a specific temperature, resulting in the colour change of the pH indicator. The first part contains the reactants, including microorganisms and colour indicator made of aqueous ink with appropriate carrier separating them, both of which are attached to a transparent plastic film; the second part comprises an activator with an aqueous adhesive layer, and it gets activated when the reactant contact with the activator.	Varlet-Grancher et al., (2006)

Types Chemical	Principle	Working	Example	Experimental	References
	Lactobacillus-based TTI	The lactic the acid generated by lactic acid bacteria to change the pH under certain conditions, which leads to a colour change to indicate the accumulation effect of time and temperature.		It is a two-dimensional code containing the colour information of the TTI that changes with mutative time and temperature; people can judge the quality of the product quickly by scanning the two-dimensional code to obtain the colour information.	Lee et al., (2013)
Others TTI	TTI based on phototonic lattice change			Based on the photonic crystal (PC-TTI), which consists of a substrate, a mesh layer, a photonic crystal patterned film and a hardener pouch. The photonic crystal in the patterned film shows the correlation between band-gap shift and temperature, thus providing a visible indication of the range of temperature.	
	TTI based on thermochromic polymer/dye blends			Based on the phase separation between dye molecules and a polymer host. Polymers with built-in threshold temperature sensors exhibit a change of their absorption characteristics in response to external heat.	Chen et al., (2013)

Existing TTIs have a limitation on flexibility at a range of reaction temperature or reaction period. Since storage characteristics of each kind of food are very different even at the same storage temperature, ideal TTIs should have flexibility corresponding to each food characteristics by adjusting the reaction rate. However, most of the existing TTIs do not have flexibility control, the variation rate in the same temperature and broaden available temperature range. TTI colour changes need to be correlated to food quality and safety changes to provide additional information on food distribution under the selected temperature conditions.

Sensors

A sensor is an instrument that is used to detect, locate or quantify a problem and then send signals to measure its physical or chemical characteristics(Muller and Schmid, 2019). A sensor has the ability to frequently detect an event or changes in the surrounding environment (Vanderroost et al., 2014). Usually sensors are made up of a receptor and transducer. The function of a receptor is to convert physical or chemical information into a form of energy, and a transducer changes the energy to an analytical signal (Biji et al., 2015). For food quality and safety assurance, it may be very important to develop intelligent food packaging systems using portable sensors to check different compounds and gas molecules, especially, H_2, O_2, NO_2 and CO_2 in modified atmosphere packaging. There are different types of sensors that investigate different parameters, such as chemical sensors, gas sensor, and biosensors.

Gas sensor

These sensors are designed to detect the presence of the gaseous analyte in the package. The progress of spoilage can be determined by the concentration of certain gases, like H_2S, $NH_{4,}$ CO_2 and volatile biogenic amines. These chemical sensors are the best alternatives to the time-consuming analytical instruments such as gas chromatography-mass spectrometer (GC-MS), which can only be applied by breaking the food package integrity (Llobet, 2013).The gas sensors make use of these properties by monitoring them. It responds quantitatively and reversibly to the presence of a gas by changing the physical parameters of the sensor (Matindoust *et al.*, 2016). Such systems are based onthe principle of luminescence quenching or absorbance changes caused by direct contact with the gas analyte (Kalpana et al., 2019).

A high sensitive optical dual sensor for the detection of oxygen using different levels of temperature was developed (Baleizao *et al.*, 2008). The sensor was composed of two light-emitting compounds, one can be used for the detection of temperature while the other for the detection of oxygen. As Ruthenium tris-1, 10-phenanthroline is highly luminescent compound: it was used as the temperature-sensitive dye. The probe used for the detection of oxygen-sensitivity was fullerene C_{70} because of its strong, thermally activated and delayed fluorescence at high temperatures. Their results confirmed that the dual sensor had the capacity of detecting temperature change ranging from 0 to 12°C and the minimum detection limits for oxygen was 50 ppmv. An opto-chemical CO_2 sensor was developed by composing of a Pt-TFPP dye and a colorimetric pH indicator a-naphthol phthalein bounded in a plastic shield combined with a tetraoctyl- or cetyltrimethyl ammonium hydroxide that acts as a phase transfer agent. After the optimization of the composition and the working conditions of the developed sensor for the measurement of CO_2 in foods stored under modified atmosphere packaging, it was concluded that the sensor could retain its sensitivity to CO_2 at 4°C for almost three weeks. On the other hand, a sensor was developed for checking the freshness of packaged cod fillets (Heising et al., 2015). The working principle of this sensor is to monitor the volatile compounds released from the fish in the storage. The sensor has the ability to monitor the freshness of the fish at temperatures as low as 0°C. Temperatures higher than 4°C, a conductivity meter should be combined with the temperature sensor to successfully check the freshness of the packaged fish.

Biosensor

Biosensors detect, record, and transmit information pertaining to biological reaction (Yam et al., 2005). They consist of a bio-receptor specific to a target analyte (enzymes, antigens, microbes, nucleic acids, hormones) and a signal transducer element (optical, colorimetric, and electrochemical) which is connected to the data acquisition and processing systems (Hogan et al., 2008). The main difference between a chemical sensor and a biosensor is in the recognition layer, at chemical sensor the receptor is a chemical compound, at biosensor it is an organic or biological material (Ghaani et al., 2016). Biosensors are able to monitor food freshness in a more specific way than freshness indicators, as they can detect the formation of degradation products and might be designed in a tailored manner according to the type of product being packed (Realini and Marcos, 2014). Important characteristics of biosensors are their specificity, sensitivity, reliability, portability, and simplicity (Yam et al., 2005).

Some commercial examples of biosensors are Toxin Guard and Food Sentinel System. Toxin Guard is a visual sensor based on antibody-antigen reactions that indicate the presence of a pathogenic bacteria. In the presence of pathogenic bacteria, the bacterial toxin is bound to the antibodies and immobilized on a thin layer of film, resulting in the color change (Ghaani et al., 2016). Food Sentinel System is a biosensor developed to detect food pathogens with a specific pathogen antibody that is attached to a membrane-forming part of a barcode. The presence of contaminating bacteria causes the formation of a localized dark bar, rendering the barcode unreadable upon scanning (Yam et al., 2005).

Radiofrequency Identification Devices (RFID)

The RFID system consists of a tag, a reader, and a central node or computer. A tag is a small electronic device composed of a microchip and antenna. Each tag has a unique identification, which may be a serial number, or product-related information (e.g., a stock number, lot or batch number, production date) to be stored electronically via the microchip (Lee *et al.*, 2014). RFID tags can be classified into three types: active, passive, and semi-passive. Passive tag has no battery and is powered by the energy emitted by the reader. Reading range is shorter, up to approximately 5m (Dobrucka et al., 2014). Active tag has its own battery for powering the microchip's circuitry and broadcasting signals to the reader, and operates at a distance of up to approximately 50m (Dobrucka et al., 2014). Semi-passive tag has battery only to power its microchip and is unable to produce return signal.

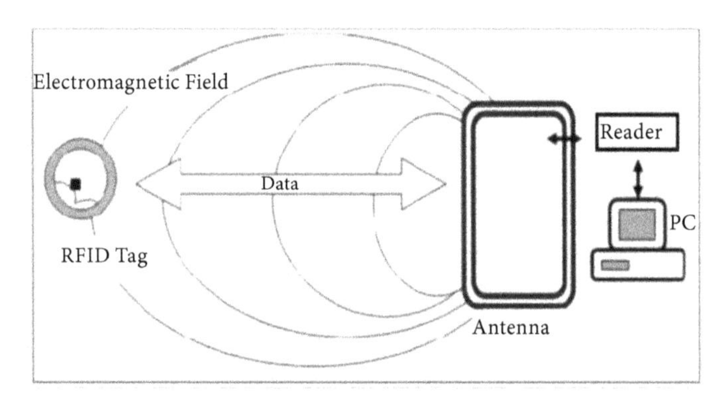

Fig. 5: RFID Tag

RFID systems allow the integration of other functions into the RFID tag, such as time-temperature indicators or biosensors, to monitor and communicate the temperature history of the product as well as to give quality information by obtaining microbial data (Guillory et al., 2012). By evaluating the environmental conditions, the quality of product and shelf life can be estimated and decisions for optimizing logistic can be supported. With including a flexible pH sensor, the wireless pH sensing system was tested for in situ monitoring of the spoilage processes in fish products (Huang et al., 2012).

Printed electronics

Printed electronics refer to a process in which printing technology is used to produce various kinds of electronic goods, such as electronic circuits, displays, sensors, RFID. It is based on organic conducting and semiconducting as well as printable inorganic materials. The area of printed electronics is very wide, which includes a large number of different materials and printing technologies (Nilsson et al., 2012). Because it can provide a new functionality to the packaging, printing electronics is seen as one of the emerging technologies for food packaging in the future.

One of the first market applications of printed electronics was RFID embedded in cartons, besides passive ID cards printed on paper used for ticketing or toys (Kleper,

2014)). Flexible lithium polymer batteries produced in a roll-to-roll process, printed antennas for RFID tags, printed electrodes for glucose test strips, printed RF-driven smart objects, and large-area organic pressure sensors for applications such as retail logistics were among first commercial products that are present on the market from around 2010 (Lupo et al., 2013). Printed passive components, such as RFID antennas, based on printable conductors and dielectrics are already well known in electronics manufacturing. Printed wireless power transmission devices, electrochromic displays, and sensors have been produced using roll-to-roll printing (Kang et al., 2014). Antitheft/forgery labels, temperature sensors, test strips, and smart labels are some of the integrated systems which are now emerging in the market. Integrated smart systems are built from different components, such as printed batteries, printed memory, active (transistors, diodes, logic circuits, display elements) or passive (resistors, capacitors, conductors) devices integrated by one process or by a combination of separately produced components.

The key challenges for integrated smart systems are related to integration of different components and especially interfacing to printed electronic circuitry. Intense development is related to incorporation of sensors into an integrated smart system. The company Thin-Film demonstrated and introduced the industry's first integrated printed electronic temperature-tracking sensor system powered solely by batteries coupled with near-field communication functionality, designed for monitoring perishable goods.

Legal and safety aspects of intelligent packaging

European Commission, 2004, appeared the first law related to materials intended to come in the contact with food. The law enabling that the component present in the packaging shall not be transferred to food in forbidden quantity. Moreover, the intelligent material used in food packaging shall not provide misleading information to the consumer. Regarding migration of the chemical or biological component European Commission, 2006 amended law that such compounds shall not be allowed in the unacceptable quantities (Dainelli et al., 2008). Specific law focusing on the intelligent packaging on food was established in 2009, by European Commission which states that it is mandatory to write the words "DO NOT EAT" on the device or component used as intelligent packaging, if possible put a specific symbol on the package. The risk for releasing substance from intelligent packaging occurs mainly when they are put inside the primary packaging. The risk causing factor could be leaching or swallowing of active components (Restuccia et al., 2010). In order to increase the acceptance of consumers toward the intelligent packaging, it is necessary to minimize such potential risks of food contamination (Lee et al., 2008).

Conclusion

Smart packaging has incredible aptitude for enhancing supply chain effectiveness and efficiency. There are many applications that could improve customer satisfaction; increase visibility; improve security and reliability of supply; prevent product diversion, counterfeit, and theft; and reduce product wastage. Further research is required to enable a wider application (Fuertes *et al.*, 2016). Customers always want better quality and more information about the products, but most of them are not really willing to pay more for that. If they were well informed about the benefits of the systems, the customers might be more willing to spend more on food with smart packaging (Vanderrost et al., 2014). In addition, consumer confidence in the safety of the systems also needs to be strengthened. Therefore, further steps should be taken to promote the technologies (Sohail et al., 2018). The manufacturer also has to realize that the use of intelligent packaging can offer them a real market advantage. If all these aspects can be accomplished, a more extensive use of smart packaging would be possible.

References

Aday, M. S., Caner, C., &Rahvalı, F. (2011). Effect of oxygen and carbon dioxide absorbers on strawberry quality. *Postharvest Biology and Technology*, 62(2), 179-187.https://doi.org/10.1021/jf001438l

Ahn, B. J., Gaikwad, K. K., & Lee, Y. S. (2016). Characterization and properties of LDPE film with gallic-acid-based oxygen scavenging system useful as a functional packaging material. *Journal of Applied Polymer Science*, *133*(43).https://doi.org/10.1002/app.44138

Akkapeddi, M. K., & Lynch, B. A. (2019). *U.S. Patent No.* 10,208,200. Washington, DC: U.S. Patent and Trademark Office.https://patents.google.com/patent/US10208200B2/en

Almenar, E., Del-Valle, V., Hernández-Muñoz, P., Lagarón, J. M., Catalá, R., &Gavara, R. (2007). Equilibrium modified atmosphere packaging of wild strawberries. *Journal of the Science of Food and Agriculture*, 87(10), 1931-1939.https://doi.org/10.1002/jsfa.2938

Anthierens, T., Ragaert, P., Verbrugghe, S., Ouchchen, A., De Geest, B. G., Noseda, B., & Du Prez, F. (2011). Use of endospore-forming bacteria as an active oxygen scavenger in plastic packaging materials. *Innovative Food Science & Emerging Technologies*, 12(4), 594-599. https://doi.org/10.1016/j.ifset.2011.06.008

Appendini, P., & Hotchkiss, J. H. (2002). Review of antimicrobial food packaging. *Innovative Food Science & Emerging Technologies*, 3(2), 113-126.https://www.researchgate.net/deref/http%3A%2F%2Fdx.doi.org%2F10.1016%2FS1466-8564(02)00012-7

Arvanitoyannis, I. S., & Stratakos, A. C. (2012). Application of modified atmosphere packaging and active/smart technologies to red meat and poultry: a review. *Food and Bioprocess Technology*, 5(5), 1423-1446. https://doi.org/10.1007/s11947-012-0803-z

Balamatsia, C. C., Patsias, A., Kontominas, M. G., &Savvaidis, I. N. (2007). Possible role of volatile amines as quality-indicating metabolites in modified atmosphere-packaged chicken fillets: Correlation with microbiological and sensory attributes. *Food Chemistry*, 104(4), 1622-1628.https://doi.org/10.1016/j.foodchem.2007.03.013

Baleizao, C., Nagl, S., Schäferling, M., Berberan-Santos, M. N., & Wolfbeis, O. S. (2008). Dual fluorescence sensor for trace oxygen and temperature with unmatched range and sensitivity. *Analytical chemistry*, 80(16), 6449-6457.https://doi.org/10.1021/ac801034p.

Baltrusaitis, J., Schuttlefield, J. D., Zeitler, E., Jensen, J. H., & Grassian, V. H. (2007). Surface Reactions of Carbon Dioxide at the Adsorbed Water– Oxide Interface. *The Journal of Physical Chemistry C*, 111(40), 14870-14880.https://doi.org/10.1021/jp051868k

Barbosa-Pereira, L., Cruz, J. M., Sendón, R., de Quirós, A. R. B., Ares, A., Castro-López, M., &Paseiro-Losada, P. (2013). Development of antioxidant active films containing tocopherols to extend the shelf life of fish. *Food Control*, 31(1), 236-243.http://dx.doi.org/10.1016%2Fj.foodcont.2012.09.036

Baughman, R. H., Hall, L. J., Kozlov, M., Smith, D. E., & Prusik, T. (2012). *U.S. Patent No. 8,269,042*. Washington, DC: U.S. Patent and Trademark Office.https://patents.google.com/patent/US8269042B2/en

Bhardwaj, A., Alam, T., & Talwar, N. (2019). Recent advances in active packaging of agri-food products: a review. *Journal of Postharvest Technology*, 7(1), 33-62.

Biji, K. B., Ravishankar, C. N., Mohan, C. O., & Gopal, T. S. (2015). Smart packaging systems for food applications: a review. *Journal of food science and technology*, *52*(10), 6125-6135. https://doi.org/10.1007/s13197-015-1766-7

Bolumar, T., Andersen, M. L., &Orlien, V. (2011). Antioxidant active packaging for chicken meat processed by high pressure treatment. *Food Chemistry*, *129*(4), 1406-1412. https://www.researchgate.net/deref/http%3A%2F%2Fdx.doi.org%2F10.1016%2Fj.foodchem.2011.05.082

Borchert, N. B., Kerry, J. P., &Papkovsky, D. B. (2013). A CO2 sensor based on Pt-porphyrin dye and FRET scheme for food packaging applications. *Sensors and Actuators B: Chemical*, 176, 157-165.https://doi.org/10.1016/j.snb.2012.09.043

Brody, A. L., Strupinsky, E. P., & Kline, L. R. (2001). *Active packaging for food applications*. CRC press.https://doi.org/10.1201/9780367801311

Busolo, M. A., &Lagaron, J. M. (2012). Oxygen scavenging polyolefin nanocomposite films containing an iron modified kaolinite of interest in active food packaging applications. *Innovative Food Science & Emerging Technologies*, *16*, 211-217.http://dx.doi.org/10.1016%2Fj.ifset.2012.06.008

Byun, Y., Whiteside, S., Cooksey, K., Darby, D., & Dawson, P. L. (2011). α-Tocopherol-loadedpolycaprolactone (PCL) nanoparticles as a heat-activated oxygen scavenger. *Journal of agricultural and food chemistry*, 59(4), 1428-1431.https://doi.org/10.1021/jf103872g

Cabrera, I., Krongauz, V., &Ringsdorf, H. (1988). Photo-and thermo-chromic liquid crystal polymers with spiropyran groups. *Molecular Crystals and Liquid Crystals*, *155*(1), 221-230.https://doi.org/10.1080/00268948808070366

CAI, H. W., REN, F. Z., ZHANG, H. T., & LIU, W. (2006). Research of Amylase-Time-Temperature Indicator [J]. *FoodScience*, *11*.https://iopscience.iop.org/article/10.1088/1757-899X/612/2/022042/pdf

Chaix, E., Guillaume, C., &Guillard, V. (2014). Oxygen and carbon dioxide solubility and diffusivity in solid food matrices: a review of past and current knowledge. *Comprehensive Reviews in Food Science and Food Safety*, 13(3), 261-286.https://doi.org/10.1111/1541-4337.12058

Charles, F., Sanchez, J., & Gontard, N. (2006). Absorption kinetics of oxygen and carbon dioxide scavengers as part of active modified atmosphere packaging. *Journal of food Engineering*, *72*(1), 1-7.https://doi.org/10.1016/j.jfoodeng.2004.11.006

Cherpinski, A., Gozutok, M., Sasmazel, H., Torres-Giner, S., &Lagaron, J. (2018). Electrospun oxygen scavenging films of poly (3-hydroxybutyrate) containing palladium nanoparticles for active packaging applications. *Nanomaterials*, *8*(7), 469.doi: 10.3390/nano8070469.

Choi, W. S., Singh, S., & Lee, Y. S. (2016). Characterization of edible film containing essential oils in hydroxypropyl methylcellulose and its effect on quality attributes of 'Formosa'plum (Prunussalicina L.). *LWT-Food Science and Technology*, 70, 213-222. https://doi.org/10.1016/j.lwt.2016.02.036

Claeys, W. L., Indrawati, A. M., & Hendrickx, M. E. (2003). are intrinsic TTIs for thermally processed milk applicable for high-pressure processing assessment?. *Innovative Food Science & Emerging Technologies*, 4(1), 1-14.https://doi.org/10.1016/S1466-8564(02)00066-8

Coles, R., McDowell, D., & Kirwan, M. J. (Eds.). (2003). *Food Packaging Technology* (Vol. 5). CRC Press.https://doi.org/10.1002/pts.655

Cooksey, K. (2005). Effectiveness of antimicrobial food packaging materials. *Food Additives and Contaminants, 22*(10), 980-987.https://doi.org/10.1080/02652030500246164

Courbat, J., Briand, D., Damon-Lacoste, J., Wöllenstein, J., & De Rooij, N. F. (2009). Evaluation of pH indicator-based colorimetric films for ammonia detection using optical waveguides. *Sensors and Actuators B: Chemical*, 143(1), 62-70.https://www.researchgate.net/deref/http%3A%2F%2Fdx.doi.org%2F10.1016%2Fj.snb.2009.08.049

Crump, J. W., Chau, C. C., McKedy, G. E., Pyne, D. S., Powers, T. H., Solovyov, S. E., & Hurley, T. J. (2015). *U.S. Patent Application No.* 14/*320,192*.

Cruz, R. S., Camilloto, G. P., & dos Santos Pires, A. C. (2012). Oxygen scavengers: an approach on food preservation. *Structure and function of food engineering, 2*.DOI: 10.5772/48453

Cruz, R. S., Soares, N. D. F. F., & Andrade, N. J. D. (2006). Evaluation of oxygen absorber on antimicrobial preservation of lasagna-type fresh pasta under vacuum packed. *Ciência e Agrotecnologia, 30*(6), 1135-1138.http://dx.doi.org/10.1590/S1413-70542006000600015

da Cruz, N. F., Simões, P., & Marques, R. C. (2012). Economic cost recovery in the recycling of packaging waste: the case of Portugal. *Journal of Cleaner Production*, 37, 8-18.http://doi.org/10.1016/j.jclepro.2012.05.043

Davidson, P. M., Cekmer, H. B., Monu, E. A., &Techathuvanan, C. (2015). The use of natural antimicrobials in food: an overview. *Handbook of natural antimicrobials for food safety and quality*, 1-27.https://doi.org/10.1016/C2013-0-16441-0

Day, B. P. (2008). Active packaging of food. *Smart packaging technologies for fast moving consumer goods, 1*. 10.1002/9780470753699

Di Maio, L., Scarfato, P., Galdi, M. R., &Incarnato, L. (2015). Development and oxygen scavenging performance of three-layer active PET films for food packaging. *Journal of Applied Polymer Science, 132*(7).https://doi.org/10.1002/app.41465

Dobrucka, R., &Cierpiszewski, R. (2014). Active and intelligent packaging food–research and development–a review. *Polish Journal of Food and Nutrition Sciences*, 64(1), 7-15.https://doi.org/10.2478/v10222-012-0091-3

Ellouze, M., & Augustin, J. C. (2010). Applicability of biological time temperature integrators as quality and safety indicators for meat products. *International journal of food microbiology, 138*(1-2), 119-129.https://doi.org/10.1016/j.ijfoodmicro.2009.12.012

Ezrielev, R. I., & Barrett, R. B. (1995). *U.S. Patent No. 5,476,792*. Washington, DC: U.S. Patent and Trademark Office.

Fava, F. (2014). *U.S. Patent No. 8,871,846*. Washington, DC: U.S. Patent and Trademark Office.

Ferrari, M. C., Carranza, S., Bonnecaze, R. T., Tung, K. K., Freeman, B. D., & Paul, D. R. (2009). Modeling of oxygen scavenging for improved barrier behavior: Blend films. *Journal of Membrane Science, 329*(1-2), 183-192.https://doi.org/10.1016/j.memsci.2008.12.030

Gaikwad, K. K., Lee, J. Y., & Lee, Y. S. (2016). Development of polyvinyl alcohol and apple pomace bio-composite film with antioxidant properties for active food packaging application. *Journal of food science and technology, 53*(3),1608-1619.https://doi.org/10.1007/s13197-015-2104-9

Gaikwad, K. K., Singh, S., & Lee, Y. S. (2017). A pyrogallol-coated modified LDPE film as an oxygen scavenging film for active packaging materials. *Progress in Organic Coatings, 111*, 186-195.https://doi.org/10.1016/j.porgcoat.2017.05.016

Gaikwad, K.K., Singh, S., & Lee, Y.S. (2018). Oxygen scavenging films in food packaging. *Environmental chemistry letters*, 16(2), 523-538.https://doi.org/10.1007/s10311-018-0705-z

Gaikwad, K. K., Singh, S., & Negi, Y. S. (2019). Ethylene scavengers for active packaging of fresh food produce. *Environmental Chemistry Letters*, 1-16.https://doi.org/10.1007/s10311-019-00938-1

Galagan, Y., & Su, W. F. (2008). Fadable ink for time–temperature control of food freshness: Novel new time–temperature indicator. *Food research international*, 41(6), 653-657.http://dx.doi.org/10.1016%2Fj.foodres.2008.04.012

Galdi, M. R., Nicolais, V., Di Maio, L., &Incarnato, L. (2008). Production of active PET films: evaluation of scavenging activity. *Packaging Technology and Science: An International Journal, 21*(5), 257-268.https://doi.org/10.1002/pts.794

Ghaani, M., Cozzolino, C.A., Castelli, G. and Farris, S., 2016. An overview of the intelligent packaging technologies in the food sector. Trends in Food Science & Technology, 51,1-11. http://dx.doi.org/10.1016%2Fj.tifs.2016.02.008

Giannakourou, M. C., Koutsoumanis, K., Nychas, G. J. E., &Taoukis, P. S. (2005). Field evaluation of the application of time temperature integrators for monitoring fish quality in the chill chain. *International journal of food microbiology, 102*(3), 323-336.https://doi.org/10.1016/j.ijfoodmicro.2004.11.037

Gibis, D., &Rieblinger, K. (2011). Oxygen scavenging films for food application. *Procedia Food Science, 1*, 229-234.https://doi.org/10.1016/j.profoo.2011.09.036

Gogou, E., Katapodis, P., Christakopoulos, P., & Taoukis, P. S. (2010). Effect of water activity on the thermal stability of Thermomyces lanuginosus xylanases for process time–temperature integration. *Journal of food engineering*, 100(4), 649-655.https://doi.org/10.1016/j.jfoodeng.2010.05.014

Gómez, A. H., Hu, G., Wang, J., & Pereira, A. G. (2006). Evaluation of tomato maturity by electronic nose. *Computers and electronics in agriculture*, 54(1), 44-52.http://dx.doi.org/10.1016/j.compag.2006.07.002

Gómez-Estaca, J., López-de-Dicastillo, C., Hernández-Muñoz, P., Catalá, R., &Gavara, R. (2014). Advances in antioxidant active food packaging. *Trends in Food Science & Technology*, 35(1), 42-51.https://doi.org/10.1021/jf201246g.

Guillory, M., & Strandhardt, G. (2012). NVC World Review on supply chain applications of RFID and sensors in packaging. *NVC Netherlands Packaging Centre, available at: http:// www. en. nvc. nl/pasteur-sensor-enabled-rfid/(accessed 10 March 2014).*

Hamilton, R. J., Kalu, C., Prisk, E., Padley, F. B., & Pierce, H. (1997). Chemistry of free radicals in lipids. *Food Chemistry, 60*(2), 193-199.https://doi.org/10.1016/S0308-8146(96)00351-2

Han, J. H. (2003). Novel food packaging techniques. *Antimicrobial food packaging. Cambridge: Woodhead Publishing Ltd.*

Han, J. H., &Floros, J. D. (1997). Casting antimicrobial packaging films and measuring their physical properties and antimicrobial activity. *Journal of Plastic Film & Sheeting, 13*(4), 287-298.https://doi.org/10.1177%2F875608799701300405

Heising, J. K., Dekker, M., Bartels, P. V., & Van Boekel, M. A. J. S. (2014). Monitoring the quality of perishable foods: opportunities for intelligent packaging. *Critical reviews in food science and nutrition*, 54(5), 645-654.10.1080/10408398.2011.600477.

Heising, J. K., Van Boekel, M. A. J. S., & Dekker, M. (2015). Simulations on the prediction of cod (Gadusmorhua) freshness from an intelligent packaging sensor concept. *Food Packaging and Shelf Life*, 3, 47-55.https://doi.org/10.1016/j.fpsl.2014.10.002

Hogan, S. A., & Kerry, J. P. (2008). Smart packaging of meat and poultry products. *Smart packaging technologies for fast moving consumer goods*, 33-54.https://doi.org/10.1016/j.nbt.2013.11.006

Holck, J., Larsen, D. M., Michalak, M., Li, H., Kjærulff, L., Kirpekar, F., & Meyer, A. S. (2014). Enzyme catalysed production of sialylated human milk oligosaccharides and galactooligosaccharides by Trypanosoma cruzi trans-sialidase. *New biotechnology*, 31(2), 156-165.https://doi.org/10.1016/j.nbt.2013.11.006

Huang, W. D., Deb, S., Seo, Y. S., Rao, S., Chiao, M., & Chiao, J. C. (2011). A passive radio-frequency pH-sensing tag for wireless food-quality monitoring. *IEEE Sensors Journal, 12*(3), 487-495.https://doi.org/10.1109/JSEN.2011.2107738

Isbitsky, R., Freedman, P. J., Solomon, I. M., & Kagan, M. L. (2007). *U.S. Patent No. 7,232,253.* Washington, DC: U.S. Patent and Trademark Office.https://patents.google.com/patent/US7232253B2/en?oq=[U.S.+Patent+7%2c232%2c253]

Jang, J. D., Hwang, Y. I., & Lee, D. S. (2000). Effect of packaging conditions on the quality changes of fermented soy paste and red pepper paste. *KOREAN JOURNAL OF PACKAGING SCIENCE & TECHNOLOGY*, *6*(1), 31-36.http://dx.doi.org/10.1016/j.jef.2015.02.002

Järnström, L., Johansson, K., Jönsson, L. J., Winestrand, S., Chatterjee, R., Nielsen, T., & Kotkamo, S. (2013). ENZYCOAT II-Enzymes embedded in barrier coatings for active packaging. *Nordic Innovation Publication*, 203, 1-96.

Johansson, K., Gillgren, T., Winestrand, S., Järnström, L., &Jönsson, L. J. (2014). Comparison of lignin derivatives as substrates for laccase-catalyzed scavenging of oxygen in coatings and films. *Journal of biological engineering*, 8(1), 1 10.1186/1754-1611-8-1

Kalpana, S., Priyadarshini, S. R., Leena, M. M., Moses, J. A., & Anandharamakrishnan, C. (2019). Intelligent packaging: Trends and applications in food systems. *Trends in Food Science & Technology*.https://doi.org/10.1016/j.tifs.2019.09.008

Kang, H., Park, H., Park, Y., Jung, M., Kim, B. C., Wallace, G., & Cho, G. (2014). Fully roll-to-roll gravure printable wireless (13.56 MHz) sensor-signage tags for smart packaging. *Scientific reports*, 4, 5387. doi:10.1038/srep05387

Kerry, J. P. (2014). New packaging technologies, materials and formats for fast-moving consumer products. In *Innovations in food packaging* (pp. 549-584). Academic Press. https://doi.org/10.1016/B978-0-12-394601-0.00023-0

Kerry, J. P., O'grady, M. N., & Hogan, S. A. (2006). Past, current and potential utilisation of active and intelligent packaging systems for meat and muscle-based products: A review. *Meat science*, 74(1), 113-130.https://doi.org/10.1016/j.meatsci.2006.04.024

Kerry, J., & Butler, P. (Eds.). (2008). *Smart packaging technologies for fast moving consumer goods*. John Wiley & Sons.https://doi.org/10.1016/j.meatsci.2013.04.050

Kim, E., Choi, D. Y., Kim, H. C., Kim, K., & Lee, S. J. (2013). Calibrations between the variables of microbial TTI response and ground pork qualities. *Meat science*, *95*(2), 362-367.https://doi.org/10.1016/j.meatsci.2013.04.050https://doi.org/10.1016/j.meatsci.2013.04.050

Kim, J. U., Ghafoor, K., Ahn, J., Shin, S., Lee, S. H., Shahbaz, H. M., & Park, J. (2016). Kinetic modeling and characterization of a diffusion-based time-temperature indicator (TTI) for monitoring microbial quality of non-pasteurized angelica juice. *LWT-Food Science and Technology*, 67, 143-150.https://doi.org/10.1016/j.lwt.2015.11.034

Kim, M. J., Jung, S. W., Park, H. R., & Lee, S. J. (2012). Selection of an optimum pH-indicator for developing lactic acid bacteria-based time–temperature integrators (TTI). *Journal of food engineering*, 113(3), 471-478.http://dx.doi.org/10.1016%2Fj.jfoodeng.2012.06.018

Kleper, M. L. (2004). *Printed electronics and the automatic identification of objects: An Investigation of the Emerging and Developing Technologies Related to the Generation Beyond Print-on-Paper.* GATF Press.

Koivukunnas, P., & Hurme, E. (2008). *U.S. Patent No. 7,430,982*. Washington, DC: U.S. Patent and Trademark Office.

Koutsoumanis, K. P., & Gougouli, M. (2015). Use of time temperature integrators in food safety management. *Trends in Food Science & Technology*, 43(2), 236-244.http://dx.doi.org/10.1016%2Fj.tifs.2015.02.008

Koutsoumanis, K., Taoukis, P. S., & Nychas, G. J. E. (2005). Development of a safety monitoring and assurance system for chilled food products. *International journal of food microbiology*, 100(1-3), 253-260-https://doi.org/10.1016/j.ijfoodmicro.2004.10.024

Kreyenschmidt, J., Christiansen, H., Hübner, A., Raab, V., & Petersen, B. (2010). A novel photochromic time–temperature indicator to support cold chain management. *International Journal of Food Science & Technology*, 45(2), 208-215.https://doi.org/10.1111/j.1365-2621.2009.02123.x

Kuswandi, B., Maryska, C., Abdullah, A., &Heng, L. Y. (2013). Real time on-package freshness indicator for guavas packaging. *Journal of food measurement and Characterization*, 7(1), 29-39.http://dx.doi.org/10.5614%2Fj.math.fund.sci.2015.47.3.2

Kuswandi, B., Oktaviana, R., Abdullah, A., &Heng, L.Y. (2014). A novel on package sticker sensor based on methyl red for real time monitoring of broiler chicken cut freshness. *Packaging Technology and Science, 27,* 69-81.https://doi.org/10.1016/bs.coac.2016.04.010

Kuswandi, B., Wicaksono, Y., Abdullah, A., Heng, L. Y., & Ahmad, M. (2011). Smart packaging: sensors for monitoring of food quality and safety. *Sensing and Instrumentation for Food Quality and Safety,* 5(3-4), 137-146.10.1007/s11694-011-9120-x

LABUZA, T. P., & Fu, B. I. N. (1995). Use of time/temperature integrators, predictive microbiology, and related technologies for assessing the extent and impact of temperature abuse on meat and poultry products. *Journal of Food Safety,* 15(3), 201-227.https://doi.org/10.1111/j.1745-4565.1995.tb00134.x

LaCoste, A., Schaich, K. M., Zumbrunnen, D., & Yam, K. L. (2005). Advancing controlled release packaging through smart blending. *Packaging Technology and Science: An International Journal,* 18(2), 77-87.https://doi.org/10.1002/pts.675

Lee, H. L., An, D. S., & Lee, D. S. (2016). Effect of initial gas flushing or vacuum packaging on the ripening dynamics and preference for kimchi, a Korean fermented vegetable. *Packaging Technology and Science,* 29(8-9), 479-485.https://doi.org/10.1002/pts.2227

Lee, J. W., Cha, D. S., Hwang, K. T., & Park, H. J. (2003). Effects of CO2 absorbent and high-pressure treatment on the shelf-life of packaged Kimchi products. *International journal of food science & technology, 38*(5), 519-524.https://doi.org/10.1046/j.1365-2621.2003.00699.x

Lee, S. J., & Rahman, A. M. (2014). Intelligent packaging for food products. In *Innovations in Food Packaging* (pp. 171-209). Academic Press.https://doi.org/10.1016/B978-0-12-394601-0.00008-4

Li, H., Tung, K. K., Paul, D. R., Freeman, B. D., Stewart, M. E., & Jenkins, J. C. (2012). Characterization of oxygen scavenging films based on 1, 4-polybutadiene. *Industrial & Engineering Chemistry Research,* 51(21), 7138-7145.https://doi.org/10.1021/ie201905j

Lim, S., Gunasekaran, S., &Imm, J. Y. (2012). Gelatin-Templated Gold Nanoparticles as Novel Time–Temperature Indicator. *Journal of food science,* 77(9), N45-N49.https://doi.org/10.1111/j.1750-3841.2012.02872.x

Llobet, E. (2013). Gas sensors using carbon nanomaterials: A review. *Sensors and Actuators B: Chemical,* 179, 32-45.https://doi.org/10.1016/j.snb.2012.11.014

Llorens, A., Lloret, E., Picouet, P. A., Trbojevich, R., & Fernandez, A. (2012). Metallic-based micro and nanocomposites in food contact materials and active food packaging. *Trends in Food Science & Technology,* 24(1), 19-29.https://doi.org/10.1016/j.tifs.2011.10.001

López-de-Dicastillo, C., Gómez-Estaca, J., Catalá, R., Gavara, R., & Hernández-Muñoz, P. (2012). Active antioxidant packaging films: development and effect on lipid stability of brined sardines. *Food Chemistry,* 131(4), 1376-1384.https://doi.org/10.1016/j.foodchem.2011.10.002

Lopez-Rubio, A., Almenar, E., Hernandez-Muñoz, P., Lagarón, J. M., Catalá, R., &Gavara, R. (2004). Overview of active polymer-based packaging technologies for food applications. *Food Reviews International,* 20(4), 357-387.https://doi.org/10.1081/FRI-200033462

Lu, L., Zheng, W., Lv, Z., & Tang, Y. (2013). Development and application of time–temperature indicators used on food during the cold chain logistics. *Packaging Technology and Science,* 26, 80-90.https://doi.org/10.1002/pts.2009

Lupo, D., Clemens, W., Breitung, S. and Hecker, K., 2013. OE-A roadmap for organic and printed electronics. In Applications of Organic and Printed Electronics 1-26. Springer, Boston, MA https://link.springer.com/book/10.1007/978-1-4614-3160-2

Mahieu, A., Terrié, C., & Youssef, B. (2015). Thermoplastic starch films and thermoplastic starch/polycaprolactone blends with oxygen-scavenging properties: Influence of water content. *Industrial Crops and Products,* 72, 192-199.https://doi.org/10.1016/j.indcrop.2014.11.037

Mangaraj, S., Goswami, T. K., & Mahajan, P. V. (2009). Applications of plastic films for modified atmosphere packaging of fruits and vegetables: a review. *Food Engineering Reviews*, 1(2), 133.http://dx.doi.org/10.1007%2Fs12393-009-9007-3

Marx, D., Joss, L., Hefti, M., Pini, R., & Mazzotti, M. (2013). The role of water in adsorption-based CO2 capture systems. *Energy Procedia*, 37, 107-114.https://doi.org/10.1002/app.33718

Matche, R. S., Sreekumar, R. K., & Raj, B. (2011). Modification of linear low-density polyethylene film using oxygen scavengers for its application in storage of bun and bread. *Journal of Applied Polymer Science*, 122(1), 55-63.https://doi.org/10.1002/app.33718

Matindoust, S., Baghaei-Nejad, M., Zou, Z., & Zheng, L. R. (2016). Food quality and safety monitoring using gas sensor array in intelligent packaging. *Sensor Review*, 36(2), 169-183. https://doi.org/10.1108/SR-07-2015-0115

Ming, X., Weber, G.H., Ayres, J.W., & Sandine, W.E. (1997). Bacteriocins applied to food packaging materials to inhibit Listeria monocytogenes on meats. *Journal of Food Science*, 62(2), 413-415.https://doi.org/10.1111/j.1365-2621.1997.tb04015.x

Müller, P., & Schmid, M. (2019). Intelligent Packaging in the Food Sector: A Brief Overview. *Foods*, 8(1), 16.https://doi.org/10.3390/foods8010016

Mustafa, F., & Andreescu, S. (2018). Chemical and biological sensors for food-quality monitoring and smart packaging. *Foods*, *7*(10), 168.https://doi.org/10.3390/foods7100168

Nilsson, H. E., Unander, T., Sidén, J., Andersson, H., Manuilskiy, A., Hummelgard, M., &Gulliksson, M. (2012). System integration of electronic functions in smart packaging applications. *IEEE Transactions on Components, Packaging and Manufacturing Technology*, 2(10), 1723-1734 10.1109/TCPMT.2012.2204056.

Nopwinyuwong, A., Trevanich, S., & Suppakul, P. (2010). Development of a novel colorimetric indicator label for monitoring freshness of intermediate-moisture dessert spoilage. *Talanta*, 81(3), 1126-1132.https://doi.org/10.1016/j.talanta.2010.02.008

Nugraha, B., Bintoro, N., & Murayama, H. (2015). Influence of CO_2 and C_2H_4 adsorbents to the Symptoms of Internal Browning on the Packaged 'Silver Bell'Pear (Pyrus communis L.). *Agriculture and Agricultural Science Procedia*, 3, 127-131.https://doi.org/10.1016/j.aaspro.2015.01.025

Nuin, M., Alfaro, B., Cruz, Z., Argarate, N., George, S., Le Marc, Y., & Pin, C. (2008). Modelling spoilage of fresh turbot and evaluation of a time–temperature integrator (TTI) label under fluctuating temperature. *International Journal of Food Microbiology*, 127(3), 193-199. doi: 10.1016/j.ijfoodmicro.2008.04.010.

Nyberg, C., &Tengstål, C. G. (1984). Adsorption and reaction of water, oxygen, and hydrogen on Pd (100): Identification of adsorbed hydroxyl and implications for the catalytic H_2–O_2 reaction. *The Journal of chemical physics*, 80(7), 3463-3468.https://doi.org/10.1063/1.447102

Ohta, M. (2008). Development of the indicator to warn deviant temperature raise in cold chain. *Nippon Shokuhin Kagaku Gakkaishi (Jpn. J. Food Chem.)*, 15, 18-22.

Otles, S., & Yalcin, B. (2008). Intelligent food packaging. *LogForum* 4, 4, 3.http://www.logforum.net/

Ozdemir, M., &Floros, J. D. (2004). Active food packaging technologies. *Critical reviews in food science and nutrition*, 44(3), 185-193.https://doi.org/10.1080/10408690490441578

Padgett, T., Han, Y., & Dawson, P. L. (2000). Effect of lauric acid addition on the antimicrobial efficacy and water permeability of corn zein films containing nisin. *Journal of Food Processing and Preservation*, 24(5), 423-432. https://doi.org/10.1111/j.1745-4549.2000.tb00429.x

Park, H. R., Kim, K., & Lee, S. J. (2013). Adjustment of Arrhenius activation energy of laccase-based time–temperature integrator (TTI) using sodium azide. *Food control*, 32(2), 615-620.

Park, H. R., Kim, K., & Lee, S. J. (2013). Adjustment of Arrhenius activation energy of laccase-based time–temperature integrator (TTI) using sodium azide. *Food control*, *32*(2), 615-620.https://doi.org/10.1016/j.foodcont.2013.01.046

Park, J. D., Lim, S., & Kim, H. (2015). Patterned silver nanowires using the gravure printing process for flexible applications. *Thin Solid Films*, 586, 70-75.https://doi.org/10.1016/j.tsf.2015.04.055

Pavelková, A. (2013). Time temperature indicators as devices intelligent packaging. *Acta Universitatis Agriculturaeet Silviculturae Mendelianae Brunensis*, 61(1), 245-251.

Pavelková, A. (2013). Time temperature indicators as devices intelligent packaging. *Acta Universitatis Agriculturae et Silviculturae Mendelianae Brunensis*, 61(1), 245-251. https://doi.org/10.11118/actaun201361010245

Pereira de Abreu, D. A., Cruz, J. M., &Paseiro Losada, P. (2012). Active and intelligent packaging for the food industry. *Food Reviews International*, *28*(2), 146-187.https://doi.org/10.1080/87559129.2011.595022

Peter, A., Nicula, C., Mihaly-Cozmuta, A., Mihaly-Cozmuta, L., & Indrea, E. (2012). Chemical and sensory changes of different dairy products during storage in packages containing nanocrystallised TiO2. *International journal of food science & technology*, 47(7), 1448-1456.https://doi.org/10.1111/j.1365-2621.2012.02992.x

Phollookin, C., Wacharasindhu, S., Ajavakom, A., Tumcharern, G., Ampornpun, S., Eaidkong, T., &Sukwattanasinitt, M. (2010). Tuning down of color transition temperature of thermochromically reversible bisdiynamidepolydiacetylenes. *Macromolecules*, *43*(18), 7540-7548.

Phollookin, C., Wacharasindhu, S., Ajavakom, A., Tumcharern, G., Ampornpun, S., Eaidkong, T., & Sukwattanasinitt, M. (2010). Tuning down of color transitiontemperature of thermochromically reversible bisdiynamidepolydiacetylenes. *Macromolecules*, 43(18), 7540-7548.https://doi.org/10.1002/pts.963

Phoopuritham P., Thongngam, M., Yoksan, R., &Suppakul, P. (2012). Antioxidant Properties of Selected Plant Extracts and Application in Packaging as Antioxidant Cellulose-Based Films for Vegetable Oil. *Packaging Technology and Science*, 25(3), 125-136.https://doi.org/10.1002/pts.963

Poyatos-Racionero, E., Ros-Lis, J. V., Vivancos, J. L., & Martínez-Máñez, R. (2018). Recent advances on intelligent packaging as tools to reduce food waste. *Journal of cleaner production*, 172, 3398-3409. https://doi.org/10.1016/j.jclepro.2017.11.075

Prinzmetal, W., Treiman, R., & Rho, S. H. (1986). How to see a reading unit. *Journal of memory and language*, 25(4), 461-475.https://doi.org/10.1016/0749-596X(86)90038-0

Puligundla, P., Jung, J., & Ko, S. (2012). Carbon dioxide sensors for intelligent food packaging applications. *Food Control*, *25*(1), 328-333.https://doi.org/10.1016/j.foodcont.2011.10.043

Quintavalla, S., & Vicini, L. (2002). Antimicrobial food packaging in meat industry. *Meat science*, 62(3), 373-380.https://doi.org/10.1016/S0309-1740(02)00121-3

Raab, V., Bruckner, S., Beierle, E., Kampmann, Y., Petersen, B., & Kreyenschmidt, J. (2008). Generic model for the prediction of remaining shelf life in support of cold chain management in pork and poultry supply chains. *Journal on Chain and Network Science*, 8(1), 59-73.https://doi.org/10.3920/JCNS2008.x089

Realini, C. E., & Marcos, B. (2014). Active and intelligent packaging systems for a modern society. *Meat science*, 98(3), 404-419.

Realini, C. E., & Marcos, B. (2014). Active and intelligent packaging systems for a modern society. *Meat science*, *98*(3), 404-419. doi: 10.1016/j.meatsci.2014.06.031

Restuccia, D., Spizzirri, U. G., Parisi, O. I., Cirillo, G., Curcio, M., Iemma, F., ... &Picci, N. (2010). New EU regulation aspects and global market of active and intelligent packaging for food industry applications. *Food control*, 21(11), 1425-1435.https://doi.org/10.1016/j.foodcont.2010.04.028

Robertson, G. L. (2005). *Food Packaging: Principles and Practice.* CRC press.
Robertson, G. L. (2006). Active and intelligent packaging. *Food packaging: Principles and practice, 2.*
Rodriguez-Aguilera, R., & Oliveira, J. C. (2009). Review of design engineering methods and applications of active and modified atmosphere packaging systems. *Food Engineering Reviews, 1*(1), 66-83.
Rokugawa, H., & Fujikawa, H. (2015). Evaluation of a new Maillard reaction type time-temperature integrator at various temperatures. *Food control, 57,* 355-361.https://doi.org/10.1016/j.foodcont.2015.05.010
Rooney, M. L. (1995). Overview of active food packaging. In *Active food packaging* (pp. 1-37). Springer, Boston, MA.
Rooney, M. L. (2005). Oxygen-scavenging packaging. In *Innovations in food packaging* (pp. 123-137). Academic Press.
Sajilata, M. G., Savitha, K., Singhal, R. S., &Kanetkar, V. R. (2007). Scalping of flavors in packaged foods. *Comprehensive reviews in food science and food safety,* 6(1), 17-35.https://doi.org/10.1111/j.1541-4337.2007.00014.x
Sakanaka, S., & Tachibana, Y. (2006). Active oxygen scavenging activity of egg-yolk protein hydrolysates and their effects on lipid oxidation in beef and tuna homogenates. *Food Chemistry,* 95(2), 243-249.https://doi.org/10.1016/j.foodchem.2004.11.056
Salman, H., Tenetov, E., Feiler, L., & Raimann, T. (2011). *U.S. Patent Application No. 12/523,110.*
Scarfato, P., Di Maio, L., &Incarnato, L. (2015). Recent advances and migration issues in biodegradable polymers from renewable sources for food packaging. *Journal of Applied Polymer Science,* 132(48).https://doi.org/10.1002/app.42597
Schaefer, D., & Cheung, W. M. (2018). Smart Packaging: Opportunities and Challenges. *Procedia CIRP, 72,* 1022-1027.
Schroeder, G. O., Schroeder, G. M., & Tsai, B. C. (2001). Oxygen Scavenger Technologies, Applications, And Markets: A Strategic Analysis for the Packaging Industry. *GeorgeO. Schroeder Associates: Appleton, WI.*
Schroeder, J. I., Allen, G. J., Hugouvieux, V., Kwak, J. M., & Waner, D. (2001). Guard cell signal transduction. *Annual review of plant biology, 52*(1), 627-658.https://doi.org/10.1146/annurev.arplant.52.1.627
Shin, D. H., Cheigh, H. S., & Lee, D. S. (2002). The use of Na2CO3-based CO2 absorbent systems to alleviate pressure buildup and volume expansion of kimchi packages. *Journal of Food Engineering,* 53(3), 229-235.
Shin, Y., Shin, J., & Lee, Y. S. (2011). Preparation and characterization of multilayer film incorporating oxygen scavenger. *Macromolecular research,* 19(9), 869.https://doi.org/10.1007/s13233-011-0912-y
Siripatrawan, U., & Harte, B. R. (2010). Physical properties and antioxidant activity of an active film from chitosan incorporated with green tea extract. *Food Hydrocolloids,* 24(8), 770-775.Siripatrawan, U., & Harte, B. R. (2010). Physical properties and antioxidant activity of an active film from chitosan incorporated with green tea extract. Food Hydrocolloids, 24(8), 770–775. doi:10.1016/j.foodhyd.2010.04.003
Sofi, S. A., Singh, J., Rafiq, S., Ashraf, U., Dar, B. N., &Nayik, G. A. (2018). A Comprehensive Review on Antimicrobial Packaging and its Use in Food Packaging. *Current Nutrition & Food Science, 14*(4), 305-312.https://doi.org/10.2174/1573401313666170609095732
Song, R., Zhu, GM., Xie, JQ. et. al., (2013). Synthesis of 2,4-hexadiyn-1,6 bis(ethylurea) and its polymerization. *Polymer Materials Science and Engineering,* 29, 1–4.
Spevacek, J. A. (2003). *U.S.* Patent No. 6,614,728. Washington, DC: U.S. Patent and Trademark Office.
Sung, S. Y., Sin, L. T., Tee, T. T., Bee, S. T., Rahmat, A. R., Rahman, W. A. W. A., &Vikhraman, M. (2013). Antimicrobial agents for food packaging applications. *Trends in Food Science & Technology,* 33(2), 110-123.

Taoukis, P. S., & Labuza, T. P. (1989). Applicability of time-temperature indicators as shelf life monitors of food products. *Journal of Food Science*, 54(4), 783-788.https://doi.org/10.1111/j.1365-2621.1989.tb07882.x

Taoukis, P. S., & Labuza, T. P. (2003). Time-temperature indicators (TTIs). *Novel Food Packaging Techniques*, 103-126.

Vaikousi, H., Biliaderis, C. G., & Koutsoumanis, K. P. (2009). Applicability of a microbial Time Temperature Indicator (TTI) for monitoring spoilage of modified atmosphere packed minced meat. *International Journal of Food Microbiology*, 133(3), 272-278.https://doi.org/10.1016/j.ijfoodmicro.2009.05.030

Vanderroost, M., Ragaert, P., Devlieghere, F., & De Meulenaer, B. (2014). Intelligent food packaging: The next generation. *Trends in Food Science & Technology*, 39(1), 47-62.https://doi.org/10.1016/j.tifs.2014.06.009

Varlet-Grancher, X. (2006). Time temperature indicator (TTI) system. *European Patent, 1725846*.https://patents.google.com/patent/EP1725846A1/en

Vermeiren,L.,Devlieghere,F.,vanBeest,M.,deKruijf,N.,&Debevere,J.(1999). Developments in the active packaging of foods. *Trends in food science & technology*, 10(3),77-86. https://doi.org/10.1016/S0924-2244(99)00032-1

Versteylen, S., Stoll, B. N., & Michaels, M. H. (2015). *U.S. Patent Application No. 14/497,973*. https://patents.google.com/patent/US20150093478A1/en

Wang, G. Y., Wang, H. H., Han, Y. W., Xing, T., Ye, K. P., Xu, X. L., & Zhou, G. H. (2017). Evaluation of the spoilage potential of bacteria isolated from chilled chicken in vitro and in situ. *Food Microbiology*, 63, 139-146.https://doi.org/10.1016/j.fm.2016.11.015

Wang, S., Liu, X., Yang, M., Zhang, Y., Xiang, K., & Tang, R. (2015). Review of time temperature indicators as quality monitors in food packaging. *Packaging Technology and Science*, 28(10), 839-867.https://doi.org/10.1002/pts.2148

Wanihsuksombat, C., Hongtrakul, V., & Suppakul, P. (2010). Development and characterization of a prototype of a lactic acid–based time–temperature indicator for monitoring food product quality. Journal of Food Engineering, 100(3), 427-434https://doi.org/10.1016/j.jfoodeng.2010.04.027

Watkins, C. B. (2000). Responses of horticultural commodities to high carbon dioxide as related to modified atmosphere packaging. *HortTechnology*, 10(3), 501-506.https://doi.org/10.21273/HORTTECH.10.3.501

Wessling, C., Nielsen, T., Leufvén, A., &Jägerstad, M. (1998). Mobility of α-tocopherol and BHT in LDPE in contact with fatty food simulants. *Food Additives & Contaminants*, 15(6), 709-715.https://doi.org/10.1080/02652039809374701

Winestrand, S., Johansson, K., Järnström, L., &Jönsson, L. J. (2013). Co-immobilization of oxalate oxidase and catalase in films for scavenging of oxygen or oxalic acid. *Biochemical engineering journal*, 72, 96-101.https://doi.org/10.1016/j.bej.2013.01.006

Wrona, M., Bentayeb, K., & Nerín, C. (2015). A novel active packaging for extending the shelf-life of fresh mushrooms (Agaricus bisporus). *Food Control, 54*, 200-207.https://doi.org/10.1016/j.foodcont.2015.02.008

Wu, D., Hou, S., Chen, J., Sun, Y., Ye, X., Liu, D., ... & Wang, Y. (2015). Development and characterization of an enzymatic time-temperature indicator (TTI) based on Aspergillus niger lipase. LWT-Food Science and Technology, 60(2), 1100-1104.https://doi.org/10.1016/j.lwt.2014.10.011

Xu, D., Xiao, P., Zhang, J., Li, G., Xiao, G., Webley, P. A., & Zhai, Y. (2013). Effects of water vapour on CO2 capture with vacuum swing adsorption using activated carbon. *Chemical engineering journal, 230*, 64-72.https://doi.org/10.1016/j.cej.2013.06.080

Yam, K. L., Takhistov, P. T., & Miltz, J. (2005). Intelligent packaging: concepts and applications.*Journalof foodscience*,70(1),R1-R10.https://doi.org/10.1111/j.1365-2621.2005.tb09052.x

Yam, V. W. W., & Wong, K. M. C. (2005). Luminescent Molecular Rods—Transition-Metal Alkynyl Complexes. In *Molecular wires and electronics* (pp. 1-32). Springer, Berlin, Heidelberg.

Yam, V. W. W., & Wong, K. M. C. (2005). Luminescent Molecular Rods—Transition-Metal Alkynyl Complexes. In *Molecular wires and electronics* (pp. 1-32). Springer, Berlin, Heidelberg. 10.1007/b136069.

Yildirim, S., Röcker, B., Pettersen, M. K., Nilsen-Nygaard, J., Ayhan, Z., Rutkaite, R., & Coma, V. (2018). Active packaging applications for food. *Comprehensive Reviews in Food Science and Food Safety*, 17(1), 165-199.

Yildirim, S., Röcker, B., Pettersen, M. K., Nilsen-Nygaard, J., Ayhan, Z., Rutkaite, R., & Coma, V. (2018). Active packaging applications for food. *Comprehensive Reviews in Food Science and Food Safety*, 17(1), 165-199.https://doi.org/10.1111/1541-4337.12322

Yu, J., Liu, R. Y. F., Poon, B., Nazarenko, S., Koloski, T., Vargo, T., & Baer, E. (2004). Polymers with palladium nanoparticles as active membrane materials. *Journal of applied polymer science*, *92*(2), 749-756.https://doi.org/10.1002/app.20013

Zeng, J., Roberts, S., & Xia, Y. (2010). Nanocrystal-based time–temperature indicators. *Chemistry–A European Journal*, *16*(42), 12559-12563.https://doi.org/10.1002/chem.201002665

Zhang, C., Yin, A. X., Jiang, R., Rong, J., Dong, L., Zhao, T., ... & Yan, C. H. (2013). Time–Temperature indicator for perishable products based on kinetically programmable Ag overgrowth on Au nanorods. *ACS Nano*, 7(5), 4561-4568.https://doi.org/10.1021/nn401266u

Zheng, G. L., Qian, J., & Feng, Q. (2013). Preparation of Time-Temperature Indicator Reaction System Based on Glucoamylase. *Food Science*, 12.

Zweig, S. E. (2005). *U.S. Patent No. 6,950,028*. Washington, DC: U.S. Patent and Trademark Office.

12

Energy Bars: A Perfect Choice of Nutrition to All

Padmashree Ananthan, Gopal Kumar Sharma **and** ***Anil Dutt Semwal***

Grain Science and Technology Division, Defence Food Research Laboratory Siddarthanagar, Mysuru 570 011,Karnataka, India

Introduction

Over the years, processed food industry has become the largest growing industry in the world. The convenience food industry with the state of the art technology has been proved very successful in reaching every household and have started replacing fresh and healthy foods with the processed ones (Srinivasan and Shende, 2016).

Demand for processed and convenience foods are constantly on rise and convenience food segment is rising by leaps and bounds. Changing life style, food habits, displacement of rural population to urban centres, insufficient time to prepare healthy food and changing socio economic needs have led to the industrialization of the food supply and resulted in the development of convenience foods and today a great variety of the same are available for sale to the consumer throughout the world (Tillotson, 2003).

These convenience foods are available readily, convenient, inexpensive with pleasing organoleptic quality, requires minimal or no preparation before consumption. Varieties of convenience foods which are available in the market play a prominent role in the food choice of today's consumer, and they include frozen and canned vegetables, bakery products, soups, sauces, prepared meat and fish, chilled and frozen dairy products as well as ready to eat and shelf stable products (Tillotson, 2003).

Rapid switchover of the consumer from the traditional dining practices to fast food centres, generally feed on quick and easily made products called fast foods, often termed as unhealthy junk foods. These foods are rich in saturated fatty acids, several refined components, high carbohydrates etc., which ultimately lead to the negative impact on physical and mental health of the consumer, contributing to several chronic diseases such as cardiovascular complications, cancer, atherosclerosis, liver disease, type 11 diabetes (Veggie et al., 2018) etc.

Therefore, focussing much about the health of the consumer, there is a great need for a healthy and a balanced diet to meet the basic needs of the human body (Mendes et al 2013) with an appealing organoleptic properties and appearance.

Though most of the convenience foods are ready to eat in nature, often are nutritionally imbalanced, containing higher amounts of carbohydrates and fats, lower in protein, dietary fibres etc. Therefore food researchers and nutritionists have come out with the solution of developing ready to eat nutritionally balanced foods having sufficient oragnopleptic properties to replace unhealthy foods. One such healthy, nutritionally balanced food is the nutritious energy bars having gained more importance and high popularity during recent years both in terms of convenience and nutritional balance occupying larger space in the consumer market (Padmashree et al., 2012).

The first energy bars were manufactured by Pillsbury Company for astronauts in late 1960's and called as Space Food Sticks'. These space food sticks were described as 'non frozen balanced energy rods' containing nutritionally balanced amounts of protein , fat and carbohydrates, having different flavours like mint, orange and pea nut butter which could easily fix in to an airtight space of astronaut helmets. In mid 1970's, Pillsbury Company extended its market to the public and dropped 'Space' from its name. The first commercially released bar was 'power bar' for athletes in 1986 for survival of long distance events without running the lack of glycogen (https://www.erinnudi.com). Later, cliff bar, balance bar, bars from muscle milk etc entered in to the market claiming as healthy and energizing (https://www.manrepeller.com).

Though, these bars were originally designed and marketed to those who involved in strong physical activity such as athletes, as a quick source of energy (Chandagare et al 2018) and nutrition, today, with the increasing attention on nutrition and healthy food habits and increasing strength of the people involved in strong physical activities, energy bars have become a great choice of quality source of energy. These bars have proved quite valuable in terms of convenience and health requirement, requires little or no preparation, no refrigeration required and offers long shelf life (Pooja Ravindra and Dhanashree sunil 2018). They are simple, hygienic, convenient well proportioned (Chandegara et al., 2018) and portable way to deliver energy in the form of calories. They can be stored in small packets or pouches, making it more convenient to carry or hold, can fit anywhere and can be eaten at any point of time.

Definition of energy bars

Energy bars, nutritious bars, snack bars are the different names referred to food bars in general, are convenient ready to eat food items in the form of tablets, usually made by compression technology (Padmashree et al., 2012), baking technology (Ramirez Jiminez et al 2018, Sunwater house et al 2009), or by the use of binders of choice (Hogan et al., 2012, Aramouni et al., 2011, Da silva et al., 2016).These bars contains nutrients which includes vitamins, minerals, sufficient amount of carbohydrates, proteins and fat to keep the body functioning. Depending on the specific needs, energy bars are focussed on proteins, carbohydrates, fats and many times serves as a complete meal with sufficient calorific load. Bars come in different flavours, usually taste sweet, more appetizing, and often look similar to baked goods. Generally energy bar weighs between 45-80 g, providing 200-300 kcal of energy, 7-15 g protein, 3-9 g fat, and 20-40 g carbohydrate (Maheshwari alla and Jitendran 2018).

Types of energy bars

Numerous types of energy bars have been developed and are referred by different names such as protein bars, cereal bars, snack bars, diet bars, granola bars, neutraceutical bars, health bars, whole food bars, fibre bars and so on (Painter and Prisecaru (2002). Each bar has different characteristic and purpose, specially designed for teenagers, athletes, adults, men, woman, children, pregnant and lactating mothers, bars with low glycemic index for sustained release of energy and the one with omega 3 fatty acids etc, containing different amounts of protein, fat, carbohydrates, minerals and vitamins from different sources, meant for providing energy conveniently, slowly or quickly depending on the physical activity and the calorific needs of the individual.

Protein rich bars

The protein deficiency is one of the main causes of malnutrition in under developed countries and the problem needs to be addressed very seriously, especially in younger sections of population. As per the data of UNICEF 2016, 22.9% children worldwide under the age of 5 had stunted growth and 50% deaths under the age of 5 are mainly attributable to malnutrition leading to the loss of 3 million young lives every year (Aravind kumar et al., 2018).

As per the data of FAO 2010, vegetable protein sources have occupied the global market (57%), followed by non vegetarian protein sources such as dairy (10%), meat (18%), fish and shell fish (6%) and other animal products (9%).

Pulses are the important sources of dietary proteins and other nutrients and have a significant role in overcoming the challenges related to protein energy malnutrition in developed and underdeveloped countries (Henchion et al. 2017). Concentrates and isolates of proteins are generally used to enhance protein content in foods. Varieties of legumes used in the development of protein isolates include soybean, peanut, canola, cashewnuts, sesame, pinto and navy beans (Seyam et al., 1983). Isolates of protein originated in US around 1950's (Jay and Micheal 2004) are playing a vital role in the formulation of various foods which includes beverages, infant foods, children milk foods, speciality foods and other protein products (Olaofe 1998).

For the past few years, demand for snack food products with improved nutrition has increased due to their convenience, functionality and desirability especially among young population (Potter et al., 2013). These products include bars, beverages, etc containing high protein body building products offering about 20 g of protein per serving. High protein bars containing 20-25 g protein per 100 g, serves as one of the ways to incorporate substantial amount of protein in the diet of children (Aravind kuamr et al. 2018). Also, lactose intolerant individual can safely consume these products (Akintayo et al. 1999) as the significant amount of lactose gets removed during the process. The bars which are claimed to be high protein ones contains protein in the range of 15-35% (Loveday et al., 2009).

Many research works have been carried out and great numbers of patents are available on the development and stability of protein bars since decades using different protein sources.

A high protein fat occluded food composition useful as a binder containing at least 15 % by weight of protein, wherein protein source was mainly from dairy and vegetable was described by Kelly et al. (1977). The food bar prepared by using the above binder was found completely free from off flavours arising from protein. A nutritional high protein athletic bar having calorific distribution of at least 10% from protein having a shelf life of at least 6-12 months was described by Michnowski et al., (1989). Sharma et al. (2006) describes a ready-to-eat high protein compressed bar consisting of defatted soy flour, ground nuts, cashew nuts, sesame seeds, bengal gram, as main sources of proteins along with other ingredients like sugar, binder, hydrogenated fat which was stable for more than 12 months. Sarma and Roger (2007) describes a protein containing food product in the form of a bar comprising of about 10-65% weight of protein particulate material, about 35-40% of lipid containing material and a small amount of emulsifier that can be used as a coating composition which helps to maintain the moisture level and prevents it from becoming hard as well as add to the nutritional value of a bar. Gowtham Akhilesh (2008) in his invention of nutritional protein bar, describes high protein nutrition bar in the form of nuggets having high levels of selected protein containing more than 50% by weight of non soy protein selected from milk, rice and pea proteins with good organoleptic properties. Dates along with dry skim milk powder and single celled protein used in the development of shelf stable date bar has enhanced protein content significantly with increased protein bioavailability and several B vitamins (Kamel and Kramer 1977). Date bar with the fortification of yeast protein and dry skim milk in different proportions has significantly improved amino acid profile with the two fold increase in the chemical score and increased significantly the calculated protein efficiency ratio from 1.36 to 2.35-2.43 as a result of protein fortification. A low cost nutritious date bar with the fortification of vetch protein isolate (VPC) and whey protein concentrate (WPC) was developed especially for school children to meet their body requirements. The addition of VPC and WPC has significantly improved nutritional profile of bars in terms of protein content, taste and texture (Nadeem et al. 2012). A novel nutritious high protein bar was developed specifically for malnourished children using spirulina, a multicellular filamentous cyanobacteria along with bengal gram dhal, ground nuts, corn flakes and other ingredients and the protein content was enhanced by 167% more than the control sample (Arvind kumar et al., 2018).

Protein rich bars usually become harder shortly after their manufacture and this makes the product unpalatable. It has been hypothesised that hardening may be due to the thiol-disulphide interchange reactions during storage which lead to protein cross linking, aggregation and network formation (Zhou et al., 2008). It has also been reported that, bar hardening during storage is favoured by more ordered protein secondary structure and lower surface hydrophobicity of protein particles (Baier et al., 2007). Moisture migration and maillard reaction between reducing sugars and reactive lysine is also thought to be responsible for the hardening of protein bars (Gerard 2002). Incorporation of hydrophilic protein in the bar formulations can activate these reactions and make bars to stay chewable, soft and flexible. Childs et al. (2007) reported that whey protein hydrolysates can effectively inhibit bar hardening. Loveday et al. (2009) prepared high protein model protein bar using a milk protein concentrate as a protein source, sugar and other low molecular weight polyhydroxy compounds and studied the physico chemical changes occurring in the bar during the storage period of 50 days at 20°C and found that the chemical changes play a minimum or no role in hardening of bars, but changes in molecular mobility and changes in microstructure driven by the moisture migration may be responsible for bar hardening. Wagner et al. (2007) describes a high protein bar consisting of protinaceous material comprising of a combination of isolated soy protein and a milk protein consisting of 10% to 90% each by weight of isolated soy protein and milk protein with an improved texture having a mechanical hardness of less than 2500 gms force and an extended shelf - life as compared to conventional bars.

Cereal bars

Cereal bars are shaped in the form of a bar, generally made from processed cereals. Dried fruits, berries and nuts are also used in the preparation of a bar which are held together by glucose syrup by adhering the ingredients (Aleksejiva et al. 2017). The use of a glucose syrup as a binding agent helps in providing rapid energy absorption (Da Silva et al., 2013). Wheat, oat, rice and soy are the commonly used ingredients in a cereal bar and is considered as a good choice for a quick meal due to its high nutritional value (Degaspari et al., 2008).Cereal bar can act as a good substitute for a less healthy snack, providing good source of energy before strong physical activity or can be used as a meal substitute. Also, when used as a food supplement or as portable nutrition, cereal bars can be a good source of carbohydrate and promotes quick recovery of energy following workouts (Brito, 2005). In 1990's Americans consumed cereal bar during mountaineering, climbing and marathon running as a quick source of energy and later it has attracted larger segments of population, especially common consumers (Palazzolo, 2003). The global consumption of cereal bar increased to 11% in 2007 and cereal bar market was estimated at around 4 billion USD and with the availability of wide varieties of it with the diversity of ingredients (Sharma et al 2014). The change in life style and people concern in getting quality and nutritive food has undoubtedly changed the eating habits and has witnessed a growth of cereal bar market by 20% per year (Lin et al., 2010).

As cereal bars are prepared using wide range of ingredients, from the nutritional point of view, these bars are classified in to 4 types such as, fibrous bars, energy bars, diet (light) bars and protein bars. Fibrous bar contains high fibre and glucose providing energy value near to 100 kcal/ unit. Energy bar contains less fibre and high calorific value of 280 kcal, are easily absorbed in to the body, generally, recommended for energy replacement after strenuous physical activity. Diet bars are sugar free, usually have 65 calories, while protein bar contains approximately 17g protein / unit with less fat, highly beneficial for hardworking people (Degaspari et al., 2008).

The nutritional quality of cereal bar largely depends on the ingredients used for its preparation. The biggest challenge is getting a cereal bar with the combination of several ingredients having specific functionality containing protein, fat, fibre, minerals, vitamins, thickening agents, sweeteners, flavourings and switching them in to a product with a good texture, flavour, taste by achieving goals of specific nutrient (Lima, 2004).

Epidemiological studies have reported that, the diet rich in whole grain cereals helps in reducing the risk factor such as diabetes, cardio vascular diseases, obesity, metabolic syndrome and certain types of cancers (Esmaillzadeh and Azadbakht 2006, Qi and Hu, 2007, Williams et al., 2008, Wu et al., 2015) .Therefore,there is a growing demand from the consumers for a variety of nutritionally balanced cereal bars which should be healthy, natural, convenient (Sharma et al., 2014) with pleasing organoleptic properties. These bars may ideally substitute meal and the complex carbohydrate starch, present in cereal aids in slow liberation of energy which can be absorbed slowly for a longer period of time (Brito 2005).They are the convenient products, would be an ideal food format for delivering nutrition in adequate amounts to the consumer. Plenty of literature is available on the development and stability of varieties of cereal bars using various ingredients. Da Silva et al. (2013) developed a cereal bar formula using a mixture of Brazilian cassava flour, hydrogenated vegetable fat, dried banana, ground nuts, cashew nuts, glucose syrup etc and the bar was found safe for consumption up to 180 days of storage. The stability and antioxidant potential of cereal bars made with oats, rice flakes, fruit peels and baru nuts was found to be stable for 120 days in different packaging materials studied Mendes et al. (2013). A snack bar recipe was developed by Silva et al. (2015) using crushes of corn starch biscuit, oat, skim milk powder, rice flakes, corn syrup and jeriva flour which is grown in Brazil, showed a significant amount of protein, vitamin C , fibre and a greater antioxidant activity. Cereal bar developed by using mixture of rice, oat flakes, dried apple, salt, cinnamon, gum acacia, inulin and maltodextrin showed a_w between 0.552 to 0.596, with good sensory acceptance and a considerable amount of soluble fibre inulin (Srebernich et al., 2016). A low calorie cereal bar developed by Jo- Su- Ah et al. (2018) using job's tears, sorghum, black rice, oat , red quinoa seeds, chick pea, linseed, and lentil along with various concentrations of saccharin was found stable for 90 days of storage period. Author's claim that the bar can be useful for those who wish to consume sugar free products. Carvalho et al (2011) successfully developed nutritional cereal bar using almonds of chicha, sapucaia and gurgueia nut and the residue of pineapple. Burrie et al. (2003) describes a ready-to-eat nutritionally valuable cereal bar containing agglomerated particles or flakes of different cereals comprising of rice flour, wheat flour and a flaked bar mainly comprising of corn flour and corn starch along with amylaceous materials of different cereal flours mixing with the binder of sugar, milk solids and a binding agent to obtain a bar. Ramakrishna and Sunki Reddy (2006) describes a process for the preparation of cereal bar based on puffed and expanded rice products mixing with jaggery or glucose or sugar syrup, modified starch, toasted pea nuts and processing the mixture in the form of a bar to have a soft chewy texture with an extended shelf – life. Mesu (2017) describes crunchy textured cereal bar comprising of dry cereal mixture and a binder made of potato starch, whey protein and blendes thereof. The crunchy texture retained throughout the bar and is produced by drying the bar at a temperature between 110-150^0 C for 10-60 min.

Fibre bars

Fibres are the groups of substances composed of structural components of plant cell walls, primarily cellulose, hemicellulose, pectins and lignins that are resistant to digestive enzymes, an extremely beneficial element in providing various health benefits to a person by decreasing the incidence of several diseases (Dhingra et al. 2012). Fibres can either be water soluble or water insoluble. Soluble part of the fibre includes fruits, vegetables, gums and pectins, beans and peas, nuts, seeds, oat, barley and rice bran whereas insoluble fibre includes wheat and corn bran, cereals, vegetables, fruit skins, nuts and so on.

The changes in milling process of cereals have considerably decreased the level of fibre content apart from reducing the other nutrients such as vitamins, minerals and antioxidant compounds (Mellen et al. (2008). Changing lifestyles and eating habits have also influenced the decreased intake of fibres in the diet. Several epidemiological studies have revealed the relationship between the decreased fibre consumption as a dietary fibre and increase in certain diseases such as diverticulitis, colonic cancer, etc. (Painter and Burkitt, 1971). Therefore, the enrichment of fibre to the diet has been receiving much focus during recent research studies. The incorporation of dietary fibre may help to protect the individual against obesity, diabetes, constipation, diverticulitis, colon cancer, duodenal cancer, stroke, hypertension, cardio vascular diseases (Otles and Ozgoz 2014) etc. The essential role of dietary fibre in the diet is related to its bulk density, hydration capacity, binding property, fermentability (Dhingra et al. 2012). The fibre consumption in the diet can be increased by the consumption of whole grains, fruits and vegetables and by incorporating fibre enriched foods in the diet (Vetter, 1984). Various fibre enriched foods have been formulated for increasing the dietary intake of fibre. Fibre rich bars are usually prepared with higher fibre content with different sources of fibre such as maltodextrin, methyl cellulose, fruit juice concentrates, cereal brans, etc. Demand has been increased considerably for the variety of fibre rich foods among the health conscious people during recent years.

Azlyn et al. (1989) formulated low calorie fibre bar by incorporating wheat and corn bran at 5 different combinations and reported that optimum level of replacement of fibre with the flour were 20% by weight either as complete wheat bran or mixture containing half of wheat and corn bran to produce sensorily acceptable low calorie fibre bars. Coleman et al. (2007) describes the preparation of low calorie nutrition bar with high soluble fibre content having fructo-oligosaccharides as a fibre source along with protein crisp, caramel, reduced sugar content with an excellent flavour and multi textured properties providing less than 110 kcal/28 g serving. Linscott (1989) describes the preparation of granola bar with the incorporation of dietary fibre in the form of compressed flakes by mixing granola ingredients and suitable binding materials along with rice flour. Dekker and Lamar (2006) describes the preparation of binder composition for bar by using partially digestible fibre such as dextrin as one of the fibre component. Wurtman et al. (2003) discloses a bar with rapidly digestible carbohydrates such as maltodextin, dextran, starch along with psyllium and bran from oat, corn, rice, barley, buck wheat and numerous other sources of carbohydrates. Sudha et al (2004) describes a high fibre biscuit composition for increasing its dietary fibre content comprising of 30 to 55% wheat flour, 5 to 40% fibre from different cereal sources and 0.1 to 3.5% emulsifier and other food grade additives. Sanchez et al (2007) investigated the effect of wheat fibre on the restructured fish product with the incorporation of 3 to 6% wheat fibre. The addition of fibre resulted in increased water binding and water absorption capacity. Products with 3% fibre were whiter in colour with less rigidity and cohesiveness and found more sensorily acceptable.

Meal replacement bars

The term meal replacement refers to pre-packaged, calorie controlled products presented in the form of a bar or beverage which replaces regular meal. In US, the term 'meal replacement' is not defined in Food and Drug Administration (FDA) regulations. They generally provide calories in the range of 200-250, often fortified with minerals and vitamins in good or excellent amounts, with low sugar and fat (Leslie Kransny 2004). The macronutrients like carbohydrates, protein and fat are present in various combinations in meal replacement products (Miller et al., 2007). Meal replacement products were initially consumed by elderly and malnourished individuals. But however, during recent years, the marketing of these products have targeted healthy adults by representing as a healthy and a convenient source of balanced nutrition and intended for specific purposes such as weight loss, weight gain and maintenance (Leslie Krasny, 2004).

Meal replacement shakes and bars are simple and effective in managing weight especially in overweight and obese individuals with diabetes and have shown advantages over self selective weight loss diets. Meal replacement bars helps the individual in managing the weight since they provide precalculated amount of food with a known calorific level (Tsai and Wadden, 2006).

As per the American Diabeties Association's (ADA), nutrition recommendation, consumption of meal replacement products once or twice a day to replace usual meal may result in a significant weight loss (ADA 2008).

Several studies also have reported the weight loss in obese individuals following the consumption of meal replacement shakes and bars. The study conducted on 100 obese individuals by prescribing a meal and snacks prepared from self selected, conventional food for one group and prescribing the similar self selected diet for the other group except that two of the three main meals which includes breakfast, lunch and dinner were replaced with one of the three meal replacement options (shakes, hot chocolate or soup) and snacks with nutrition bars for 3 months, indicated that participants who had meal replacement options and bars lost 7.8% of their total initial body weight, while the group eating conventional food lost 1.5% of total initial body weight (Ditschuneit et al (1999). Clinical studies have shown that, the consumption of meal replacement bar curb the appetite, will aid in a sustained and a safe weight loss (Rothacker and Wattenberg, 2004).

Owoc (2008) describes a high protein meal replacement bar with delicious taste and texture, with a composition consisting of various ingredients in suitable proportions delivering requisite amount of proteins needed for energy, endurance and muscle development. Kemeny (2003) describes a meal equivalent sweet and savory nutrition bar which can be eaten singly or as segments of meal equivalent food bar with balanced amount of carbohydrates, protein and fat with an improved flavour by minimizing sugar and sugar substitutes and unhealthy forms of fats. Kemeny (2004) describes an improved meal substitute bar or meal equivalent bar equivalent to conventional meal, containing functional appetite stimulants, main course of major nutrition ingredients with functional appetite depressants, having predetermined calories and providing a meal with improved physiological and psychological value to a consumer. Ryan (2009) described a nutritionally balanced beverage or bar composition for human consumption, which can be used as a partial or total meal replacement containing relative ratios of proteins, fat and saturated fat, sugar, carbohydrate, fibre and one or more nutritionally balanced vitamins or minerals.

Health bars

Health foods are those foods which are specially claimed to be beneficial to human health. No single food provides a body with the entire vital nutrient it requires; therefore it becomes necessary to consume a variety of health foods to derive the nutritional requirement of the body.

Health bars are one of the health foods that have specific favourable effects on health. These bars are loaded with great amount of nutrients in terms of carbohydrates, fats and proteins with added vitamins and minerals, nutraceuticals and other dietary supplements to suit specific needs of the body.

Painter and Prisecaru (2002) studied the effect of various bars on the blood glucose levels of human and indicated that commercial formulations containing glucose or high fructose corn syrup had higher glycemic index, intended for the quick energy release and some ingredients such as barley, oats, fructose, etc., had the lower glycemix index intended for sustained release of energy. Portman (2000) describes a nutritional composition in dry powder form consisting of carbohydrate to protein ratio of 2.8 to 4.2 to 1 part respectively, wherein carbohydrate was used as an energy source during exercise and proteins for stimulating the release of insulin during exercise and repair of muscle cells after exercise. Dimichele (2002) describes a solid nutritional composition in the form of a bar containing immuno nutritional products such as combination of antioxidant and structured glyceride component useful in reducing the immunological suppression which results from severe stress. King et al. (2003) describes a high energy multi-saccharide food bar containing galactose as an ergogenic aid particularly used by sports people during exercises and also to alleviate hunger and fatigue in general population. Manning et al. (2003) describes food bars for consumption especially by pregnant and lactating mothers with fortification of one or more vitamins and minerals up to 2.5 to 50% by weight of food bar, DHA 0.05 to 3% by weight, calcium in an amount of 1000-3000 mg consisting of anti-constipation and regularity agents. Petty et al. (2008) discloses a novel low allergenic meal replacement snack or the food bar to alleviate hunger and fatigue in individuals who are diagnosed of having food allergy, degenerative or autoimmune disease and also to use as an ergogenic aid for sport persons consisting of a protein source selected from spelt, quinoa, amaranth, buckwheat, rice, garbanzo flour and carbohydrates selected from pea, buckwheat, rice, tapioca, sorghum, fava and garbanzo flour and dietary supplement selected from a group consisting of creatine, choline, leucine and μ-lipoic acid. Hayashi et al. (1989) provides a low moisture long shelf - stable snack bar consisting of all the minerals and vitamins that meets 25% of recommended daily allowances as prescribed by US Dept. of Agriculture. Chimel et al (2008) describes a process for the preparation of ready-to-eat chocolate cereal granola health bar containing a mixture of high antioxidant with coco solids and having a high procyanidin content of 50-80 mg along with soy crisp, dry cereal ingredients, lecithin and sterol esters. Martin and Borr (2004) describes a food bar specifically used during musculo-skeletal disorders such as arthritic conditions of all type consisting of about 250 mg to 2500 mg of amino-2-deoxyglucose sulphate, 2 amino deoxyglucose sulfate hydrochloride, or acetyl 2 amino-2-deoxyglucose sulfate and combinations of these consisting of vitamin B, C, E, fibre, protein, fat and flavourings. Cicci et al. (2006) describes a preparation of nutrition bar fortified with vitamin A and minerals like Ca, Mg and Fe to meet RDA requirements of teenagers.

Prosise (2004) discloses a nutritional food bar with a balanced mix of amino acids, fat and carbohydrates and their process of making, which provides an alternative to unhealthy foods. Gilles et al. (2001) discloses a solid matrix material specially designed for diabetic persons, includes a source of fructose in combination with at least one non absorbent carbohydrate as the two component carbohydrate system is said to slow down the post prandial carbohydrate response. Nielson and Jacobson (2009) prepared energy bars supplemented with the fish oil at the level of 5% since fish oil is enriched with omega-3 fatty acids provide multiple health benefits to the consumer. Nemeth et al (2008) describes a fruit filled bar fortified with omega-3 fatty acids in the form of powder containing DHA & EPA and their mixtures. The product was found stable for 12 weeks at 85^{0}C without the development of fishy taste or aroma over a storage period.

Various other types of bars have also been formulated to suit the specific needs of the individual such as diet or low carbohydrate bars, that can be eaten by diet conscious people who are intended to maintain weight and also whole food bars containing raw or minimally processed cereals and nuts compressed together to have a bar form.

Binder composition used in the preparation of bars

Food binders are the complicated mixtures usually used to adhere solid particles of food, mostly consists of mixtures of protein, carbohydrates, fats and oils, water, some leavening agents and some minute components such as minerals, vitamins and flavours. In order to give a perfect shape to the food product, a variety of methods and products as binders have been used. Food binders play a major role in giving textural characteristics to the product, which in turn depends on its solid content and sugar type. To obtain a better texture in the food product, food binders should preferably have low moisture content and a high soluble solid content (76-85%) that should disperse properly with the food product at the time of mixing. The binders having higher content of moisture and low soluble solid is of poor quality, have a limited shelf - life with an unacceptable texture and more susceptible to microbial attack (Pandey et al., 2007).

During the preparation of the product using binders, binders are first or during preparation are subjected to a wide range of thermal treatments such as baking, boiling, steaming, freezing and also mechanical treatments such as kneading, mixing, extruding, etc., and all these treatments results in a good quality product (Pandey et al., 2007). One of the main properties associated with the binder is that, it should not affect the taste and flavour of the final product and should not introduce any undesirable flavour or taste to the product.

The role of binders is of significant importance in the preparation of bars. The desired consistency and the texture of bars depend upon the binders used in their preparation. Various binders like honey, sugar syrup, rice syrup, corn syrup, high fructose concentrates, grain dextrins and their combinations, hydrocolloids such as gum arabic, agar-agar, gum tragacanth, guar gum, propyl glycol alginate, psyllium, xanthan gum, pectin, etc. can be used in the preparation of bars. Non hydrocolloid binder composition includes carbohydrate, glucose, fructose, galactose, raffinose, stachyose, starch, inulin, fructo oligosaccharides, polydextrins and others. Sugar alcohol includes sorbitol, xylitol, mannitol, erythritol, lactitol, glycerol etc., (Pandey et al., 2007). Various types of food binders used in the preparation of bars by various authors are described below .

Murray et al. (1981) discloses a food composition containing egg white as a binder, useful in the preparation and development of a soggy product with limited shelf - life. Persson (1987) discloses a binder composition for biscuits having 8-30 parts by weight of sucrose and / or mixtures of glucose and its polymers. Budd et al (1990) disclosed a binder comprising of sugar and a sweetness suppressor which exhibits water activity (a_w) in the range of 0.65-0.90, used mainly for adhering seasoning topping. Pandey et al (2007)discloses a savoury food binder which can be used for the preparation of savoury nutritious bars, consisting of less than to about 10% by weight of fat, between 5 and 50 % by weight of glycerol and between 10 to 80% of non hydrocolloid carbohydrate. Hitchner (1987) describes a low water activity binder comprising of two binder materials such as the crispy binder material comprising of whipped protein foam and a protein foam stabilizer and a chewy binder material comprising of gelled solution of gelatine in glycerol. Kelly et al. (1977) describes a fat occluded food composition as a binder which includes fat, protein and carbohydrate ingredients in which the protein ingredient is non fat milk solid or sodium caseinate or a vegetable protein product and was prepared by blending carbohydrate and protein with a molten fat at an elevated temperature. Coleman et al. (2007) describes low calorie whole grain cereal bar, a binder with carbohydrate composition consisting of brown rice syrup, caramel, cane juice, oligo fructose, inulin and mixtures binding them in different composition. Dekker and Lamar (2006) teaches a process for the preparation of binders containing few or more mono and disaccharides, a saturated fatty acid and a hydrocolloid such as gelatine and pectin to improve the binding capacity of the binder with one of the component as a partially digestible fibre such as dextrin. Froseth et al (2003) discloses a layered cereal bar with glycerol, corn syrup, sugar, maltodextrin, fructose, calcium, sorbitol as the binding ingredients. Funk et al. (2004) describes a layered cereal bar wherein binder may be a complex carbohydrate made from soy protein, fat, sweeteners, water, gelatine and conventional maltodextrin. Mihalos et al. (2003) describes the preparation of binding paste consisting primarily 6 x sugar (52-56%), filler fat (30-35%), lecithin (0-1%), salt (0-1%), flavour (0-2%), coating fat (0-13%) and blending of the above ingredients with proper mixing, agitated and stored at 140^0 F. Nana et al (2004) describes the use of maltodextrin as a binder in the preparation of protinacious food composition wherein, binder is made from maize starch, corn starch and potato starch. Coleman et al (2004) discloses binder components in cereal bar consisting of glycerol and sorbitol in combination with carbohydrate based binders such as corn syrup, corn syrup solids, honey, molasses etc.

Table 1: Dry and wet ingredients utilization in the preparation of bars

Sl No	Dry ingredients	Dry/Wet ingredients for binding	Reference
1	Dates, walnuts, dry skim milk powder, single cell protein	Dates	Kamel and Kramer 1978
2	Rice crispies, oat flakes, almond , pineapple peel powder, salt	Glucose syrup, brown sugar, vegetable fat	Carvalho et al 2011
3	Roasted baru nuts, dry fruit peels (Apple and papaya), rice flakes, oat and salt	Commercial sucrose, soy lecithin, glucose syrup, vegetable fat, water, gum acasia	Mendes et al 2013
4	Crushed corn starch biscuit, oats, skim milk powder, rice flakes, corn syrup, jeriva flour	Corn glucose and honey	Silva et al 2016
5	Rice, oat flakes, dried apple, cinnamon, salt and aspertame	Sorbitol syrup, acasia gum, inulin, maltodextrin	Srebernich et al 2016
6	Sorghum, black rice, oat, red quinoa seeds, chickpeas, linseed and lentils	Sacharin and gelatine solution	Jo- Su-Ah et al 2017
7	Sorghum flour, peanut butter, cocoa butter, cocoa powder, defatted soy flour, roasted gram flour, water melon seeds	Sugar, honey, corn syrup, glucose syrup and water	Saritha verma et al 2018
8	Oats, chocolate, nuts (almond and cashew), seeds (sunflower and watermelon)	Date syrup and honey	Chandegara et al 2018
9	Ground nuts, bengal gram, dry coconut, puffed rice, spirulina and corn flakes	Ghee, jaggery and liquid glucose	Arvind kumar et al 2018)
10	Rice crispies, quick cooked rolled oats, whey protein concentrate, citric acid	Glucose syrup, honey, vegetable oil, glycerol, maltodextrin, sugar, pectin and apple puree	Sun-Waterhouse et al 2009
11	Puffed rice, white oats, defatted soy flour, milk powder	Jaggery and honey	Ankush giri and Mridula 2016
12	Ground nuts, bengal gram dhal, dried coconut, puffed rice, corn flakes, spirulina	Liquid glucose, ghee and jaggery	Arvind kumar et al 2018

Methods of preparation of bars

Several authors have mentioned the different methods of preparation of different types of bars. But, generally nutritious energy bars are prepared by mixing different ingredients (roasted or raw) of choice in a desired proportion followed by coating them with some syrup or using a suitable binder and compressing them using hydraulic tablet making machine or heating the mixed ingredients to a desired temperature to get the dry product, mixing them with a suitable binder, compressing, spreading or cutting them to a desired shape. Bars are also prepared either by pressing them or passing through a series of rollers by compression to form a slab or passing through the extruder and cutting them to a required shape and finally packing them in a suitable packaging material with or without vacuum. Bars can also be flushed with an inert gas to eliminate substantially the oxygen content in the sealed package. Here are some of the examples which describe different methods of preparation of bars.

Farajzadeh and Golmakani (2011) and Ankush giri and Mridula (2016) reported the mixing of dry and wet ingredients of energy bar in a food processor at different speed levels and moulding the obtained mixture in a manual moulder and cutting and weighing them to the desired shape and size. In another study, snack bar was prepared by blending both wet and dry ingredients and heating the mixed ingredients for a certain period of time (1hr: 15min) to obtain thick consistency. The paste was allowed to cool and then rolled in to a strip and cut in to a strip of size 70 mmx20mmx5mm followed by spreading of banana puree on the strip in layers (Ho et al 2016). In the preparation of cassava flour based bars, bars were pressed mechanically (200kgf/m^2) in an appropriate aluminium mould (10cmx3cmx1cm) and drying partially for 1 hr at 65^0 C in an air circulation oven to preserve its characteristic crunchiness (Silva et al 2013). The cereal bar was prepared by mixing the roasted and cooked ingredients with agglutination syrup and spreading them in a rectangular shaped steel container, cooling and cutting them to rectangular pieces of 10g (Carvalho et al 2011).The snack bar was prepared by mixing the dry ingredients containing corn starch, skim milk powder, rice flakes and oat flakes with a binder syrup (Mixture of corn glucose and honey), homogenizing and laminating them to a size of 10cmx2cm approximately weighing 25 g each (Silva et al 2016). Veggie et al (2018) prepared protein bar by homogenizing dry ingredients first and then mixing with solid and semi solid ingredients and baking them in a PVC baking tin and covering with diet milk chocolate weighing 20g of final mass. Chandegara et al (2018) followed a method of roasting dried ingredients at 125^0 C for 10 min, mixing them with the syrup, honey and chocolate, transferring to mould, pressing them to make a sheet of 2 cm thickness using roller pin and roasting again at 130 ^{0}C for 10 min, cooling at room temperature and cutting to 10mmx3mmx2 mm thickness. In the preparation of energy bars, Srebernich et al 2016, Jo- Su- Ah et al 2017, Saritha verma et al 2018 used different dry and wet ingredients and the method followed was, the mixing of dry ingredients with a binder syrup made from different binding ingredients, putting them in a mould, cooling and rolling them to a desirable thickness and finally cutting them to a required shape. Mihalos (2003) describes the preparation of cold formed food bar by mixing the ingredients like cookie pieces, sugar, filler fat, puffed wheat, puffed rice, etc. and rolling the ingredients in a compression roll into a slab and slicing the slab into ropes and cutting the ropes into conventionally sized pieces. Nana et al (2004) describes a method of preparation of a protinaceous food bar by mixing dry ingredients in a Hobart mixer to get a homogenous blend followed by the extrusion and finally sheeting to a thickness of one and a half to three fourth inch and cutting into bars of desired shape. Azlyn et al (1989) reported preparation of low calorie food bar by thoroughly mixing the bar ingredients first and the mixture was shaped and placed in an aluminium cookie sheet and the sheet was kept in a preheated oven set at a temp of 177°C and baked for 10 min. The bars were cooled, cut and sealed in 2.7 mm zip lock. Sharma et al (2006) describes a process for the preparation of ready - to - eat protein rich compressed bar by mixing the dry ingredients along with binders in a planetary mixer and conditioning mixed ingredients at room temperature for 1 hr and compressing them by using a pressure of 15000- 16000 pounds with the help of hydraulic press into a square shaped tablets of size (65 x 65 x 11 mm) and immediately packing them in a skin tight cellophane. Gautam Akhilesh et al (2004) describe a high protein nutrition bar wherein the protein nuggets were prepared by extrusion process and extrusion conducted at temperatures from 60 – 140°C. Nuggets were dried using belt or conveyor or fluidized bed drier, also extruded at lower extrusion temperature up to 90°C and one or more supercritical fluids were injected prior to extrusion to form a puffed product.

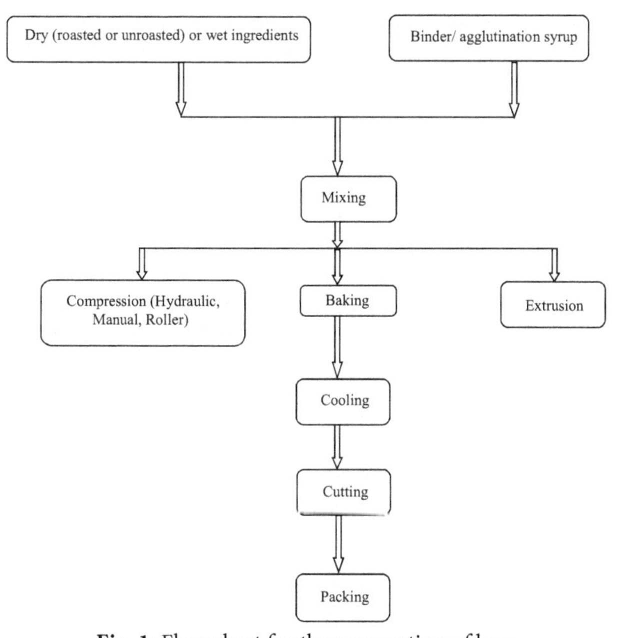

Fig. 1: Flow chart for the preparation of bars

Packaging material used in the preparation of energy bars

Food packaging plays an important role in predicting the shelf life of foods as they act as a moisture barrier and loss or gain of moisture in foods. Selection of a suitable packaging material is of utmost importance for packing a food product, as it not only gives aesthetic look to the product, but also enhances product stability during storage by preventing the lipid oxidation (Seacheol and Zhang 2005).

Energy bars can be packed in different packaging materials based on the requirement of the stability, which largely depends on the nature of the ingredients used in its preparation. Use of many packaging materials has been reported for packing of energy bars. Aluminium foil and LDPE were the commonly used packaging materials by most of the researchers for packing energy bars. Energy bars developed for soldiers was packed in nitrogen gas controlled atmosphere in an aluminium foil (Zarnab Golestan: Iran) which was covered on both the sides by low density polyethylene (LDPE) (Faijzadeh and Golmakani 2011). Packaging material like polyethylene- aluminium-polyethylene (PVC/Aluminium/PVC) was used for packing cassava flour based bar to preserve it for 210 days (Silva et al 2013). Carvalho et al (2011) reported the use of bioriented polypropylene metallised flexible packing for packing cereal bars prepared by using almonds of chichi, sapucaia and gurguei nuts. Polyethylene film (Srebernich et al 2016, Paula et al 2013), aluminium foil (Kamel and Kramer 1978, Ankush giri and Mridula 2016, Sharma and Mridula 2015) and LDPE pouches (75μ) (Chandegara et al 2018) were also used for the packing of energy bars. Padmashree et al (2013) packed energy bars in different packaging materials viz metallised polyester, paper- aluminium foil- polyethylene and polypropylene pouches to establish its stability.

Growth of energy bar market across the globe

Due to the increased interest for the consumption of energy rich, convenience form of products, packed with nutrition and energy boosting components from the health conscious consumers across the world, global energy bar market is expected to grow at a CAGR of 4.9% during the forecast period of 2019-2024. The demand for energy based products like energy drinks and bars from the consumer of US has made the United States to dominate the sale of energy bars globally. In the last few years, global energy bar product launches have nearly doubled with the dominance of European market also, because Europeans and Americans have prominent eating patterns and prefer to have ready to eat health foods like energy snack bars. Super markets, hypermarkets, nutrition stores, convenience stores, vending machines, on line stores etc, are the major marketing channels for energy bars. The growing interest on health and wellness of Asia Pacific population is also providing an opportunity for the growth of energy bar market in these countries. Manufacturers of energy bars are attracting consumers with ample options of taste, flavours and health benefits. Claims on the product has also accelerated the growth of the market with various label claims such as 'no artificial additive', high or added fibre, 'reduced sugar' in the past 5 years. Kellogg's is the prominent global player in the energy bar market followed by General Mills and PepsiCo. Associated British Food, Glaxo SmithKline, PLC and Abbott Laboratories Inc are the other market players in the snack bar market with global presence. Product innovation was a key strategy adapted by the market players to attract the attention of consumers. Larger groups of younger populations in emerging countries like India and China are also targeted by energy bar market (http://www.mordorintelligence.com).

As far as the market for the protein bar is concerned, the global market is expected to have a CAGR of 4.23% during the forecast period of 2019-2024, as the consumers prefer to buy protein bars as they offer ideal protein balance required by the body, essentially needed for weight management, improving muscle mass, increasing energy etc. North America has reached saturated phase in protein bar market, while Asia pacific and Europe are picking up pace in production and consumption. Protein bar market is also increasing in Asian countries with increasing potential in product innovation (http://www.mordorintelligence.com).

The cereal bar market is also growing at a fastest rate globally and is expected to reach dollar 16.9 million by 2025, with a CAGR of 6.17% during the forecast period of 2019-2025. Asia pacific is one of the prominent market for cereal bars, China and India being two major countries, which hold highest shares in the cereal bar market. Growing middle class income, increased economy and increased health conscious are the triggering factors for the increased cereal bar market in Asia pacific region (https://www.globenewswire.com) In North America, US hold largest market for cereal bars (https://www.persistentmarketresearch.com).

Also, meal replacement shakes and bars are playing a wide market globally as consumers find it a healthy alternative to the unhealthy junk foods. The global market for the meal replacement products is expected to grow at CAGR of 6.83% during the forecast period of 2017-2021(http://www.researchandmarkets.com). The best meal replacement products offered by the market are meal replacement powders, shakes and bars. Meal replacement products are opted, as they can replace the meal with correct amount of nutrients, and can be eaten in less amount of time. The hectic life style, increased health conscious, rising number of fitness centres and health clubs is fuelling the demand for meal replacement products. The key market holders are Abbott, Herbalife, Kellogs, Nestle and Slimfast (https://www.presswire.com).

Fig. 2: Global snack food market in 2018

Growth of energy bar market in india

As per the report published by TechSci Research, 'India Nutrition bar Market, Forecast and Opportunities 2020', the nutrition bar market in India is expected to grow at a CAGR of more than 29% during 2015-2020. This increase is attributed to the change in life style, increasing number of working population with insufficient time for traditional cooking, growing incidence of life style diseases with the rising number of obesity, diabetes, asthma, cardiovascular and other diseases in the country. Though the nutrition bar market in India is still at nascent stage, but in the years to come, it may become popular among the common consumer. The manufacturers are playing a major role in creating awareness about the healthy food habits by organizing camps where consumers can seek advice from dieticians and nutrition experts. In the year 2018, India's nutrition market was stood at Dollar 9.4 million, anticipated to grow at a CAGR of over 30% to reach dollar 46.2 million in the year 2024. Nutrition bars are attracting largely the young consumers as it can be consumed easily at office space, during travel, meetings etc and is trying to become a part of their daily diet plans.

Due to the increasing awareness about the benefits of consuming multigrain or cereals, cereal bar market segment has dominated India's nutrition bar market. Besides this, protein bar also attracting consumers especially the one's suffering from life style diseases and also the weight conscious people. If region wise market for the nutrition bar is concerned, South India has dominated the nutrition bar market in 2014, and expected to maintain its dominance during 2015-2020. The major companies dealing with nutrition bar market in India are Naturell, Nouveau Medicament (P) ltd. Xterra Nutrition, General Mills etc (https://www.techsciresearch.com).

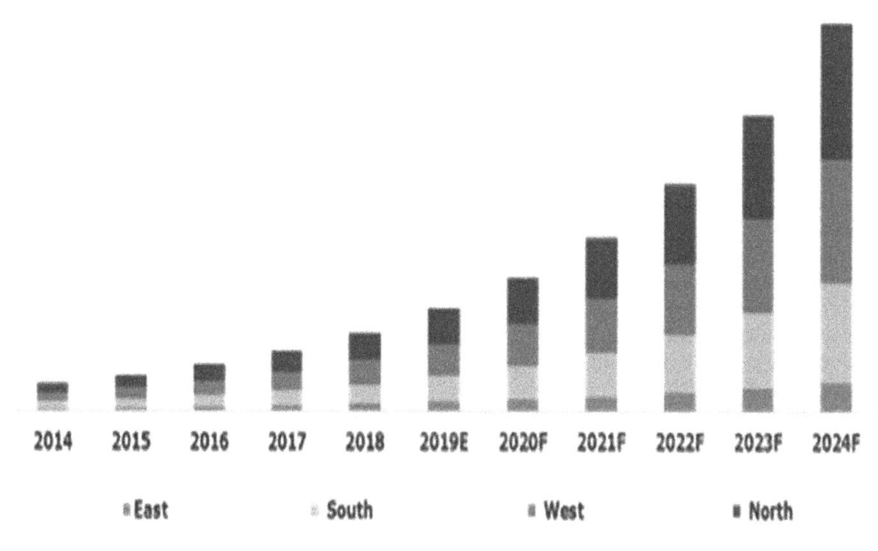

Fig. 4: Nutritional bar market size in India by region, from 2014-2024

Conclusions

Energy bars, once developed to meet the nutritional requirement of athletes and other physically active people to enhance their performance, has now attracted health conscious consumers across the world and the market is flooded with the varieties of bars claiming high energy values. Though USA and Europe have occupied larger market for energy bars, but consumption in the developing nations is still at infant stage. This may be due to the high cost of the bar with the use of cost effective ingredients in their preparation and also the lack of awareness about its nutritional and health benefits among the consumers. If the bars are made available at cheaper cost and awareness is brought at the consumer level by nutrition experts through health camps about the benefits of consumption, nutritional deficiencies can be addressed especially among younger population. Still, there is a wide scope for the further nutritional enrichment of the bar through the incorporation of nutrient rich ingredients (both micro and macro) which are less cost effective to ease consumer affordability. Despite a growing increase in the consumer need, a market research on consumer expectation is still needed so as to enhance its consumption. Introduction of these energy bars in mid day meal programme, will help in alleviating malnutrition in school children to a larger extent. Also, pregnant and lactating mothers of weaker sections can be benefitted with its consumption, if it is made available in government hospitals and dispensaries to overcome under nutrition.

References

Akintayo, E.T., Oshodi, A.A. & Esuoso, K.A.(1999). Effect of ionic strength and pH on the foaming and gelation of pigeon pea (*Cajanas cajan*) protein concentrate. *Food Chem. J.* 66: 51-56.

Aleksejiva, S., Siksna, I. & Rinkule, S. (2017). Composition of cereal bar. *J. Health. Sci.* 5: 139-145.

American Diabetic Associatio. ADA, 2008. Nutrition recommendations and interventions for diabetes (position statement). *Diabetes Care.* 31: S61-S78.

Ankush Giri, N. and Mridula, D. (2016). Development of energy bar utilizing potato extrudates. Asian. *J. Dairy Res.* 35: 241-246.

Aramouni, F.M. and Abu-Ghoush, M.H. (2011). Physico chemical and sensory characteristics of no bake wheat –soy snack bars. *J. Sci. Food. Agric.* 91: 44-51.

Aravind Kumar, Vedshree Mohanty, Yashaswini, P. (2018). Development of high protein nutrition bar enriched with *Spirulina phantesis* for under nourished children. *Curr. Res. Nutr. Food. Sci*: 835-844.

Asif Ahamad, Uroosa Irfan, Rai Mohammad, Amir Kashif, Sarfraz Abbasi (2017). Development of high energy cereal and granola nut bar. *Int. J. Agric. Biol. Sci.* 13-30 Sep- Oct.

Azlyn, K.L., Toma, R.B., Koval, J.E. and Christopher, S. (1989). Formulation and sensory evaluation of low calorie fibre bar. *J. Food. Sci.* 54: 727-729.

Baier, S.K., Guthrie, B.D., Elmore, D.L., Smith, S.A., Lendon, C.A., Muroski, A.R., Aimutis, W. (2007). Influence of extrusion on protein conformation and shelf life extension nutrition bars. Paper presented at the international symposium on the properties of water. X, 2-7 September, Bangkok. Thailand.

Brito, S.R. (2005). Barras energeticas, Disponivel em :<http://www adventuremag.com.be/ dicas/ EPAFypullkcnHOQO Fr.php>. Acessoem: 15 maio

Budd, D.L., Curtis, D.L., Dowdie, O.G. and Mehta, R.S. (1990). Topped savoury snack foods. US Patent No: 491031. Hitchner E. 1987. Composite food product. US Patent No: 4689238.

Burri, J., Daenzer-Alloncle, M., Desjardins, J., Neidlinger, S. (2003). Cereal bar and method of making. US Patent No: 6607760.

Carvalho, M.G., Costa, JMCD, Rodriguez, MCP, Sousa PHM, Clemente E. (2011). Formulation and sensory acceptance of cereal bars made with almonds of chichi, sapucaia and gurgueia nuts. *The Open Food Sci.* J.5: 26-30.

Chandegara, M., Chatterjee, B. and Sewani, N. (2018). Development of novel chocolate energy bar by using nuts. *Int. J. Food. Fer. Technol.* 8: 93-97.

Childs, J.L., Yates, M.D. and Drake, MA. (2007). Sensory properties of meal replacement bars and beverages made from whey and soy proteins. *J. Food Sci.* 72: S425-S434.

Chimel, M.J., Hammerstone, J.F., Johnson, J.C., Myers, M.E., Snyder, R.M. and Whitacre, E.J. (2008). Bars and confectioneries containing cocoa solids having a high polyphenol content and sterol/ stanol esters and process for the preparation. US Patent Application Publication No: 20050069625A1.

Cicci, MA. (2006). Nutrition bar for adolescent consumers. US Patent Application Publication No: 20060024408.

Coleman, E.C., Birney, S.L. and Brander, R.W. (2004). Cereal bar and their method of manufacture. US Patent Application Publication No: 20040126477.

Coleman, E.C., Schmid, A.H., Katz, M.C. and Birney, S. (2007). Low calorie whole grain cereal bar. US Patent Application Publication No: 20070104853.

Da Silva, E.P.D., Siqueira, H.H., Damiani, C, Boas, V. and De Barros, EV. (2016). Physico chemical and sensory characteristics of snack bars added of Jeriva flour (*Syagrus romanzoffiana). Food Sci. Technol.* (Campinas). 36: 421-425.

Dasilva, E.C., Sobrinho, V.S. and Cereda, M.P. (2013). Stability of cassava flour based food bars. *Food Sci. Technol.* (Campinas). 33: 192-198. Jan –Mar.

Dasilva, E.P., Siqueira, H.H., Do Lago, R.C., Rosell, C.M. and Boas, V., De Barros, E.V. (2014). Developing fruit based nutritious snack bars. *J. Sci. Food Agric.* 94: 52-56.
Degaspari, C.H., Blinder, E.W., Motten, F. (2008). Perfil nutricional do consumidor de barras de cereais. *Visao Academica.* 9: 49-61.
Dekker, R., Lamar, E. (2006). Food bar. US Patent Publication Application No: 20060088628 A1.
Dhingra D, Michael M, Rajput H, Patil RT. 2012. Dietary fibre in foods: A Review. *J. Food Sci. Technol.* 49:.235-266.
Dimichele, S.J., McEwen, J.W., Wood, S.M. (2002). Product and method to reduce stress induced immune suppression. US Patent No: 6444700.
Ditschuneit, H.H., Flechtner-mors M., Johnso, T.D. and Adler G. (1999). Metabolic and weightloss effects of dietary intervention in obese patients. *Am. J. Clinical Nutr.* 69: 198-204.
Esmaillzadeh A, Azadbakht L. (2006). Whole-grain intake, metabolic syndrome, and mortality in older adults. *Am. J. Clin. Nutr.* 83:1439–1440.
FAO (2010). The state of food insecurity in the world, addressing food security in protracted crises. Rome, Italy.
Farajzadeh, D., Golmakhani, M.T. (2011). Formulation and experimental production of energy bar and evaluating its shelf life and qualitative properties. *Iranian. J. Military Sci.* 13: 181-187.
Froseth, B.R., Funk, D.F., Strehlow, D. (2003). Layered cereal bar and their mthods of manufacture. US Patent Application Publication No: 20030091697.
Funk, D.F., Harrison, R.J., and Smith, D.J. (2004). Layered cereal bar and their methods of manufacture. US Patent Application Publication No: 20040013771.
Gautam Akhilesh, Albert Johan, Z., Mark Edward, J. (2008). Nutritional bar and components. US Patent No: 20080020098.
Gerard, J.A. (2002). Protein-protein cross linking in food; methods, consequences and applications. *Trends. Food. Sci. Technol.* 13: 391-399.
Gilles, S.M., Wolf, B.W., Zinker, B.A., Garleb, K.A.,Walton, J.E., Nicholson, S.E. (2001). Diabetic nutritionals and method of using. US Patent No: 6248375.
Henchion, M., Hayes, M., Muller, A.M., Fenelon, M., Tiwari, B. (2017). *Foods.* 53: 1-21.
Hitchner, E. 1987. Composite food product. US Patent N0: 4689238.
Ho, L.H., Tang, J.Y.H., Mazaitul Akma, S., Mohammed Aiman, H., Roslan, A. (2016). Development of novel energy snack bar by utilizing local Malaysian ingredients. *Int. Food. Res J.* 23: 2280-2285.
Hoffman, J.R., Flavo, M.J. (2004). Micronutrient utilization during exercise: implications for performance and supplementation. *Sports. Sci. Modern. J.*3: 118-130.
Hogan, S.A., Chaurin, V., O'Kennedy, B.T., Kelly, P.M. Influence of dietary proteins on textural changes in high protein bars. *Int. Dairy. Sci.* 26: 58-65.
Jay, R.H., Michael, J.F. (2004). Protein-which is best?. J Sports Med. 3: 118-130.
Jo-Su- Ah, Maruf Ahmad and Jong- Bang Eun. (2018). Physico chemical characteristics, textural properties and sensory attributes of low calorie cereal bar enhanced with different levels of saccharin during storage. J. Food Proc. Preservation. http://doi.org.10.1111/jfpp.13486
Kamel, B.S. and Kramer, A. (1977). Development of high protein date bar and their stability at different temperatures. *J. Food Quality.* 1: 359-371.
Kelly Ray G, Pruitt sr Kenneth R and Kershman Alvin L. (1977). Food bar and process of preparing same. US Patent No: 4055669.
Kemeny, E.S. (2003). Sweet and Savoury food bars for meal equivalent nutrition segments. US Patent Application Publication No: 20030087004.
Kemeny, E.S. (2004). Meal equivalent food bar. US Patent No: 6808727.
Khan, M.A., Semwal, A.D., Sharma, G.K., Yadav, D.N. and Srihari, K.A. 2008. Studies on the development and shelf stability of ground nuts (*Arachis hypogea*) burfi. *J. Food Quality.* 31: 612-626.

King, R.F.G.J., Gale, R.W. and Lester, S.E.G. (2003). Energy Bar. US Patent No 6585999.

Leslie Krasny (2004). Meal replacements- Convenience or wellness. Wellness Food Regulatory Issues. Pp. 14-15.

Lima, A.C. (2004). Estudo para a agregacao de calor aos produtos de caju: elaboracao de ormulacoes de frutas ecastanha em barras. PhD Thesis. UNICAMP, Campinas-SP. Brazil.

Lin, P.H., Miwa, S., Li, Y.J., Wang, Y., Levy, E., Lastor, K.C. and Champagne, C. (2010). Factors influencing dietary protein sources in the premier trial population. J. American. Dietetic. Association. 110: 291-295. http://dx.Doi.org/10.1016/j.joda.2009.10.041.PMid: 20102859.

Linscott, S.E. (1989). Granola bar with supplemental dietary fibre and method. US Patent No: 4871557

Loveday, S.M., Hindmarsh, J.P., Creamer, L.K., Harjinder, Singh. (2009). Physico-chemical changes in model protein bar during storage. 42: 798-806.

Maheshwari Alla, G.V., Jitendran, L. (2018). Development and Analysis of nutri bar enriched with zinc for sports athletes. *Int. J. Adv. Res. Sci. Engg. Technol.* 5: 5558-5570.

Manning, P.B., Schramm, J.H., McGrowth James, W.J.R. (2003). Food bars containing nutritional supplements. US Patent Application Publication No: 20030108594.

Martin, K.A., Barr, T.L. (2004). Food bar for treating musculoskeletal disorders. US Patent Application Publication No: 20040253296.

Mellen, P., Walsh, T., Herrington, D. (2008). Whole grain intake and cardio vascular disease; a meta analysis. *Nutr. Metab. Cardiovasc.* Dis. 18: 283-290.

Mendes, N.R., Gomez- Ruffi, C.R., Lage, M.E., Becker, F.S., Melo, A.A.M., Silva, F.A., Damiani C. (2013). Oxidative stability of cereal bars made with fruit peels and baru nuts packaged in different types of packaging. *Food. Sci. Technol.* (Campinas). 33:730-736. Oct-Dec.

Messina, M.J. (1997). Ch-10. In due K, Editor, Soybean chemistry, technology and utilization, Newyork, Chapman and hall. Pp: 442-466.

Mesu. B. (2017). Cereal bar having a crunchy texture. US Patent No 9603381.

Michnowski, J. (1984). Nutritional athletic bar. US Patent No: 4859475.

Mihalos, M.N., Schwartzberg, J. (2003). Cold formed food bars containing fragile baked inclusions. US Patent Publication Application No: 2003170348.

Miller, W.M., Nori Janosz, K.E., Zalesin, K.C., Mccullough, P.A. (2007). Nutraceutical meal replacements: More effective than all food diets in the treatment of obesity. *Therapy*. 4: 623-629.

Murray, E.D., Maurice, T.J., Berker, L.D. (1981). Protein binder in food composition. US Patent No: 4247573.

Nadeem, M., Rehman, S., Anjum, F.M., Murtaza, M.A., Ghulam Mueen–Uddin. (2012). Development, characterization and optimization of protein level in date bars using RSM. *The. Scientific. World. J.* Article ID: 518702. Pp 1-10.

Nana, R., Schwetlik, G.R. (2004). Food composition and food bars. US Patent Application Publication No: 20040241313.

Nemeth, K., Bello, A., Floyd, C. (2008). Baked fruit filled bar fortified with omega- 3- fatty acids and process for making same.US Patent No: 20080050473.

Nielsen, N.S., Jacobson, C. (2009). Methods for reducing lipid oxidation in fish oil enriched energy bars. *Int. J. Food. Sci. Technol.* 44: 1536-1546.

Olaofe, O., Arogundad, LA., Adeyeye, E.I., Falusi, OM. (1998). Composition and food of the variegated grasshopper. *Tropical. Sci.* 38: 233-237.

Otles, S., Ozgoz, S. (2014). Health effects of dietary fibre. *Acta Sci. Pol. Technol. Aliment.* 13: 191-202.

Owoc, J.H. (2008). Neutral meal replacement high protein bar, method of making the bar and improving the flavour and texture thereof. US Patent Application Publication No: 2008/0187640.

Padmashree, A., Sharma, G.K., Govindaraj, T. (2013). Development and evaluation of shelf stability of flaxoat nutty bar in different packaging materials. *Food. Nutr. Sci.* 4: 538-546.

Padmashree A, Sharma GK, Srihari KA, Bawa AS (2012). Development of shelf stable protein rich composite cereal bar. *J. Food. Sci. Technol.* 49:335-341

Painter, J.E., Prisecaru, V.I. (2002). The effects of various protein and carbohydrate ingredients in energy bars on blood glucose level in humans. *Cereal. Food. World.* 47: 236-241.

Painter, N.S., Burkitt, D.P. (1971). Diverticular disease of the colon: a deficiency disease of western civilization. *British. Med. J.* 2: 450-454.

Palazzolo, G. (2003). Cereal bars: they are not just for breakfast anymore. *Cereal. Food. World.* 48: 70-75.

Pandey, P.K., Guerrero, J.M., Ciaston, M. (2007). Non sweet binder for savoury food product. US Patent Publication Application No: 2007/0063557.

Persson, T., Duc, H., Buhler, M. (1987). Biscuit process. US Patent No: 4650685.

Petty HT, McCoy SC, Kristinsson HG. (2008). Novel low allergenic food bar. US Patent Application Publication No: 20080085343.

Pooja Ravindra, M., Dhanashree Sunil, M. (2018). Development and quality evaluation of puffed cereal bar. *Int. J. Pure. Appl. Biosci.* 6: 930-836.

Portman, R. (2000). Composition for optimising muscle performance during excercise. US Patent No: 6051236.

Potter, K., Stojceska, V., Plunkett, A. (2013). The use of fruit powders in extruded snacks suitable for children diets. *LWT-Food. Sci. Technol. May.* 51: 537-544.

Prosise, R.L., Beharry, C.R., Elsen, J.J., Helmers, R.L. JR, Kearney, T.J., Kester, J.J., Murphy, B.K., Niehoff RL, Noble KH, Reinhart RN JR, Sarama KJ, Tsai L, Waimin Siu SR, Wehmeier TJ, Wong VYL. (2004). Ready to eat nutritionally balanced food compositions having superior taste systems.US Patent Application Publication No: 20040166202.

Qi, L., Hu, F.B. (2007). Dietary glycemic load, whole grains, and systemic inflammation in diabetes: the epidemiological evidence. *Curr. Opin. Lipidol.* 18:3–8.

Ramakrishna, C., Sunki Reddy, Y.R. (2006). Cereal bar formulation and process thereof. EP No: 1699300B1

Ramirez-Jimenez, A.K., Gaytan-Martinez, M., Morales-Sanchi, Z.E., Loarca-Pina, G. (2018). Functional properties and sensory value of snack bars added with common bean flour as a source of bioactive compounds. *LWT- Food. Sci. Technol.* 89:674-680.

Rothacker, D.Q., Watemberg, S. (2004). Short term hunger intensity changes following ingestion of a meal replacement bar for weight control. *Int. J. Food. Sci. Nutr.* 55: 223-226.

Ryan, G. (2009). Nutritionally balanced food or beverage product. US Patent Application Publication No: 2009/ 0220649A1.

Sanchez, I., Haji Maleki, R., Borderias, A.J. (2007). Wheat fibre as a final ingredient in restructured fish products. *Food. Chem.* 100: 1037-1043.

Sarita Verma, Neelam Kaetrapaul, Vandana Verma. (2018). Development and standardization of protein rich sorghum based cereal bar. 7: 2842-2849.

Sarma, M., Roger, L.D. (2007). Protein containing food product and coating for a food product and method of making same. US Patent No: 20070087085.

Seacheol, M., Zhang, Q.H. (2005). Packaging for non thermal food processing. In: JH Haan, Ed., Innovation in Food Packaging. Pp: 482.

Seyam, A.A., Banank, O.J., Breen, M.D. (1983). Protein isolates from navy and pinto beans, their uses in macaroni products. *J. Agric. Food.Chem.* 31: 499-502.

Sharma, C., Kaur, A., Agarwal, P., Singh, B. (2014). Cereal bars: A healthful choice, a review. Carpathian. *J. Food. Sci. Technol.* 6: 29-36.

Sharma, G.K., Padmashree, A., Roopa, N. and Bawa, A.S. (2006). Storage stability of protein rich compressed bar. *J. Food. Sci. Technol.* 43: 404-406.

Sharma, M., Mridula, D. (2015). Development and quality evaluation of maize based fortified nutritious bar. Agric. Res. DOI: http://10.1007/s40003-014-0140-8.

Silva, E.C.D., Sobrinho, V.S., Cereda, M.P. (2013). Stability of cassava flour based food bars. *Food. Sci. Technol.* (Campinas). 33: 192-198.

Silva, E.P., Siqueria, H.H., Damiani, C., Vilas Boas, E.V.B. (2016). Physico chemical and sensory characteristics of snack bars added of Jeriva flour (*Syagrus romanzoffiana*). Food. Sci. Technol. DOI: http:// dx.doi.org/10.1590/1678-457x.08115.

Srebernich, S.M., Goncalves, G.M.S., Ormenese, R.C.S.C., Ruffi, C.R.G. (2016). Physico chemical, sensory and nutritional characteristics of cereal bars with addition of acasia gum, inulin and sorbitol. Food. Sci. Technol. DOI: http://dx.doi.org/10.2090/1678-457x.05416.

Sudha ML, Vetrimani R, Leelavathi KK (2004). High fibre biscuit composition and process for preparing the same.US Patent Application Publication No: 2004/ 0191393.

Sun Waterhouse, D., Teoh, A., Massarotto, C., Wibisoka, R., Wadhwa, S. (2010). Comparative analysis of fruit based functional snack bars. *Food Chem.* 119: 1369-1379.

Sylvester-Bradley, R., Folks, B.F. (1976). Cereal grains: their protein components and nutritional quality. *Sci. Prog. Oxf.* 63: 241-263.

Tillotson, S.E. (2003). Convenience foods. Encyclopaedia of food science and nutrition. 2nd ed. Pp 1616-1222.

Tsai, A.G., Wadden, TA. (2006). The revolution of very low calorie diets: an update and meta analysis. *Obesity* 14: 1283-1293.

Umar Garba and Sawinder Kaur (2014). Protein isolation, production, functional properties and applications. *Int. J. Current. Res. Rev.* 6: 35-45.

Veggie, N., Voltarelli, F.A., Periera, J.M.N., Silva, W.C., Novalta, J.W., Cavenaghi, D.F.L., Borros WM (2018). Quality of high protein diet bar plus chia (*Salvia hispanica* L) grain evaluated sensorilly by untrained testers. *Food. Sci. Technol.* (Campinas). 38 (Suppl 1): 306-312.

Vetter, J.L. (1984). Fibre as a food ingredient. *Food. Technol.* 38: 64-68.

Wagner, T.J., Bopp, K. and Cho, M.J. (2007). High protein food bars comprising sugar alcohol and having improved texture and shelf life. US Patent Application Publication No: 20070042106.

Williams PG, Grafenauer SJ, O'Shea JE. (2008). Cereal grains, legumes, and weight management: a comprehensive review of the scientific evidence. *Nutr. Rev.* 66:171–182.

Wu, H., Flint, A.J., Qi, Q., van Dam, R.M., Sampson, L.A., Rimm, E.B., Holmes, M.D., Willett, W.C., Hu, F.B. and Sun Q. (2015). Association between dietary whole grain intake and risk of mortality: two large prospective studies in US men and women. *JAMA Intern Med.*175:373–384.

Wurtsman, J.J., Wurtsman, R.J. (2003). Weight loss compositions and methods for individuals who may have gastroacidity. US Patent Application Publication No: 2003/0039739.

Zhou, P., Lieu, X., Labuza, T.P. (2008). Effect of moisture induced whey protein aggregation on protein conformation, the state of water molecule and the microstructure and texture of high protein containing matrix. *J. Agric. Food. Chem.* 56: 4534-4540.

Lightning Source UK Ltd.
Milton Keynes UK
UKHW020105230722
406270UK00004B/243

9 781032 157429